Writing Computer and Information History

Writing Computer and Information History

Approaches, Connections, and Reflections

Edited by William Aspray

ROWMAN & LITTLEFIELD
Lanham • Boulder • New York • London

Published by Rowman & Littlefield
An imprint of The Rowman & Littlefield Publishing Group, Inc.
4501 Forbes Boulevard, Suite 200, Lanham, Maryland 20706
www.rowman.com

86-90 Paul Street, London EC2A 4NE

Selection and editorial matter © Rowman & Littlefield 2024
Copyright in individual chapters is held by the respective chapter authors.

British Library Cataloguing in Publication Information Available

Library of Congress Cataloging-in-Publication Data
Names: Aspray, William, editor.
Title: Writing computer and information history : approaches, connections, and reflections / William Aspray.
Description: Lanham : Rowman & Littlefield, [2024] | Includes bibliographical references and index.
Identifiers: LCCN 2023055532 (print) | LCCN 2023055533 (ebook) | ISBN 9781538183816 (cloth) | ISBN 9781538183823 (ebook)
Subjects: LCSH: Computer science—Historiography. | Information technology—Historiography. | Technical writing—Historiography. | Electronic data processing—History—Sources.
Classification: LCC QA76.16 .W75 2024 (print) | LCC QA76.16 (ebook) | DDC 004.072/2—dc23/eng/20240212
LC record available at https://lccn.loc.gov/2023055532
LC ebook record available at https://lccn.loc.gov/20)

Contents

Preface

While the reader may gain considerable knowledge about the history of computing and the history of information from reading this book, that is not its main purpose. The goal here instead is to reflect upon the practice of researching and writing these types of history. Thus, our goal here is historiographical rather than historical. When we are teaching our students about the field of computer and information history, or we are mentoring our junior colleagues on how to be more effective scholars and teachers, we often point them to a model piece of scholarship—a book or an article—and tell them to learn from their reading. But there are several problems with this implicit approach. We often do not give them complete directions about what to look for in these publications. We also know that the formal presentation of results in books and papers often hides the process by which the topic was selected, boundaries were drawn, evidence was selected, analytic approach was chosen and applied, and results were presented; and also how this work fits into a larger body of scholarship, the implicit goals and biases of the author, and many other similar issues. This collection of reflective essays by eighteen leading scholars, discussing the craft that they practice, enriches this process of learning about the various ways to carry out computer history or information history.

A second important goal of this book is to build closer ties between the practices of computer history and information history. This book concerns both of these topics. The first scholarship in computer history by professionally trained historians began to appear in print in the 1970s, so we are approaching a half century of research and publication in this area. These early scholars have produced numerous pieces of exemplary scholarship from a variety of perspectives including intellectual history of individual technologies and academic disciplines, business histories of firms, economic histories of market sectors, and externalist histories of funding and professionalization, among others. However, the field continues to evolve, especially as computing and communication technologies have drawn together in the context of the Internet and social media; and with this trend a new set of scholars are being drawn to the study of computer and information history. They come not only from the history of science and technology field, but also from communication and media studies. Powerful theories, approaches, and frameworks are being increasingly drawn more widely from both the humanities and the social sciences to inform the practice of computer history. Thus, the practice of writing the history of computing, although it has been around for some years now, is still evolving.

While there are subdisciplines within information history, such as library history and book/publishing history, that are as old or older than computing history, the efforts to forge a unified discipline of information history is of a more recent vintage—no more than twenty-five years old. Thus, there is much more ambiguity today about what this field is, what its acceptable methods are, and where its boundaries are drawn—when compared to computer history. We know that information history includes the histories of various institutions (e.g., libraries, museums, scientific research centers), professions (e.g., archivists, computer scientists, management consultants), and technologies (e.g., computers, televisions, filing cabinets). But it is much more than that. It includes the development of various means of collecting, organizing, analyzing, and disseminating information. This takes place in formal structures such as libraries and universities, but also in informal settings such as when the photocopier repairmen meet over coffee to figure out how to work together to share written and tacit knowledge to carry out their jobs effectively, or the housewives of a rural Virginia community identify means of sharing information vital to overcoming the isolation of rural life.[1]

In an earlier article, this editor argued that information history is the sum of work produced by at least six communities, each of which developed its own conferences and journals, but which did not interact much with one another: computer historians, communication historians, book and publishing historians, library historians, archival historians, and information science historians.[2] However, there are even more players today. Internet studies and media studies share some overlap with communication studies, but they have also developed their own distinct professional cultures and their own bodies of historical scholarship. Scholars in the digital humanities have become increasingly interested in computer and information history. And mainstream historians—even those studying times much earlier than the second half of the twentieth century (when "information" became part of common parlance)—have discovered value in analyzing events from an information perspective. Thus, these scholarly communities are still finding new ways to study information. It is my longer goal to bring these various groups of historians together, and this book is a step in that direction.

The chapters in this book might occasionally cite exemplary pieces of literature, but the main purpose is not to offer literature reviews. Similarly, the goal is not to make a needs and opportunity statement of the traditional sort: that topics A, B, and C still need further study, while topics X, Y, and Z have already received adequate coverage. Instead, these chapters are intended to be reflective essays, by leading scholars, about the nature, advantages, joys, frustrations, and challenges of writing computer or information history in a particular way.

These essays address many different aspects of the writing of computer and information history. Some chapters discuss how a particular type of history (e.g., business history) complements more traditional accounts of computing history and shows the kinds of questions a new approach can reach that are not so readily answered when studied in other ways. Some chapters discuss the units of analysis (e.g., individuals, ideas, firms, industries, societies, etc.) of computer and information history; or the key concepts that arise when looking at computers or information in a particular way. Our authors discuss the kinds of evidence they use and the challenges that arise in the procurement or manipulation of this evidence; and in some chapters, they discuss how their approach came about and how it has evolved over time. Another topic covered in several of these chapters is how scholarship employing a particular approach to computer or information history (e.g., business history of computing) contributes to other scholarly areas (e.g., business history of technology, innovation studies).

These and many other issues are raised in the chapters of this book as the authors reflect on what their discipline is and how to practice it. As one of our authors, Martin Campbell-Kelly, said: "Every chapter suggests another way of seeing [in the writing of computer and information history]."[3] All the authors are either experienced in writing computer or information history, or work in an adjacent scholarly field that provides them with a useful perspective on our topic. A few of the authors regard themselves primarily as practitioners of other academic disciplines—such as information studies, business and management studies, or philosophy—but even these authors hold long and deep connections to computing and information history. Brief biographies of each of the authors can be found in the back of the book.

The authors were encouraged to develop their own style for these essays, and the styles vary considerably from chapter to chapter. Some are autobiographical, e.g., a story is told in one chapter by a traditionally trained historian of modern Europe about how he became familiar with the literature in information history and developed new concepts and frameworks to strengthen this nascent field. Other chapters are presented in a more traditional academic manner, e.g., identifying themes, topics, and approaches that are central to a particular approach and identifying ways in which a particular historical lens can be used to understand phenomena about computing or information more broadly—without discussing anything about their personal paths through this academic subject area.

The primary audiences for this book are scholars (and their graduate and advanced undergraduate students) who are either already interested in computer history or information history, or who are trying to determine their interest. There is also a wider potential audience among general historians, communication scholars, media studies scholars, digital humanities scholars, computer scientists, and library and information science scholars who see a value in computer and information history to their own work. This is a jargon-free book, and it is entirely accessible to a member of the educated public.

The book is organized in four sections. The first section provides an overview of computer history and information history: what is the nature and extent of the field, what are the primary approaches and concepts, when did the field emerge, and how has it evolved over time. Toni Weller accomplishes this in a partly autobiographical essay—appropriate since she is one of the leading early scholars of information history. She argues that there has been an "information turn" in history, involving "an acknowledgement by historians that information is a category of historical enquiry in its own right and that historical processes and methods should be applied when exploring notions or practices of information in the past."[4] Weller discusses how the field of information history has emerged in the context of the sociological study of the information society and Shannon's theory of information, which has been so important to communication engineers, but how it has become much broader by incorporating the story of information institutions such as libraries and archives, and by appropriating concepts and frameworks from the humanities.

Thomas Haigh presents "the big picture" in computing history through reflections about the writing of his recent book with Paul Ceruzzi, *A New History of Modern Computing*, which is a leading "master narrative"[5] of computing history that has reconceptualized the way in which the story is told.[6] He recounts the origins of the study of the history of computing in the 1970s, the change over time in the topics studied in computing history (partly changed by the emergence of important new technologies such as the Internet and mobile computing), and a new approach to presenting the computer in its various guises, e.g., as a calculating device, a communication device, and a shopping platform.

Part II contains three essays elaborating on the emergence of information history as a subject for study. In the first essay, James Cortada discusses his journey from being a traditional European historian to being an active early participant in information history. He discusses the frameworks, theories, and approaches that were new to him—and to which he extensively added as the field began to emerge—such as information ecologies and information infrastructures. In particular, he discusses how he crafted his two main books on information history: *All the Facts: A History of Information in the United States Since 1870* and *Birth of Modern Facts: How the Information Revolution Transformed Academic Research, Governments, and Businesses.*[7]

The second essay, by Laura Skouvig, displays the connections between communication studies and information history. Her chapter examines how historians understand information, history, and information history; and how rhetorical genre theory—developed as part of communication theory—can be used as a theoretical lens to do so. She uses a case study approach, drawing from her scholarship on the relationship between law and information from late 17th- to early 19th-century Denmark.

The final essay of this section is written by Geoffrey Bowker and John Leslie King, who have been leading figures in the creation of the information school movement over the past quarter century. This educational movement led to the transformation of some existing computer science and many library and information science departments into so-called "information schools" or "iSchools."[8] This movement was driven in part by an effort to study information, technology, and people in an interdisciplinary way; but also in part by an effort to create an alternative professional education for workers in the information fields that is not as singularly technical as a computer science education is. This chapter provides valuable reflections on writing the intellectual history of information, the history of professionalization, and the history of education.

Part III discusses the way in which the history of computing was originally practiced and various means that have been employed to enhance its study. In the first essay, David Gugerli and Daniela Zetti provide a problem-oriented perspective on the early years of computing, covering the time from the first writings in the 1970s up through the end of the 20th century. They take a critical look at the origin stories of the computer, the involvement of computer practitioners and journalists in telling these stories, the search for antecedents to connect these stories to, the role of early historical research organizations such as the Charles Babbage Institute in organizing this era of historical scholarship, and the insularity of computing history from the themes and practitioners of history of technology during its first twenty years.

While the earliest writings in computing history were about the intellectual origins of the technologies, the next major approach to appear was the business history of computing, with scholars studying both the structure of the industries (mainframe, minicomputer, software, semiconductor industries, etc.) that produced computing technologies and the strategies of firms within these industries. In the second essay of this section, Shane Greenstein discusses a key business history topic of product launches in both the computer and Internet industries and gives a behind-the-scenes analysis of how to craft this kind of historical study. He does so in a way that is particularly helpful to students or computer scientists not conversant with business history. He illustrates his approach using iconic developments such as IBM's System/360 and Personal Computer, the Netscape Web Browser, the Google Search Engine, Facebook, and the Apple iPhone.

There are many similarities and interactions between business history and economic history, but also significant differences in their approaches and their topics of study. In the third essay in this section, Daniel Garcia-Swartz and Martin Campbell-Kelly address writing the economic history of computing. They argue that the critical aspects in writing the economic history of computing are the use of economic models; quantification connected to these models; the roles of institutions, organizations, and policies; and comparative "outcomes across industries, regions, and countries to identify causes and effects." The economic themes they cover to illustrate the nature of economic history of computing include "increasing returns to scale and transportations costs, the role of policies and institutions, knowledge spillovers, and the differences between closed and open standards." They draw on well-known examples from computer history: computer timesharing, the early years of the international computer industry, the beginnings of information-technology geographic clusters, and several open and proprietary standards in the mobile-phone industry.

The final essay in this section, by Philip Doty and William Aspray, discusses "the importance of law and public policy in the writing of the story of the creation, use, maintenance, and spread of information and computing since 1930." Law and public policy almost always provide a context for the development and use of computing and information technologies, they argue. In many instances, the regulators, lawmakers, judges, and various interest groups engaged in the political process are important shapers of the way computing technologies are made and used. However, many of the existing accounts of computer history use the law and policy as merely background for the center-stage story of technology development and dissemination. The authors identify multiple ways in which law and public policy shapes computer technology development and use, and they present examples of scholarship that integrates the policy and technological aspects to provide a richer history. The authors conclude by illustrating the breadth of scholarship involving policy and law by summarizing four radically different approaches to the study of privacy and information technology as used by a legal scholar, a sociologist, a lawyer-philosopher, and a literary scholar.

Part IV focuses on ways in which computer and information history can be broadened and enriched by the addition of topics that have not traditionally been a major part of their scholarship. There are five chapters in this section. In the first chapter of this section, Valerie Schafer discusses the history of telecommunications and data networks, and their relations to computer and information history. This is a multifaceted topic, and Schafer brings to her discussion a sense of its many dimensions, such as the history of the ARPANET, the Internet, and packet switching networks and the ways in which the scholarship on these topics has evolved from the purely technical to the sociotechnical. She notes that the world of networks is much larger than these well-known topics, including, for example, national and proprietary networks including those with specialized applications such as computerized networks for banking and online information providers of legal information. Other topics that she introduces include issues related to law and policy such as net neutrality and privacy; the user experience with various kinds of technology that have enabled people to connect with others; and the approaches and topics brought to this study area by media studies and media archeology scholars.

In the second chapter in this section, Matthew Jones discusses what it means to write data into the histories of computing and information. He considers what it means to give front stage to data in these historical accounts and how, given that data is human-created, this history necessarily draws in many different kinds of academic study to treat it adequately—not only history of computing and history of science and technology, but also military history, policy study, and privacy and ethics, among others. Jones briefly surveys a number of important topics related to data history: the use of large-scale data by the military beginning in the 1950s, the rise of concern over government databases in the 1960s, the creation of database management systems as part of information science and computer science, the glut of data in society, the ways in which data mining methods have transformed various kinds of scientific enterprises, the popularity of search engines (notably Google's) in everyday life, ethical problems over the classification of data and its use in making everyday decisions such as creditworthiness of an individual, and other ethical and privacy considerations related to the use of data in society.

In the third chapter in this section, Howard Rosenbaum recounts issues that he and various co-authors have faced as they write about social informatics, and as they use its approach in their research. Social informatics is an approach, originating in the 1980s, that integrates the discussion of people and context with the study of information technologies. It has a similarity to the social turn that occurred in general history in the second half of the 20th century, and it serves as a kind of unofficial ethos for the interdisciplinary iSchool movement, which is discussed in chapter 5. The theory of scientific and intellectual movements, proffered by sociologists Scott Frickel and Neil Gross, is used as an organizing frame to describe the ways in which writing about social informatics has changed over time.[9]

The fourth chapter in this section, written by Greg Downey, provides a "personal tour of the intersections between history, geography, and the study of computing and information." Downey begins with a discussion of "the laboratory as a backstage area within which the contingency and contradiction of scientific exploration . . . could be confronted and clarified before becoming packaged as 'knowledge.'" He expands from individual laboratories to consideration of entire geographic regions, such as Silicon Valley, where scientific discovery and technological development have historically been concentrated. He next raises the question "How does technology shape a place?" as he considers networked cities and homes. He then considers sociotechnical systems, networks, and infrastructures that work across space rather than in a single place. In the final sections of the chapter, he discusses how human geographers analyze socio-spatial places, and the social implications that geographical information systems have for the study on computer history.

The final chapter beautifully describes the career path of a scholar, Liesbeth De Mol, trained originally as an analytic philosopher who finds herself today as an outsider to both the philosophical and history of computing communities, but who draws upon both traditions in her personal style of research and writing. She provides an interesting story of how logic, programming, analytic philosophy, Continental philosophy, and history of science and technology have all had shaping influences on her work but how she nevertheless stands outside of each of these traditions and is subject to resistance from each of these academic communities because she does not buy wholesale into their agendas and approaches. Her account shows how standing outside of these disciplines has provided her with new insights and a fresh approach to topics that have been studied repeatedly by historians of computing, such as the early work of Turing and the development of ENIAC, or later episodes in programming practice. She describes the need to find non-traditional homes for some of her scholarship and finding at least a partial solution in the founding of the History and Philosophy of Computing Conferences, which admit (and even encourage) multiple views and approaches.

While there probably will be some readers who read from cover to cover—and the book is organized to enhance that reading strategy, the expectation is that most readers will pick and choose among these chapters. There are no barriers to doing so, e.g., there are no chapters that are built on the material covered in earlier chapters. While we imagined when we conceived this volume that our principal audience would be graduate students and young faculty members who are just beginning to explore or establish a place in the history of computing or history of information, my experience from editing this book informs me that even very senior people in the field will find value in reading these essays. Although I have been practicing history of computing since the late 1970s and history of information from the early 2000s, I was surprised by how much I learned from every chapter in this book; and I am deeply appreciative that all these distinguished scholars have taken time to capture their reflections on their professional practices here.

There are far too many relevant topics for this book to achieve comprehensiveness. Most of the topics listed in this paragraph are discussed in passing in one or more of the chapters, but none of the chapters focuses directly on them. One important topic is gender and race. Chapter authors were encouraged from the beginning of the project to discuss these issues if it made sense to do so in their essays, and several of the chapters do just that. Another important topic not directly discussed in this book are the practices of creating computer science and technology. We might have included, for example, chapters on computing as a mathematical discipline, an engineering discipline, or a scientific discipline. There are also no chapters directly focused on writing labor history, media history, or game history as they pertain to computing and information. While there is one chapter on the iSchool movement, there are many other topics not covered here concerning the history of education in computer science, computer engineering, operations research, information science, management information systems, library and information studies, etc. Similarly, there is little coverage on the maintenance and upkeep of computing and information artifacts; the embodiment of computing and information into various kinds of work systems or application areas; and the ways in which information is used as an asset in government, business, and nonprofit organizations or by individuals in their everyday lives.

I want to give my personal thanks to Michael Sean Mahoney and W. Boyd Rayward. They were among the earliest scholars to address issues about the historiography of computing and information, respectively. I learned a great deal from them both, in discussions going back more than a quarter century.

Thanks to Charles Harmon, Lauren Moynihan, Dominique McIndoe, and others at Rowman & Littlefield who have improved this book's quality and made this publication possible. Thanks also to Daniel McNaughton, our professional indexer. This is my seventh book with Rowman & Littlefield and its imprints, and I continue to value the opportunity to work with them.

Notes

1. For these two cases, see Julian E. Orr, *Talking About Machines* (Ithaca, NY: ILR Press, 1996) and James Cortada, *Building Blocks of Society* (Lanham, MD: Rowman & Littlefield, 2020), respectively.

2. William Aspray. "The Many Histories of Information." *Information & Culture*, 50.1 (2015): 1–23.

3. Personal communication to editor, August 20, 2023. The term "ways of seeing" was popularized in a 1972 BBC documentary that resulted in the book: John Berger, *Ways of Seeing* (London: Penguin, 1972).

4. Here, and throughout the remainder of this preface, direct quotations are taken from the chapter themselves and are not specifically cited.

5. The notion of the "master narrative" in historical presentation is problematical. There is a positive value of the master narrative when it concisely presents the lay of the land; but there is also a negative connotation because master narratives have been used to impart views of society favored by the powerful and leave out other aspects, e.g., narratives about race, gender and sexual orientation, power, and poverty. We, of course, mean here only the positive aspect of getting an overview of computer and information history.

6. Thomas Haigh and Paul Ceruzzi, *A New History of Modern Computing* (Cambridge, MA: MIT Press, 2021).

7. James Cortada, *All the Facts: A History of Information in the United States Since 1870* (Oxford: Oxford University Press, 2016) and *Birth of Modern Facts: How the Information Revolution Transformed Academic Research, Governments, and Businesses* (Lanham, MD: Rowman & Littlefield, 2023).

8. These iSchools, incidentally, have been the home to many of the most active scholars writing on computing and information history, including William Aspray, Alistair Black, Geoffrey Bowker, Michael Buckland, Nathan Ensmenger, Thomas Haigh, Bonnie Mak, and Boyd Rayward.

9. Scott Frickel and Neil Gross. "A General Theory of Scientific/Intellectual Movements." *American Sociological Review* 70, no. 2 (2005): 204–232. Quote from p. 206.

Part I

The Master Narratives

Chapter 1

Historians and the Information Turn in History

Toni Weller

It is an exciting moment to be an information historian. Academic and public engagement with the concept of information is more compelling than ever before, from studies of misinformation perpetuated by ancient Greek and Roman historians to the notion of "information weapons" in light of the Russia-Ukraine conflict.[1] Our contemporary world is saturated with discussion of all things informational; precipitating a vast interest—both scholarly and popular—in the precedents, context, and history of information themes. When it comes to research, an increasing number of historians have adopted an information focus, exploring sources in new ways to reshape our understanding of historical societies, events, and themes ranging from statecraft and economics to technology and imperialism.[2]

But how are scholars studying these topics? What sources are they using and what methodologies? How are they defining information and how has information history changed since its early inception? These are important questions to ask about any body of scholarship, but for "information" these questions have raised a great deal of attention over the past four decades because of both the abstract nature of information itself and the influence of the "information" world in which we are currently living. Apropos to these ideas, this chapter uses information historiography, including my own information scholarship, as anchors to discuss how historical methodology and process have been applied to the idea of information, establishing a recognizable information turn in historical thinking dating from the last decade of the twentieth century.

One can understood a "turn" in history as recognition of a change of emphasis in humanities discourse, acknowledging the importance of historical context and processes towards a particular theme or subject; in this case, information. Significantly, such recognition can be witnessed outside of the academic field of history itself, where the application of historical methodology towards understanding information in past societies has been picked up by scholars working in other disciplines. This is certainly evident, with work in the history of information published by those working in the fields of science, philosophy, economics, computer science, and media and communication studies, to name but a few. The interests and approaches of these groups can be markedly different from those with backgrounds in "pure" history.

I have been fortunate to range across a variety of information themes over the course of my research career: the origins of the information age, nineteenth-century information culture, the history of the surveillance information state, women and information, race and information, and the impact of digital information on historical thinking. I was also privileged to act as editor for *Library and Information History* for a number of years—the first academic journal which explicitly recognized and encouraged information history.[3] Having a background in both history and information science, I have argued throughout my work about the importance of historical methodology in the discussion of information in the past. I have written and spoken consistently about the emergent, although one should argue now established, historiography of information.

Historiography and the Information Turn

Much has been written about how, from the 1960s, those in the world of computing took up the notion of information, the history of information technology, and discussion of an "electric age of information," to use Manuel Castells' terminology.[4] Claude Shannon's "communication theory," including his idea of information as a measurable "bit," instigated much of this discourse. The direct antecedents for Shannon's communication theory were the works of Ralph Hartley and Harry Nyquist, who were working at Bell Labs in the United States during the 1920s.[5] Shannon's ideas were further developed during the 1940s following his work on wartime cryptography.[6] His communication theory moved from the world of engineering into that of computer processing, and in so doing, created a new method to conceptualize information as an entity. This allowed a way to consider information in "measureable and quantifiable" terms, as James Gleick has argued (although this separation of information from content had historical precedent itself, as we will come to later).[7]

Others have dated an interest in the history of information slightly later, from the 1970s.[8] This latter chronology is true of fields such as sociology, which began to explore the origins of the so-called modern "information society" during the second half of the twentieth century. Works such as those by James Beniger, Manuel Castells, Frank Webster, and Michael Buckland helped develop theories of the information society and its history (or histories).[9] These often focused on the history of an institution or technology, or they explored the teleological origins of our modern information age. This set the tone for historians to move into these arenas, but history as a discipline was slower to realize the potential of information themes in the past. Concepts of information culture or of information processes and impact in past societies perhaps first began to emerge to some extent from those searching for a disciplinary history in the fields of library and information science.

Much of this debate was centered in British and European scholarship and it was into this world that I began taking my first information history steps in the early 2000s. Others had already begun to ask questions about the links between library history, information science, and the concept of an information history, and had started theorizing and historicizing established information themes during the latter 1990s.[10] Numerous conferences and articles emerged discussing these concepts, centered within the small community of English and Nordic scholars and led by prominent information history thinkers such as Alistair Black, W. Boyd

Rayward, Ilkka Mäkinen, Magnus Torstensson, and Laura Skouvig. In 2008, in an attempt to bridge the gap between thinking in information science and history, I published *Information History: An Introduction*, the first book, I believe, to explicitly feature "information history" in the title.[11] American scholar Pamela Spence Richards, writing as part of the library and information science community during the late 1980s and early 1990s, was prescient when using the term "information historians" and "information history" in her journal publications.[12] Other U.S. scholars, including William Aspray, Wayne A. Wiegand, and James Cortada, some with backgrounds in library or library and information science, others coming from computing, continued to unequivocally theorize the idea of information in history during the 2010s.[13]

The early decades of the twenty-first century demonstrated a significant movement to place the concept of information at the forefront of much historical thinking. What was initially a small number of scholars turning their attention to the questions of information began to evolve into a mainstream fascination with the subject, for which a lively theoretical debate continues.[14] In its earliest conception, "historical informatics" and "informational history" were also used until the more general term "information history" was adopted as a more universal term.[15]

In 2010 I wrote a review piece for the journal *Library & Information History* that gave an overview of the key trends and themes in information history emerging over the first decade of the new millennium. I argued that the modern information age had led many scholars to "explore information ideas historically."[16] I suggested that, by 2010, there were some distinct themes emerging from the information literature: an awareness of information as a distinct category of historical inquiry, growth of an information historiography, a dominance of technology, state, and politics in information thinking, and an emergence of more socio-cultural information themes, alongside a greater geographic and temporal scope. It was clear when I was writing that there was beginning to be a noticeable change in historical thinking, although it would take another decade for the mainstream academe to catch up. When I wrote that overview piece in 2010, and for much of my earlier work on information, there remained a sense of having to defend the concept of information history to conventional historians, to justify the purpose and point of information research. Writing this chapter almost fifteen years later, "information" has been validated as a critical topic of historical investigation.

It is mainstream now to suggest, as each of these historians did in the first decade of the 2000s, that information "has always been there" that "it pervaded our ancestors' world too," as James Gleick noted.[17] W. Boyd Rayward and Alistair Black, in their respective histories of the information society, argued that modern fascination with information was "an intensification of processes long underway," that the "epochal nature" of information was a myth.[18] Leonard Dudley even argued that the history of the West could be re-examined as a history of information revolutions, suggesting that "each of the major turning points in the history of the West since the fall of Rome has coincided with the introduction of a radically new information technology."[19] In the early 2000s more than one scholar noted that the theme of information in history had become a "fashionable topic for historical investigation";[20] twenty years later it has become conventional enough within the academy to demand several substantial encyclopedic reference works, as well as numerous conferences and journals.[21] Moreover, the field of information history has begun to collectively self-define itself, creating research tools specific to information that are similar to those in mature fields, such as economic or political history. These include, for example, bibliographies and anthologies.[22] The content of these historical tools indirectly helps to form and define the scope of the field as much as more direct empirical information research.

There were certainly historians writing about information in history before the turn of the millennium, although they did not use the term "historians of information" or "information history" or engage with the notion of a broader movement towards the historical study of information; the field was only in embryonic stages until the first decade of the 2000s. What is evident is that historians began writing on informational themes earlier than they considered an "information history" field or concept. This perhaps is the difference between what has been termed "witting and unwitting" information history.[23] David Eastwood, for example, writing of the English state in the early 1800s discussed "information flows" but in broad inclusive terms of bureaucracy and statistics.[24] One historian writing in 2000 suggested that "historians of the future may well refer to the period around 2000 as the 'age' of information."[25] What may have been more accurate would have been to term this period the age of information scholarship. That is, arguably there have been many ages of information, but this was the first time scholars began to mindfully and deliberately explore information themes in the past. In 2011, I edited a collection of information historiography in which I argued that the turn of the millennium had witnessed a new exploration of the multifarious roles, uses, and meanings of information in past societies, influenced by contemporary fascination with all things informational. This was, I suggested, an "information turn in scholarly thinking."[26]

Prior to the 2010s, some philosophers, most notably Luciano Floridi, had used the concept of an information turn as it applied to philosophical thinking.[27] Writing in 2002, Floridi used "information turn" as a philosophical synonym for the computer revolution, although others applied it more specifically to computing.[28] The interlacing between information and computer technology was prevalent in Floridi's thinking, but this again obscured the link between the history of information technology and the growth of non-technologically driven information themes being recognized and explored by historians during this period.

The idea of an information turn is an acknowledgement by historians that information is a category of historical inquiry in its own right and that historical processes and methods should be applied when exploring notions or practices of information in the past. Others also began to use and acknowledge the idea of an "information turn" in history. William Aspray concluded in 2015 that scholars may find it useful "to examine information issues in the context of their traditional historical lenses, but by giving them an information turn."[29] By 2018, the historian Peter Burke was endorsing similar ideas in the mainstream *History Today*. The idea of an "informational turn," he wrote, is "worth reflecting on."[30] In other words, the concept of information dominating scholarly thought and thinking was evident in the late twentieth century, but it was not until the early decades of the 2000s that historians themselves began to jump on the information bandwagon and recognize the emergent historiography of information.

The notion of an information "turn" in history should not be confused with the idea that the world we are living in is a distinct break from the past; i.e., the information age or information society as a distinct shift from any other epoch. To take this view would be, as Alistair Black, David Muddiman, and Helen Plant have argued, to miss "a more careful recognition of the continuities of history."[31] Nor is "information" a new feature or phenomenon of the modern age. It has existed in some form, in some conceptual understanding or process, for millennia.[32] Neither is our awareness of information, its technologies, processes, impacts, and importance unique to our own era; much historical work referenced in this chapter has shown how contemporaries collected, disseminated, used, hid, misrepresented, or preserved information in the past. Therefore, the idea of an information turn in history is not to suggest that we are the first to live in an age of information, but instead to recognize the change in questions we, as historians, are asking about the past. The informational emphasis on past societies, experiences, and processes, explicitly recognized by historians to be a valid line of inquiry, *this* is recognizably new in the last few decades.

The late twentieth and especially early twenty-first centuries was the first time at which scholars began to deliberately and consciously explore information as it has been understood in the past. The concept of information took on a "special salience" in scholarly thinking across academic disciplines.[33] Within the field of history, the decades around the turn of the millennium, particularly the early 2000s, witnessed a body of scholarship building up, a historiography of information, which employed the contexts, processes, and methodologies of the discipline of history, but focused on information as a conceptual theme. It is this shift in historiography that can be recognized as an information turn, as I termed it over a decade ago.[34]

There is, however, a distinction to be made between information as a subject of historical inquiry and information historiography, in the sense of a body of historical literature about information. "Information" did not suddenly come into being during the last decade of the twentieth century, but historians realizing that it was a subject that had a history can be dated to around this period. As more scholars began to explore the history of information and consider its nuances and subtleties, they began to inadvertently create an information discourse, each perpetuated by the other.

The adage that the present informs the past holds true here, just as it ever has. When I was writing in the first decade of the 2000s, this "turn" was in its infancy, and was, to a great degree, influenced by the new information *zeitgeist* in public consciousness that had been building from the first decade of the millennium. Since then there has been an explosion of information-related history, as well as a growing mainstream acceptance of information as a driving force in society. Over the past three decades, scholars have discussed a broad range of information chronologies ranging from the ancient world to the very modern.[35] Information geographies were at first dominated by histories of the West, but this is becoming more balanced with scholarship emerging on Eastern Europe and the Islamic world.[36] Information has been explored alongside a huge variety of themes from the traditional historical discourses of power, government, and control through to the more diverse culture, race, computing, and even filing cabinets, based in universities from across the world.[37] More precisely, these themes have been considered with an information focus, that is, asking how and why information was processed, stored, collected, preserved, disseminated, protected, omitted, and used in various historical contexts.[38]

More broadly, turns in history have precedent. From the 1970s through to the 1990s, there was a movement away from objective narratives and institutional focus, instead focusing on the way in which events in the past were *experienced*. The same source material was re-examined, but read in a different way, challenging meanings, omissions, and practices. This became known as what was called the *cultural turn*; a refocusing of the practice and focus of history.[39] Included in this was what Hayden White termed "metahistory," in essence, the history of historical writing, the nature of history, the meaning of history, and the cause and significance of historical change.[40] These are notions that underpin this volume. This phenomenon of the late twentieth century can be represented as a significant shift in society, reflecting the social and cultural context of the period, alongside an analytical reframing within academia. This cultural shift built on the philosophical linguistic ideas of Ludwig Wittgenstein and Ferdinand de Saussure, whose work established links between language, language users, and the world.[41] Together, these movements helped change focus from meaning to experience and became known as "post-modernism."[42] Such postmodern views can be hostile to traditional historicism since they deny that there is any absolute view, instead suggesting historical reality is a discursive construct. This view certainly has its critics, some of them highly vocal. The most notable of these are perhaps Richard J. Evans and Arthur Marwick, who have fiercely criticized postmodernist approaches to the field of historical study.[43] In postmodern terms, an "age of information"—or "revolutions," "reform," or anything else—represents, in the simplest of forms, readings of the past. However, one does not need to critique postmodernism to acknowledge that the notion of "ages of information" exists only as subjective and arbitrary measures of time imposed by historians to make sense of the enormity of the past. However, there is also an argument that postmodernism and the linguistic turn have allowed for a more fluid and contingent interpretation of historical analyses.[44] Arguably, alongside a growing interest in all things informational in our modern world, this has led to something as essentially conditional as information to become a tool of heuristic inquiry.

Historians of information tend to study the *experiences* of information; the processes, technologies, impacts, omissions.[45] The publication of encyclopedic volumes of historic information are testament to the mainstream recognition of information as a historical subject, reflecting just how disparate historical views of information can be between historians.[46] These volumes continue the tradition of historical scholars studying information in the past, originating during the late twentieth and early twenty-first centuries and all contributing to a body of information historiography. This mass of information discourse alongside the "sudden rise of interest at the start of the century in the history of information" must, as Peter Burke noted, "is surely connected to current debates about the kind of 'information society' in which we are living."[47] Others agree that digital information technology has caused information to "become a separate object of historical enquiry."[48] But where information scholarship can become complicated is in both our own and contemporary historical understandings and definitions of information. As Philip Mirowski and Edward Nik-Khah eloquently express it, "information" is "an elusive something which we somehow possess while not quite understanding it."[49] This can pose a significant challenge for the information historian; what is one's own understanding of information as a subject of the modern world, and how does one interpret contemporary historical understandings of information? In other words, "is it the historian's notion or that of the subjects of historical study" that we are defining?[50] To do the latter without imposing anachronistic restrictions on one's scholarship requires a rigorous and mindful methodology.

Methodological Approaches to Information [in] History

The notion of information, as with many words and concepts, has changed over time and continues to present challenges to historians. Moreover, as Daniel Brauer has shown, even "the term 'history' has many meanings . . . history, rather than having a specific object of study, is simply one approach to any object, perceived through its changes over time."[51] To add even more complexity, depending on your disciplinary approach and background, the concept of information has different historical meanings. Historians of computing, for example, may understand different things by "information" than a historian of books or of feminism. Historians more generally tend to view information and information culture in anthropological terms, whereas, as Dan Schiller noted, postmodernist theorists such as Daniel Bell—and the computer and information scientists of the later twentieth century, most notably Claude Shannon's identification of a "bit" as a measure of information—applied the notion of information as scientifically abstracted from lived experience.[52] Poignantly, this abstraction has come almost full circle from the detachment that occurred during the nineteenth century, allowing information to be considered as separate to rhetoric and considered as a concept in its own right, as I discuss below.

Moreover, the meaning of information can also vary across applications; are we talking about the history of information content, processes, frameworks, technologies, impact, or any other host of related themes? Moreover, information meant different things to past societies; often even different words were used to describe what we might understand as information today. Up until, and into, the nineteenth century it was common for the words "knowledge" and "information" to be used interchangeably; sometimes "news" and "education" could also be used to mean similar things in different contexts.[53] Of course, a rigorous historical methodology, based on evidence, analysis, and interpretation, should still form the substance of information—or of any—history, but it is also true that interpretations change depending on the historian and the context and influences in which they are writing. Daniel Headrick noted how throughout "the course of history there have been periods of sharp acceleration . . . in the amount of information that people had access to and in the creation of information systems to deal with it."[54] One of these periods was the Victorian era when there was not only an increase in information systems—and "systems" could include everything from statistics, graphs, and maps to new postal and telegraph communication technologies—but also a change in the way in which information began to be conceived by contemporaries. Information was increasingly abstracted from context, removed from the rhetoric of the previous century. During the Enlightenment, information was not separated from a specific purpose. It was essentially a "rhetorical instrument," as Neil Postman has argued.[55] And although its semantic ground could move, depending on the context in which the term was being used, from the Victorian period onwards there was a "gradual social realisation" that "information" was increasingly "more explicitly recognized as an independent phenomenon."[56] Arguably, this was one of the most notable shifts between conceptual understandings of information in the pre-modern and modern worlds, and certainly predates the digital era.[57]

Such shifts in conceptual and semantic understanding have been fundamental in framing the modern information world, but they also demonstrate the complexities of definition. But since "information" as a category of historical inquiry does not have a stable meaning, how, then, does one study something that cannot be neatly explained, nor have a fixed temporal understanding?

Some have addressed this issue by offering more structured frameworks from which information history could be studied. Paul Sturges offered an early information history model in his 2011 analysis of the civil conflict in Uganda during the 1980s.[58] It was built upon a threefold pattern of information and communication elements: the first identified domains of information *activity*, the second divided information activities into *types*, and the third further divided these activities into overt or covert aspects. Using his information model, Sturges was able to identify that patterns common in similar resistance situations were not evident in the Ugandan conflict. He concluded that the same model could be applied elsewhere, specifically "wherever there is an issue of conflict."[59] For him, the potency of using an information model as an analytical approach to his historical study was clear; it distinguished the primacy of oral communication, the role of secrecy, and, with a degree of prescience, the role of mis/disinformation in politics, business, and society. Since 2016, James Cortada has developed what he calls "information ecosystems" and "information infrastructures" as tools to facilitate historical research on information.[60] For Cortada, who has always advocated specificity of definition over the potentially nebulous context-specific approach, these methodological structures enable a degree of control and unambiguity over both the scope and fluidity of information concepts. As with Sturges, Cortada argues that such graphical representations of information can be applied to multiple research scenarios—Cortada's background in computing is evident in his suggestion that information ecosystems can be applied to industry, but he also demonstrates its practicable use in case studies ranging from diplomacy in nineteenth-century Spain to genealogy to American homemakers in the latter half of the twentieth century.[61]

Of course, one might equally argue that in general historical practice one defines according to the context and the sources, so why should "information" be privileged with a different methodology from any other historical subject? Some "unwitting" information scholars—defined as those historians who study information in the past but who do not term themselves as specifically "information" historians[62]—are less engaged in the broader discourses about definition and methodology because they consider themselves simply to be applying the methodological approaches of the historian to their subject matter, as they would with any other research theme. David Eastwood's aforementioned discussion of information flows in the nineteenth-century English state might be considered to be one example of this.

Likewise, one might posit that studying an abstract notion such as information is no different from the historical study of humor, smell, or fear; with those, one can understand in a general sense of course that they are things that make us laugh or feel afraid—although what exactly constitutes these things in differing contexts or across discrete societies is a different matter and therefore requires contextual analysis—the fundamental tool of the historian. As Vic Gatrell noted in his epic history of laughter, it can be a methodological challenge to write about an open-ended subject, "with no simple chronology or precedent to follow."[63] A recent overview of the emergent field of sensory history discussed similar challenges, including not only the issue of understanding and definition between time and place, but also the challenges inherent with determinism and presentism.[64] These latter are vitally relevant to the historical study of information. "Information" is such a *zeitgeist* in the modern world that one must be vigilant about ensuring there is no anachronism in any historical studies of information in the past. Placing current values on the past can lead to forced comparisons which promote a Whiggish view of the world when sound historical methodology is not applied to the study of information themes.[65]

However, this is where distinctions between meaning and experience can be useful in offering some methodological structure. The historical meaning of information might be fluid and ambiguous (although as historians we should certainly still try to explore these nuances), but the historical experience of information can be a useful tool around which to form a methodology. Historians can use frameworks to anchor the fluid concept of information to a more solid semantic idea and empirical source material—a methodology based in fundamental historical practice. In his studies of the history of personal information collection and the information state in England, for example, Edward Higgs uses identification documents, local and national forms, and census data to ground his exploration of information.[66] These types of source materials have long been used by historians, but it is the information agenda and questions that mark out his study as information history. Higgs is not alone in his scholarship on the history of identifying the individual under the aegis of information history, nor in his use of the census as source material.[67] Others, such as Daniel Headrick, have used historical information systems, in the broadest sense, as a way of exploring how information might have been processed or stored. Headrick includes science, statistics, maps, graphs, dictionaries, encyclopedias, and postal and telegraphic systems under the broad banner of information systems.[68] Others have anchored information through exploration of archives in the early modern world.[69] Jillian M. Hess has suggested that eighteenth- and nineteenth-century commonplace books, scrapbooks, and albums can be understood as ways of information management through material culture.[70] In my study of Victorian information culture, I used etiquette books to show how different types of information could be disseminated.[71] Using focused case studies, Richard Ovendon scrutinized the methods and motives of those who have sought to destroy texts through which the history of the human race has been preserved in events dating from the ancient world to the very modern.[72] Essentially, historians writing on informational themes situate their subject matter "in the context of information's history"; whether or not they call themselves information historians, their methodological approach is the same.[73] Most fundamentally of all perhaps, historians should "simply tell the story" (or stories) of the development of the subject, "and let readers make up their own minds."[74]

My own approach has long been to argue that defining information in context is more vital than a universal definition. This idea was echoed by the editors of the 2021 volume *Information: A Historical Companion* in their decision to allow contributors to use "their own understandings of information rather than subordinating themselves to a single definition."[75] Fundamental historical methodology teaches us that any source material should be judged within its own contemporary context.

To encourage context-specific understandings of information enables a richer opportunity for understanding information as contemporaries did, in all its complex ambiguity. As Sandra Braman suggested in relation to information in policy, "multiple definitions apply concurrently."[76] Context is vital to how we understand information in any given situation, which will be different across geographies, temporalities, gender, class, age, race, and a whole host of other criteria. Definition, then, comes from each unique case, each historical focus; and so, inevitably and correctly, there will be multiple, numerous definitions of what information is, or was understood to be, in different societies, places, and periods.

Of course, historians are a product of their own time, and questions of the past are in flux since, as E. H. Carr noted back in 1961, "history is a constantly moving process, with the historian moving within it."[77] In other words, the lens of history and the historian changes over time. But good historical methodology should ensure that we explore the past on its own terms. To historicize the past, we must explore the informational source material, and this can be challenging and multifarious. In 2000, Paul Duguid wrote about his experience watching another historian detach sensory information in the form of a vinegar scent from an archive of handwritten letters, in order to track outbreaks of cholera from a particular town during the eighteenth century.[78] As Seeley Brown and Duguid note, this challenged his historical assumptions about "where information begins and carrier ends, about dividing text from context."[79] This has echoes of Shannon's information "bit," and even more so of the nineteenth-century disconnection of information from pre-Victorian rhetoric, but what is most striking is that such a methodological distinction really was not conceived of by historians much before the late 1990s. It is, one might argue, an empirical product of the information turn in history.

Conclusion

Before the onset of postmodernist thought, Geoffrey Elton suggested in 1967 that the writing of any history amounted to a dialogue between the historian and his materials; that the past was true and independent and that the evidence, or sources, alone dictated the questions asked of the past.[80] Yet, most, if not all information history describes or constructs processes or experiences rather than offers narrative events. Moreover, the complexity of definition and methodological information discourses means that there can never be a sole, definitive history of information in the sense that Elton thought of past events when writing in the late 1960s. Nonetheless, it is vital that a rigorous historical methodology, based on source material, analysis, and interpretation, should still form the basis of information history. As with any historical study, this can be approached in a multitude of ways, if the subject matter is contextualized.

As this chapter has tried to explore, "historiography itself is also situated in a historical context, which it tries to understand simultaneously with shaping its concepts."[81] The historiography of information, therefore, is situated within a specific era of modern informational focus—the late decades of the twentieth century onwards—while also being concerned with historical ideas of information throughout the past. These are big concepts; and meta-histories, such as those in this volume, can be useful in situating the development of a field and identifying the bigger picture of approaches, reflections, and connections—as the title of this book promises.

Notes

1. Michael Grant, *Greek and Roman Historians: Information and Misinformation* (London: Routledge, 2004); Timothy L. Thomas, "Information Weapons: Russia's Nonnuclear Strategic Weapons of Choice," The Cyber Defense Review 5, 2 (2020), 125–144.

2. For a small selection of these themes see Edward Higgs, *The Information State in England: The Central Collection of Information on Citizens Since 1500* (Basingstoke: Palgrave, 2004); Paul Slack, *The Invention of Improvement: Information and Material Progress in Seventeenth-Century England* (Oxford: Oxford University Press, 2014); Philip Mirowski & Edward Nik-Khah, *The Knowledge We Have Lost in Information: The History of Information in Modern Economics* (Oxford: Oxford University Press, 2017); Martin Campbell-Kelly et al. (Eds.), *Computer: A History of the Information Machine* (4th ed., London and New York: Routledge, 2023); and Zoë Laidlaw, *Colonial Connections, 1815–45: Patronage, the Information Revolution and Colonial Government* (Manchester: Manchester University Press, 2013).

3. Established in 1969, the journal changed its name from *Library History* to *Library & Information History* as I took over the editorship in 2008. The change was instigated by my predecessor, the prominent information historian Alistair Black.

4. For quote see Marshall McLuhan, *Understanding Media: The Extensions of Man* (London: Routledge & Kegan Paul, 1964), 343.

5. See Harry Nyquist (1924), "Certain Factors Affecting Telegraph Speed," *Bell System Technical Journal* (3:2), April, 324–346; Ralph Hartley (1928), "Transmission of Information," *Bell System Technical Journal* (7:3), July, 535–563.

6. For more on this see Ronald Kline, *The Cybernetics Moment: Or Why We Call Our Age the Information Age* (Baltimore: Johns Hopkins University Press, 2015).

7. Quote from James Gleick, *The Information: A History, A Theory, A Flood* (London: Fourth Estate, 2011), 4. For discussion of information theory and its development see Dan Schiller, *How to Think about Information* (Urbana: University of Illinois Press, 2007), 17–35. See also Claude Shannon and Warren Weaver, *The Mathematical Theory of Communication* (Urbana: University of Illinois Press, 1949).

8. Ann Blair et al. (Eds.), *Information: A Historical Companion* (Princeton: Princeton University Press, 2021), vii.

9. James Beniger, *The Control Revolution: Technological and Economic Origins of the Information Society* (Cambridge, Massachusetts: Harvard University Press, 1986); Manuel Castells, *The Rise of the Network Society, The Information Age: Economy, Society and Culture,* Vol. I (Oxford: Oxford University Press, 1996)—and Vols. II and III; Frank Webster and Raimo Blom (Eds.), *The Information Society Reader* (London: Psychology Press, 2004); and Michael Buckland, Information and Society (Cambridge, Massachusetts: MIT Press, 2017).

10. For more on this see Michael Buckland's seminal, if controversial, article "Information as Thing," *Journal of the American Society for Information Science* 42, 5 (1991), 351–360. Alistair Black, "New Methodologies in Library History: A Manifesto for the 'New' Library History," *Library History* 11 (1996), 76–85; Alistair Black, "Lost Worlds of Culture: Victorian Libraries, Library History and Prospects for a History of Information," *Journal of Victorian Culture* 2, 1 (1997), 124–141; Alistair Black, "Information and Modernity: The History of Information and the Eclipse of Library History," *Library History* 14, 1 (1998), 39–45; Alistair Black, "Information History and the Information Professional," *Library History* 20 (2004), 3–6; W. Boyd Rayward, "The History and Historiography of Information Science: Some Reflections," in Trudi Hahn and Michael Buckland (Eds.), *Historical Studies in Information Science* (New Jersey: Information Today Inc., 1998), 7–21; Boris F. Volodin, "History of Librarianship, Library History, or Information History: A View from Russia," *The Library Quarterly* 70, 4 (2000), 446–467; Ilkka Mäkinen, Alistair Black, Miha Kovac, Laura Skouvig & Magnus Torstensson, "Information and Libraries in an Historical Perspective: From Library History to Library and Information History," in Leif Kajberg & Leif Lørring (Eds.), *European Curriculum Reflections on Library and Information Science Education* (Copenhagen: The Royal School of Library and Information Science, 2005), 172–191; Toni Weller, "A New Approach: The Arrival of Informational History," *Proceedings of the of the XVI International Conference of the Association for History and Computing: Humanities, Computers and Cultural Heritage* (Amsterdam: Royal Netherlands Academy of Arts and Sciences, 2005); Alistair Black, "Information History," in Blaise Cronin (Ed.), *Annual Review of Information Science and Technology*, Vol. 40 (New Jersey: Information Today Inc., 2006); Toni Weller, "Information History: Its Importance, Relevance, and Future," *Aslib Proceedings* 59, 4/5 (2007), 437–448.

11. Toni Weller, *Information History—An Introduction: Exploring an Emergent Field*, (Oxford: Chandos, 2008).

12. Pamela Spence Richards, "Of CD-ROMs and History," *The Electronic Library* 7, 2 (1989), 74–75; Pamela Spence Richards, "The Quest for Enemy Scientific Information 1939–1945: Information History as Part of Library History," *Library History* 9, 1–2 (1991), 5–14.

13. William Aspray, "The Many Histories of Information," *Information & Culture 50,* 1 (2015), 1–23; Wayne A. Wiegand, "Libraries and the Invention of Information," in Simon Eliot and Jonathan Rose (Eds.), *A Companion to the History of the Book* (Oxford: Blackwell, 2019), 531–544; James W. Cortada, "Shaping Information History as an Intellectual Discipline," *Information & Culture* 54, 1 (2019), 102–126.

14. A small selection of these include Laura Skouvig and J. Andersen, "Understanding Information History from a Genre-Theoretical Perspective," *Journal of the Association for Information Science and Technology* 66, 10 (2015), 2061–2070; Laura Skouvig, "Present and Past: The Relevance of Information History," *Information & Culture* 58, 1 (2023), 1–16; Alistair Black, Bonnie Mak, Laura Skouvig & Toni Weller, "Situating Information History," *A Handbook of Information History* (London: Routledge, forthcoming).

15. See for example, Laszlo Karvalics, "The Claims, Pre-history and Programme of Historical Informatics," in *Periodica Polytechnica Social and Management Sciences 2,* 1 (1994), 19–30; Alistair Black, "New Methodologies in Library History: A Manifesto for the 'New' Library History," *Library History* 1 (1995), 76–85; and Toni Weller, "A New Approach: The Arrival of Informational History," *Proceedings of the XVI International Conference of the Association for History and Computing*, September 14–17, 2005, Royal Netherlands Academy of Arts and Sciences, 273–278. Among the earliest to use the specific term "information history" were Pamela Spence Richards, Alistair Black, Laszlo Karvalics, and Toni Weller.

16. Toni Weller, "An Information History Decade: A Review of the Literature and Concepts, 2000–2009," *Library and Information History* 26, 1 (2010), 83–97. Quote from 84.

17. James Gleick, *The Information: A History, A Theory, A Flood* (London: Fourth Estate, 2011), 12.

18. W. Boyd Rayward, *European Modernism and the Information Society: Informing the Present, Understanding the Past* (Aldershot: Ashgate, 2008), 9; Alistair Black et al., *The Early Information Society: Information Management in Britain before the Computer* (Aldershot: Ashgate, 2007).

19. Leonard Dudley, *Information Revolutions in the History of the West* (Cheltenham: Edward Elgar, 2008), 7.

20. See for example, Paul Slack, "Government and Information in Seventeenth-Century England," *Past & Present*, 184 (August 2004), 33–68 (see p. 33 for quote), and Alistair Black et al, *The Early Information Society: Information Management in Britain before the Computer* (Aldershot: Ashgate, 2007), (p. 8 for quote)—both refer to information as a "fashionable topic for historical investigation."

21. Encyclopedic information history collections include Ann Blair et al. (Eds.), *Information: A Historical Companion* (Princeton: Princeton University Press, 2021) and the

companion volumes Michele Kennerly et al. (Eds.), *Information: Keywords* (New York: Columbia University Press, 2021) and Eric Hayot et al. (Eds.), *Information: A Reader* (New York: Columbia University Press, 2021). See also the forthcoming Toni Weller, Alistair Black, Bonnie Mak & Laura Skouvig (Eds.), *Handbook of Information History* (London: Routledge, forthcoming). Journals dedicated to information history include the aforementioned *Library & Information History* and *Information & Culture: A Journal of History*. Conferences have abounded but a brief international selection from the last two decades include Making Connections Between Library, Book, Reading and Information History (Leeds Metropolitan University, UK, 2007); International Conference on Integrated Information (Kos Island, Greece, 2011); Information and Power in History (Huygens ING, part of the Royal Netherlands Academy for the Arts and Sciences, Free University of Amsterdam, 2017); Information and its Communication in Wartime (Institute of Historical Research, London, 2017); 14th Annual Conference for the SHOT Special Interest Group for Computing, Information, and Society (via Zoom, 2003).

22. For example, the journal *Library & Information History* has featured an information history bibliography at the back of every issue since 2008; anthologies include, among others, the many publications featured in the notes of this chapter, as well as the edited volume in which this chapter sits.

23. See Alistair Black, Bonnie Mak, Laura Skouvig & Toni Weller, "Situating Information History," *A Handbook of Information History* (London: Routledge, forthcoming).

24. David Eastwood, "Amplifying the Province of the Legislature": The Flow of Information and the English State in the Early Nineteenth Century," *Historical Research* 62, 149 (1989), 276–294.

25. Peter Burke, *A Social History of Knowledge: From Gutenberg to Diderot* (Oxford: Polity, 2000), 1.

26. Toni Weller (Ed.), *Information History in the Modern World: Histories of the Information Age* (Basingstoke: Palgrave Macmillan, 2011), 2.

27. Luciano Floridi, "Artificial Intelligence's New Frontier: Artificial Companions and the Fourth Revolution," *Metaphilosophy*, 39 (October 2008), 4–5; Luciano Floridi, "The Information Society and Its Philosophy: Introduction to the Special Issue on the Philosophy of Information, Its Nature, and Future Developments," Information Society, 25, 3 (2009), 153–158; Liang Wang, "Possibilities and Changes: In-depth Study on the Information Turn of Philosophy," *Journal of Northeastern University (Social Science)*, 18, 6 (2016), 557–563.

28. Luciano Floridi, "What Is the Philosophy of Information?" *Metaphilosophy*, 33, 1/2 (January 2002), 123–145. Chris Bissell, "The Information Turn in Modeling People and Society: Early German Work," *IEEE Technology and Society Magazine*, 36, 4 (2017), 80–87; Wei He, "'Computing' and 'Information': Two Paths in 'Information Turn,'" Proceedings of the 2021 Summit of the International Society for the Study of Information. Proceedings 81, 1 (2022), 115.

29. William Aspray, "The Many Histories of Information," *Information & Culture*, 50, 1 (2015), 1–23, quote p. 14.

30. Peter Burke, "The History of Knowledge," *History Today*, 68, 11 (November 2018).

31. Alistair Black et al., *The Early Information Society: Information Management in Britain before the Computer* (Aldershot: Ashgate, 2007), 7–8.

32. See for example, A. D. Lee, *Information and Frontiers: Roman Foreign Relations in Late Antiquity* (Cambridge: Cambridge University Press, 2006) and Rebecca Robinson, "Big Data in Early China: Networks of Surveillance in the early Chinese Empires," in Laura Skouvig and Andreas Marklund (Eds.), *Histories of Surveillance From Antiquity to the Digital Age: The Eyes and Ears of Power* (London: Routledge, 2021), 20–36.

33. Frank Webster, *Theories of the Information Society* (2nd ed., London: Routledge, 2002), 2.

34. Toni Weller, *Information History in the Modern World* (Basingstoke: Palgrave Macmillan, 2011), 1–12; Toni Weller, "The Information State: An Historical Perspective on Surveillance," *Routledge Handbook of Surveillance Studies* (London: Routledge, 2012), 57.

35. See Toni Weller, Alistair Black, Bonnie Mak & Laura Skouvig (Eds.), A *Handbook of Information History* (London: Routledge, forthcoming).

36. See Marton Gergő Vér, "The Role of Information and Disinformation in the Establishment of the Mongolian Empire—A Re-examination of the 13th century Mongolian History from the Viewpoint of Information History," *International Conference on Integrated Information*, 2011; Katherine Bowers & Simon Franklin (Eds.), *Information and Empire: Mechanisms of Communication in Russia, 1600–1854* (Cambridge: Open Book Publishers, 2017); Elias Muhanna, "Realms of Information in the Medieval Islamic World," in Ann Blair et al. (Eds.), *Information: A Historical Companion* (Princeton: Princeton University Press, 2021), 21–37.

37. A brief selection of such works includes Filippo de Vivo, *Information and Communication in Venice: Rethinking Early Modern Politics* (Oxford: Oxford University Press, 2007); Jacob Soll, *The Information Master: Jean-Baptiste Colbert's Secret State Intelligence System* (Ann Arbor: University of Michigan Press, 2009); Alfred D. Chandler Jr. & James Cortada, *A Nation Transformed by Information: How Information Has Shaped the United States from Colonial Times to the Present* (Oxford: Oxford University Press, 2010); Amy E. Earhart, "Can Information Be Unfettered?: Race and the New Digital Humanities Canon," in Matthew K. Gold (Ed.), *Debates in the Digital Humanities* (Minneapolis: University of Minnesota Press, 2012); Laszlo Karvalics (Ed.), *Information History of the First World War* (Budapest: L'Harmattan, 2015); Jon Agar, *The Government Machine: A Revolutionary History of the Computer* (London: MIT Press, 2016); Colin Koopman, *How We Became Our Data: A Genealogy of the Informational Person* (Chicago: University of Chicago Press, 2019); Ida Nijenhuis, Marijke van Faassen, Ronald Sluijter, Joris Gijsenbergh & Wim de Jong (Eds.), *Information and Power in History: Towards a Global Approach* (London: Routledge, 2020); Craig Robertson, *A Vertical History of Information* (Minneapolis: University of Minnesota Press, 2021).

38. The following is a brief selection (in addition to the rest of the references in this chapter). See for example, Alex Wright, *Glut: Mastering Information Through the Ages* (Washington, DC: Joseph Henry Press, 2007); Lisa Gitelman, *"Raw Data" Is an Oxymoron* (Cambridge, Massachusetts; MIT Press, 2013); Kenneth Cmiel & John Durham Peters, *Promiscuous Knowledge: Information, Image, and Other Truth Games in History*

(Chicago: University of Chicago Press, 2020); Ann Blair, *Too Much To Know: Managing Scholarly Information before the Modern Age* (London: Yale University Press, 2011).

39. One of the earliest works in which the phrase "cultural turn" can be found is Jeffrey C. Alexander, "The New Theoretical Movement," in Neil Smelser (Ed.), *Handbook of Sociology* (London: Sage Publications, 1988), 77–101. The foundations for the cultural turn can be seen in several key works of the 1970s: Hayden White, *Metahistory: The Historical Imagination in Nineteenth-Century Europe* (London: Johns Hopkins University Press, 1973); Clifford Geertz, *The Interpretation of Cultures: Selected Essays* (London: Hutchinson, 1973); Michel Foucault, *Discipline and Punish* (New York: Pantheon Books, 1977); and Pierre Bourdieu, *Outline of a Theory of Practice* (Cambridge: Cambridge University Press, 1977).

40. Hayden White, *Metahistory: The Historical Imagination in Nineteenth-Century Europe* (London: Johns Hopkins University Press, 1973).

41. The classic texts on the linguistic turn include Ferdinand de Saussure, *Course in General Linguistics* (London: Duckworth, 1995) and Ludwig Wittgenstein, *Tractatus Logico-Philosophicus* (London: Routledge, 2001).

42. Victoria E. Bonnell & Lynn Hunt (Eds.), *Beyond the Cultural Turn: New Directions in the Study of Society and Culture* (Berkeley: University of California Press, 1999).

43. For a forceful critique of postmodernism in history see Richard J. Evans (1997), *In Defence of History* (Cambridge: Granta Books, 1997) and Arthur Marwick, *The New Nature of History: Knowledge, Evidence, Language* (Basingstoke: Palgrave, 2001).

44. John Tosh (Ed.), *Historians on History* (London: Longman, 2000), 1–15.

45. *Meaning* is more related to the philosophy of knowledge, which is a related, but separate field—although it shares a philosophy of placing the digital revolution in context; or, as Peter Burke notes, the "perspective of changes over the long term"—quote from Peter Burke, *What is the History of Knowledge?* (Cambridge: Polity, 2016), 1. For more on the history of knowledge see Peter Burke, *A Social History of Knowledge from Gutenberg to Diderot* (Cambridge: Polity, 2000); Peter Burke, *A Social History of Knowledge from the Encyclopédie to Wikipedia* (Cambridge: Polity, 2012); Johan Östling, David Larsson Heidenblad & Anna Nilsson Hammar (Eds.), *Forms of Knowledge: Developing the History of Knowledge* (Lund: Nordic Academic Press, 2020).

46. See for example, Ann Blair, Paul Duguid, Anja-Silvia Goeing & Anthony Grafton (Eds.), *Information: A Historical Companion* (Princeton: Princeton University Press, 2021).

47. Peter Burke, "The History of Knowledge," *History Today*, 68, 11 (November 2018).

48. Ida Nijenhuis, Marijke van Faassen, Ronald Sluijter, Joris Gijsenbergh & Wim de Jong, "Information and Power in History: A New Historiographical Approach?," in Ida Nijenhuis et al. (Eds.), *Information and Power in History: Towards a Global Approach* (London: Routledge, 2020), 274.

49. Philip Mirowski and Edward Nik-Khah, *The Knowledge We Have Lost in Information: The History of Information in Modern Economics* (Oxford: Oxford University Press, 2017), 13.

50. Paul Duguid, "Introduction," in Ann Blair, Paul Duguid, Anja-Silvia Goeing & Anthony Grafton (Eds.), *Information: A Historical Companion* (Princeton: Princeton University Press, 2021), x.

51. Daniel Brauer, "Theory and Practice of Writing History in Times of Globalisation," in Concha Roldán, Daniel Brauer & Johannes Rohbeck (Eds.), *Philosophy of Globalization* (Berlin: De Gruyter, 2018), 397–408.

52. Dan Schiller, *How to Think about Information* (Urbana: University of Illinois Press, 2007), 17–35. See also Claude Shannon and Warren Weaver, *The Mathematical Theory of Communication* (Urbana: University of Illinois Press, 1949).

53. Toni Weller, *The Victorians and Information: A Social and Cultural History* (Saarbrücken: Verlag, 2009); Toni Weller, "Preserving Knowledge Through Popular Victorian Periodicals: An Examination of *The Penny Magazine* and the *Illustrated London News*, 1842–1843," *Library History*, 24, 3 (2008), 200–208.

54. Daniel Headrick, *When Information Came of Age: Technologies of Knowledge in the Age of Reason and Revolution, 1700–1850* (Oxford: Oxford University Press, 2000), 8.

55. Neil Postman, *Building a Bridge to the 18th Century: How the Past Can Improve Our Future* (New York: Vintage, 1999), 87.

56. Toni Weller, *The Victorians and Information: A Social and Cultural History* (Saarbrücken: Verlag, 2009), 174, 156. See also Toni Weller, Alistair Black, Bonnie Mak & Laura Skouvig (Eds.), "Situating Information History," in *A Handbook of Information History* (London: Routledge, forthcoming).

57. Neil Postman, *Building a Bridge to the 18th Century: How the Past Can Improve Our Future* (New York: Vintage, 1999), 82–98; Toni Weller, *The Victorians and Information: A Social and Cultural History* (Saarbrücken: Verlag, 2009); Ronald Day, *The Modern Invention of Information: Discourse, History, and Power* (Carbondale: Southern Illinois University Press, 2001); Toni Weller (Ed.), *Information History in the Modern World: Histories of the Information Age* (Basingstoke: Macmillan, 2010); James Gleick, *The Information: A History, a Theory, a Flood* (London: Fourth Estate, 2011).

58. Paul Sturges, "Modelling Recent Information History: The 'Banditry' of the Lord's Resistance Army in Uganda," in Toni Weller (Ed.), *Information History in the Modern World* (Basingstoke: Palgrave, 2011), 155–174.

59. Sturges suggested the history of information-based competition between business corporations as a potential example, 172.

60. James Cortada, "New Approaches to the History of Information: Ecosystems, Infrastructures, and Graphical Representations of Information," *Library & Information History* 32, 3 (2016), 179–202; James Cortada, "A Framework for Understanding Information Ecosystems in Firms and Industries," *Libraries & the Cultural Record*, 51, 2 (2016), 133–163; James Cortada, *Building Blocks of Modern Society: History, Information Ecosystems, and Infrastructures* (Lanham, Maryland: Rowman & Littlefield, 2020).

61. James Cortada, *Building Blocks of Modern Society: History, Information Ecosystems, and Infrastructures* (Lanham, Maryland: Rowman & Littlefield, 2020).

62. See Alistair Black, Bonnie Mak, Laura Skouvig & Toni Weller, "Situating Information History," *A Handbook of Information History* (London: Routledge, forthcoming), for discussion of this term.

63. Vic Gatrell, *City of Laughter: Sex and Satire in Eighteenth-Century London* (London: Atlantic Books, 2006), xxi.

64. William Tullet, "State of the Field: Sensory History," *History: The Journal of the Historical Society*, 106, 373 (2021), 804–820.

65. Examples of such might be Blaise Cronin and Yvonne Rogers, "From Victorian Visiting Card to vCard: The Evolution of a Communicative Genre," *Journal of Information Science*, 29, 1 (1999), 65–68, or David Bawden and Lynn Robinson, "A Distant Mirror? The Internet and the Printing Press," *Aslib Proceedings*, 52, 2 (2000), 51–57, which suggest a teleological progression without a meaningful sense of the historical context.

66. See, for example, Edward Higgs, "The Rise of the Information State: The Development of Central State Surveillance of the Citizen in England, 1500–2000," *Journal of Historical Sociology*, 14, 2 (2001), 175–197; Edward Higgs, *Identifying the English: A History of Personal Identification 1500 to the Present* (London: Continuum, 2011).

67. See Dan Bouk, *How Our Days Became Numbered* (Chicago: University of Chicago Press, 2015); Margo Anderson, *The American Census: A Social History* (London: Yale University Press, 2015); Deborah Lupton, *Data Selves* & *Quantified Self* (Cambridge: Polity, 2016); Sarah Igo, *The Averaged American: Surveys, Citizens, the Making of a Mass Public* (Cambridge, Massachusetts: Harvard University Press, 2017); Paul Schor, *Counting Americans* (Oxford: Oxford University Press, 2017); and Andrew Whitby, *The Sum of the People: How the Census Has Shaped Nations, from the Ancient World to the Modern Age* (New York: Basic Books, 2020).

68. Daniel Headrick, *When Information Came of Age: Technologies of Knowledge in the Age of Reason and Revolution, 1700–1850* (Oxford: Oxford University Press, 2000).

69. Kate Peters, Alexandra Walsham & Liesbeth Corens (Eds.), *Archives & Information in the Early Modern World* (Oxford: Oxford University Press, 2018).

70. Jillian M. Hess, *How Romantics and Victorians Organized Information: Commonplace Books, Scrapbooks and Albums* (Oxford: Oxford University Press, 2022).

71. Toni Weller, "The Puffery and Practicality of Etiquette Books: A New Take on Victorian Information Culture," *Library Trends*, 62, 3 (2014), 663–680.

72. Richard Ovendon, Burning the Books: A History of Knowledge Under Attack (London: John Murray, 2020).

73. Quote from James Cortada, *All the Facts: A History of Information in the United States Since 1870* (Oxford: Oxford University Press, 2016), xvii. Other examples of inadvertent historical information scholarship include Oz Frankel, *States of Enquiry: Social Investigations and Print Culture in Nineteenth-Century Britain and the United States* (London: Johns Hopkins University Press, 2006); Joanita Vroom, "Kütahya between the Lines: Post-Medieval Ceramics as Historical Information," *Hesperia Supplements, Vol. 40; Between Venice and Istanbul: Colonial Landscapes in Early Modern Greece*, (2007), 71–93; Jacob Soll, "From Note-Taking to Data Banks: Personal and Institutional Information Management in Early Modern Europe," *Intellectual History Review*, 20, 3 (2010), 355–375.

74. Edward Higgs, *The Information State in England: The Central Collection of Information on Citizens Since 1500* (Basingstoke: Palgrave, 2004), 8.

75. Paul Duguid, "Introduction," in *Information: A Historical Companion* (New Jersey: Princeton University Press. 2021), x. For examples of my arguments on contextual definition see Toni Weller and David Bawden, "Individual Perceptions: A New

Chapter on Victorian Information History," *Library History*, 22, 2 (2006), 137–156; Toni Weller, "Information History: Its Importance, Relevance, and Future," *Aslib Proceedings*, 59, 4/5 (2007), 437–448; Toni Weller, *Information History—An Introduction: Exploring an Emergent Field* (Oxford: Chandos, 2008); and Toni Weller, "A New Approach: The Arrival of Informational History," *Proceedings of the XVI International Conference of the Association for History and Computing*, September 14–17, Royal Netherlands Academy of Arts and Sciences, 2005, 273–278.

76. Sandra Braman, "Defining Information: An Approach to Policymakers," *Telecommunications Policy*, 13, 3 (1989), 234–235.

77. E. H. Carr, *What Is History?* (London: Penguin, 1961), 133.

78. John Seeley Brown & Paul Duguid, *The Social Life of Information* (Cambridge, Massachusetts: Harvard Business School Press, 2000), 174.

79. John Seeley Brown & Paul Duguid, *The Social Life of Information* (Cambridge, Massachusetts: Harvard Business School Press, 2000), 174.

80. Geoffrey Elton, *The Practice of History* (London: Fontana Books, 1967).

81. Daniel Brauer, "Theory and Practice of Writing History in Times of Globalisation," in Concha Roldán, Daniel Brauer & Johannes Rohbeck (Eds.), *Philosophy of Globalization* (Berlin: De Gruyter, 2018), 397.

Chapter 2

Writing the Big Story

Thomas Haigh

In 2016 I began work with Paul Ceruzzi to update his book *A History of Modern Computing*, then twenty years old and firmly established as the most cited overview history of the electronic computer. Five years later *A New History of Modern Computing* was published, with me as the lead author.

It is neither a revised edition of the old book nor an entirely new one. I see it as a substantially new book. The chapter structure is completely different, but we incorporated fragments of the old text just as a new house might reuse elements of the building that was demolished to make room for it. It tells the story in a different way, or perhaps it tells a different story. In this chapter I will be describing what we did and why.

Why Tell the Big Story?

Shortly after our book was published, a fellow historian of computing suggested that this was an important but thankless job. Two years later, it is apparent that he was right on both counts. Overview histories of computing do not generate buzz or win awards. They are not written by the faculty of elite history programs; in fact, few of the people writing them are employed by history departments in any capacity. There is something a little suspect about writing a history of technology that centers on a technology. Attempting this is not quite reactionary but it is unmistakably old-fashioned. The history of modern computing is long and broad;

given the limited volume that can be accommodated in a single book it is a matter of physics that the narrative must in some sense be shallow.

So why bother? One case is the one we articulate in the book's introduction. In little more than 75 years the programmable electronic computer has gone from ENIAC, a one-off electronic monstrosity used by scientists and engineers to crunch numbers, to today's plethora of smartphones, servers, and embedded devices. Their material form, cost, social significance, mode of operation, and user groups have changed utterly and repeatedly in a way that those of other technologies, such as cars, have not. In that respect the trajectory of the computer has been truly exceptional. Looking at that arc only one segment at a time makes the big story easy to miss.

We are a narrative-weaving species. Anyone who thinks or writes about the history of computing is going to construct some kind of framing story in their head to join whatever fragments they learn about. Without the context and scaffolding of a big picture story they are likely to join the pieces together haphazardly. Scholarship of the history of computing is increasingly migrating to information schools and media programs, where students and practitioners are unlikely to receive broad historical training. Readers and writers are likely to look at historical events in isolation from broader historical contexts and frame them directly against the concerns of the present day. They may miss the deeper historical roots of apparently new concepts like cloud computing, or jump over decades of intervening history to make misleading connections straight from famous historical figures such as the women of ENIAC to the modern world. It is important that people with a deep and broad knowledge of the history of computing, and of the history of technology more generally, occasionally attempt to put the pieces together into a coherent and comprehensive story.

Early Narratives

What is the history of computing the history of? This question, above all others, must be answered by anyone attempting to write the big story of computing. Historian Michael S. Mahoney, who posed that question in one form or another in many of his papers, observed that "nothing is really unprecedented. Faced with a new situation, people liken it to familiar ones and shape their responses on the basis of the perceived similarities."[1] The search for precedent is a search for a narrative frame. Humans make sense of the world by telling stories to themselves and each other, and so to understand and explain the first electronic computers it was necessary to package them inside a story.[2] The computer's history has been told for as long as there have been computers. As historical work necessarily takes place sometime after the events concerned, we can trace the spread of computer technology in the changing narratives of the history of computing.

The early computer was most obviously a calculating machine, literally something performing computations. It was seen as a natural evolution of earlier calculating devices such as the hand-cranked calculators widely used in business.[3] The most important early edited volume on the history of computing, *A History of Computing in the Twentieth Century,* was prepared by members of the Los Alamos scientific computing staff.[4] The first textbook for computer history, *The Computer from Pascal to von Neumann*, was a mixture of historical research and memoir from a close collaborator of von Neumann.[5]

Another answer is that the computer has always been an "information machine." This narrative was first adopted by Edmund Callis Berkeley's 1949 book *Giant Brains, or Machines That Think*. This was the first popular treatment of the new technology, giving a generation of impressionable youngsters an introduction to the hitherto obscure world of computing. It described in numbing detail computers of the 1940s such as the ENIAC, MIT's differential analyzer, and the series of machines built by Harvard and Bell Labs. But it also wove them into a grand narrative that stretched back into prehistory and forward to a future of highly intelligent machines. Berkeley called the computer a giant brain not because it could handle tasks of the kind later associated with artificial intelligence but because it "can handle information with great skill and great speed."[6] In this narrative the digital computer was preceded by earlier machines and systems for handling information, such as nerve cells, cave paintings, beads on strings, and human language.[7] But in "a deep break from the past," it could transfer "information from one part of the machine to another [with] flexible control over the sequence of its operation."[8] In other words, it could execute a program.

That framing of the computer was endorsed in the subtitle of *Computer: A History of the Information Machine* by Martin Campbell-Kelly and William Aspray.[9] First published in 1996, this remains one of the two standard histories relied on by readers and instructors looking for a well-balanced, comprehensive, and reliable overview. By the 1990s, historians began to explore the computer's administrative use and situate it in the context of earlier technologies such as punched-card machines and typewriters. *Computer* embodies this shift. On its publication, the book's most novel feature was its insistence on the computer as primarily a tool for administrative coordination rather than scientific calculation. (In other words, its framing had caught up with the world of the 1960s rather than remaining in the 1940s.) Although accessible, it is not simplistic and nicely summarizes and connects key insights and stories from the secondary literature as it had developed to the early 1990s.

Computer has a finely crafted narrative structure that exemplifies the tendency of overview histories of computing to produce coherent narratives by exploring only one kind of computer platform in each time period, typically: mainframes up to 1965, minicomputers from about 1965 to 1975, personal computers from 1975 to the mid-1990s, and the Internet since then. This simplifies the narrative, but at a cost: IBM still earns a significant share of its profits from its mainframe business, and minicomputers were much more important in the early 1980s, after personal computers were invented, than in the late 1960s. It's also an approach that is hard to extend to recent decades. Smartphones did not make other platforms go away, and the Internet can be traced further back in time than the IBM PC.

The other major overview history of computing was Ceruzzi's *A History of Modern Computing*, written in parallel with *Computer*.[10] Some years ago, when comparing the two books, I wrote that "Ceruzzi's book has less to say about applications, and skips the digital computer's forebears completely to launch the story in the mid-1940s. Ceruzzi provides more detail on the architectural development of computers and better coverage of the minicomputer, which he argues for persuasively as the source of today's personal computing technologies. His book focuses more on technical history and has a somewhat more episodic structure."[11] *A History of Modern Computing* offers some well-researched case studies of computer use, but at its heart is the story of the emergence of the modern computer as a packaged product and its subsequent evolution into new forms.

Both books were updated—*A History of Modern Computing* in 2003 and *Computer* in 2004 and 2014 (with the addition of Nathan Ensmenger and Jeffrey Yost as coauthors). These updates focused, as one would expect, on adding new material to better cover the Internet and the Web. Neither book fully assimilated newer conceptions of the computer—which today is experienced more often as a media player, communication device, or control system than as a business tool or scientific calculator. Neither had much to say on the evolution of the personal computer after the mid-1980s, on video games, or on mobile computing. Both books cloistered software in a chapter of its own rather than integrating it into the main narrative.

Much changed in the twenty years after these books appeared, but there was no serious attempt to challenge them. Walter Isaacson's startlingly old-fashioned but inarguably popular chronicle of brilliant inventors, *The Innovators*, is outselling both by several orders of magnitude but is not a serious history.[12] Strong overview business histories have appeared of specific sectors, most notably Martin Campbell-Kelly's history of the software industry and Jeffrey Yost's books on the computer industry and the computer services industry.[13]

There have been a couple of highly-compressed histories, most notably Ceruzzi's own *Computing: A Concise History* and a more recent *The History of Computing: A Very Short Introduction* by Doron Swade, as well as a few others written to be accessible to high school students.[14] David Gugerli's *How the World Got into the Computer* takes a thoughtful and application-oriented view of computers and their affordances, but like Swade's book pays close attention only to the first decades of electronic computing.[15] Books continue to appear about the very early history of electronic computing, including George Dyson's engagingly and frustratingly eccentric *Turing's Cathedral*, and a shelf full of recent books that are (unlike Dyson's) actually about Turing.[16] Dyson squares his desire to make grand claims with the reality that he wrote a closely observed study of a single early computer project by rhetorically positioning the Institute for Advanced Studies computer as a "point source" for our "digital universe." By focusing on origins and first steps, most of these narratives imply that we can best understand today's computers by looking at the relatively distant past.

In the realm of full-length synthetic history, the duopoly of *Computer* and *A History of Modern Computing* endured for more than twenty years. This was a testimony to their considerable strengths and made the prospect of supplanting one of them a daunting challenge.

The Histories of Computing(s)

So, what's *new* about *A New History of Modern Computing*? Its expanded content reflects both the many things that have happened in the past two decades and the burgeoning secondary literature that has helped us think about old events in new ways. But if you place the book next to its predecessor, the most obvious change is the narrative structure.

As we looked through Paul's earlier book, we realized that its chapter structure was not holding up as well as Paul's analysis and selection of topics. We were not always sure why the topics in a chapter were grouped together or why one topic followed another. In some places the chronology was confusing. Several things were dealt with in two separate places. We decided to impose an entirely new chapter structure on the existing material.

A strict chronological organization with five years covered in each chapter would be the easiest way to structure a book like this. Looking up the date on which something occurred would tell us which chapter it was part of. But to update the reader on what happened within each area between, say, 1975 and 1980, we would have had to constantly cycle between areas such as minicomputing, scientific supercomputing, business data processing, and personal computing. That structure would also have made it very hard to engage with computer users and use practices, which have become increasingly central to scholarly work on the history of computing during the past twenty years.

Here's what we did instead. Within each chapter one moves forward in time, anywhere between ten and thirty years from beginning to end. But the stories told in each chapter overlap in time with those told in other chapters.

There are three main sections to the book. Chapters 2 to 6 look at the development of the computer from the 1950s to the 1970s. Within this section, a three-chapter sequence explores the new capabilities of real-time operation, interactive use, and interpersonal communication. Chapters 7 to 9 focus on personal computing in the 1970s and 1980s. The year 1990 serves as a kind of pivot point with only one chapter crossing that divide. The post-1990 chapters focus on the further development of personal computing.

Every narrative needs a protagonist. Ours, as the table of contents makes clear, is *the computer*. There is not, and has never been, any such entity. Computers are very different from each other. As well as the machines themselves differing hugely, from video game consoles to supercomputers to smartphones, their users, producers, and associated social practices and cultural meanings are utterly different. From one chapter to the next, we see profound discontinuities in users and applications. Even the kinds of users we are writing about shift. Some are specific institutional users with the ability to specify their needs, such as the Department of Defense. Others are communities of enthusiasts, individuals, or small businesses who exert influence through consumption patterns or bottom-up creativity. Some users are figments of the imagination of the designers of technologies that never find broad use.[17]

Our protagonist has a rich career full of adventures. In each chapter something happens to *the computer* and in most of them it "becomes" something. The very idea that one can structure a meaningful story around *the computer* might seem a ridiculous conceit. Yet, on the level of technology and architecture, there are very real continuities between the chapters. In our story, *the computer* is not just a piece of hardware but a cluster of technologies and techniques—hardware, software, architectural features, programming languages, ideas, and practices. This assemblage moves from one chapter to another, accreting capabilities as it goes.

Table 2.1: Chapter titles and temporal coverage. Most titles start with "The Computer Becomes a . . . ," which is shortened here to TCBA.

<table>
<tr><th>1940s</th><th>1950s</th><th>1960s</th><th>1970s</th><th>1980s</th><th>1990s</th><th>2000s</th><th>2010s</th><th>2020</th></tr>
<tr><td colspan="2">1: Inventing the Computer</td><td colspan="7"></td></tr>
<tr><td></td><td colspan="3">2: TCBA Scientific Supertool</td><td colspan="5"></td></tr>
<tr><td></td><td colspan="3">3: TCBA Data Processing Device</td><td colspan="5"></td></tr>
<tr><td></td><td colspan="4">4: TCBA Real-Time Control System</td><td colspan="4"></td></tr>
<tr><td colspan="2"></td><td colspan="2">5: TCBA Interactive Tool</td><td colspan="5"></td></tr>
<tr><td colspan="2"></td><td colspan="3">6: TCBA Communications Platform</td><td colspan="4"></td></tr>
<tr><td colspan="3"></td><td colspan="2">7: TCBA Personal Plaything</td><td colspan="4"></td></tr>
<tr><td colspan="3"></td><td colspan="2">8: TCB Office Equipment</td><td colspan="4"></td></tr>
<tr><td colspan="3"></td><td colspan="2">9: TCBA Graphical Tool</td><td colspan="4"></td></tr>
<tr><td colspan="5"></td><td>10: The PC Becomes a Minicomputer</td><td colspan="3"></td></tr>
<tr><td colspan="4"></td><td colspan="3">11: TCBA Universal Media Device</td><td colspan="2"></td></tr>
<tr><td colspan="5"></td><td colspan="2">12: TCBA Publishing Platform</td><td colspan="2"></td></tr>
<tr><td colspan="5"></td><td colspan="3">13: TCBA Network</td><td></td></tr>
<tr><td colspan="5"></td><td colspan="4">14: The Computer is Everywhere and Nowhere</td></tr>
<tr><td colspan="8"></td><td>15: Epilogue</td></tr>
</table>

In the first chapter *the computer* is born, or rather invented. Each subsequent chapter engages with a different domain of use within which it is remade. It enters one realm of human existence after another. In each realm it becomes an essential and transformative part of technical practice, making possible things that would not otherwise be possible. As it transforms practices it is itself transformed. *The computer* is taught new tricks and given new capabilities to meet the specific demands of its new environment.

THE HISTORIES OF COMPUTING(S)

There's a particular reason this project daunted me. I have repeatedly questioned whether it is possible to reconcile the urgent need for the history of computing to focus on users and applications with the creation of a single coherent history of computing narrative. In 2001 I wrote:

> The use of computer technology in a particular social space (such as the laboratory, office, or factory) cannot be addressed without also studying the earlier history of this setting, the people in it, and the objectives to which the machine is put. So, while coherent one-volume histories of the computer hardware industry and its technologies can be written, it seems unlikely that we can produce a single coherent narrative about the use of computers.[18]

Returning to the theme in a 2011 review essay, I observed that "a one volume synthetic history that takes the use, work, and social dimensions of information technology seriously is probably impossible given the proliferation of computer technology in recent decades."[19] Thus, I started work on the new history backed into an intellectual corner.

My path out of that corner was inspired by Michael Mahoney's discussion of a "communities of computing" approach in his classic paper "The Histories of Computing(s)."[20] Mahoney responded to my point about the impossibility of writing one volume on computer use with a provocative claim:

> As [James W.] Cortada, Haigh, and [Jon] Agar suggest, the histories and continuing experience of the various communities show that they

> wanted and expected different things from the computer. They encountered different problems and levels of difficulty in fitting their practice to it. As a result, they created different computers or (if we may make the singular plural) computings.[21]

Mahoney's declared interest was in the programming work needed to transform what he called the "protean machine" of the modern computer into a useful tool within each community. Our goals were broader, and so we took the idea that "different computers" were created for and by different communities somewhat more literally. We were interested in how both hardware and software were reshaped to meet the needs of particular communities, and in how those innovations then became part of computing as experienced in other communities. Weaving those separate stories into a coherent overall narrative meant doubling down on a distinctive feature of Paul's earlier book: its engagement with computer architecture. Architectural features originally developed to meet the needs of one kind of user, for example, Los Alamos, eventually show up in smartphones even though Cray supercomputers, with their hand-wired circuits and exotic cooling systems, would never fit in your handbag.

Our chapters overlap a great deal in time. For example, the second, third, and fourth chapters cover much the same period. In that sense, they are like the parallel arrows Mahoney drew (figure 1) to represent the histories of technology and practice in areas such as management, data processing, and military communications as parallel stories that are all eventually transformed by the introduction of computer technology.

Much of what happens in each chapter is specific to the needs of a particular community. For example, chapter 4 explains how the specific needs of the SAGE air defense project led to computers so large that they filled the entire floor of a building, and how NASA's hopes that the space shuttle would be a reliable "space truck" led to the installation of five redundant flight computers in each orbiter, any one of which could have landed it. Neither approach would translate directly into other contexts. Yet many of the technologies developed for these real-time aerospace projects quickly became part of *the computer.* High reliability electronics, silicon chips, prioritized interrupts for real-time operations,

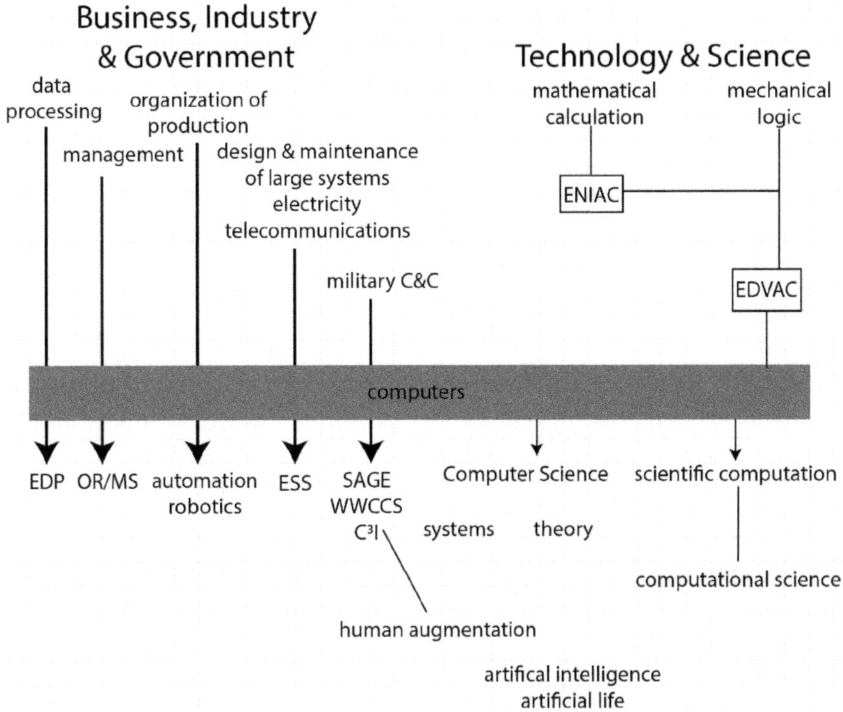

Figure 2.1.Mahoney's visual depiction of the "communities of computing" conception of the core history of computing story, in which stories of the adoption of computer technology unfold in parallel and largely independently. *Source*: Courtesy of Taylor & Francis.

and computer graphics all made the transition from exotic novelties into core features of *the computer* as later introduced into homes and offices.

We have not managed to square the circle by figuring out how to write a satisfactory one-volume history of computer use. I still think that is impossible. The social and cultural histories of the computer have so many strong discontinuities between eras, social spaces, and nations that there is little to be gained from trying to torture them all into a single volume. Only technologically grounded history can create enough of a substantive narrative connection to justify the exercise.

Each chapter in our book centers on a story that could, and indeed should, be told elsewhere as a book of its own. This would permit a different kind of narrative, focused throughout on a coherent set of users and applications. Such a book could explore the setting before the introduction of computer technology and look at a succession of technologies applied in that single space. Think, for example, of JoAnne Yates's *Structuring the Information Age*, which looked at the life insurance industry through many decades and several technological eras.[22]

What we have managed to design, I hope, is a one-volume history of computing that takes the interplay of use and innovation seriously by exploring the process in which the computer is successively remade through its encounters with different communities of users. In other words, *A New History of Modern Computing* does not aspire to comprehensively explain how "the computer changed the world."[23] It does aim to explain how specific aspects of the world changed *the computer*.

Inching Toward Universality

Any overview history of computing faces the daunting challenge of shifts over time in scale and perspective. In the 1940s, perhaps a few hundred people worked directly with programmable electronic computers. Today, most of the people in the world use smartphones daily. Their cost, physical form, modes of use, social and economic significance, and cultural meaning have changed beyond all recognition.

Many histories of computing written by computer scientists or mathematicians have positioned actual computers as physical instantiations of the "universal machine" described by Alan Turing in the 1930s. To them, almost any machine able to branch could, with sufficient time and storage, duplicate the work of any other computer. In that sense, even ENIAC was universal back in 1946. From the viewpoint of practice, however, things look very different. ENIAC had a writable memory of only 200 decimal digits, was challenging to reprogram, filled around 2,000 square feet of floor space, and cost (allowing for inflation) many millions of dollars. It could output results only by flashing lights and punching cards. It ran numerical computations and simulations, mostly for military and atomic projects. In any meaningful sense, it was a highly specialized machine.

Today, more than half of the people in the world use smartphones. They are universal in a practical sense because we use them for almost everything: step counting, chatting with friends, watching movies, listening to music, banking, finding sexual partners, shopping, and interacting with co-workers. Other computers, from supercomputers and cloud systems to embedded microcontrollers, collectively carry out a still broader range of tasks. Our book tells the story of that big transformation as the result of a series of smaller transformations. There was nothing natural about this. As several observers have noticed, the seemingly scientific nature of Moore's Law hides the vast amounts of money, human labor, and incremental innovation that were needed to keep increasing transistor densities.[24] Now that heat dissipation issues, physical challenges, and the enormous cost of constructing next-generation chip manufacturing facilities have slowed that advance, we can better appreciate how exceptional this era was. The rapid proliferation of ever more powerful computers was not just an inevitable playing out of the universality inherent in a Turing machine.

We wanted to look systematically at the development of different modes of computing over time, not just at their moment of invention. That meant, for example, talking about how the IBM PC and Macintosh platforms evolved from the 1980s onward rather than just describing their initial introductions in 1981 and 1984, respectively. The fixation on invention over adoption and use is seen most clearly in Walter Isaacson's *The Innovators*, which despite its titular claim to be about innovation, the broad process by which new technologies become successful, actually has a narrow focus on invention.[25] Isaacson loses interest in technologies long before they move into the world and become consequential. He follows the development of programmable computers only into the early 1950s, the development of the microprocessor only as far as the Intel 4004 in 1971, the development of what becomes the Internet as far as 1973, and the development of Web search up to 1998. His discussion of personal computing focuses overwhelmingly on the 1970s with brief asides on the IBM PC and the first Macintosh.

Back in 1988, Mahoney complained that the emerging scholarly literature "stops at the point where computing becomes a significant presence in science, technology, and society." Yet somehow popular histories of computing technologies stop the story of each technology well before that point.

Beginnings and Endings

As Pulp's Jarvis Cocker so memorably sang, "I don't know why, but I had to start it somewhere, so it started [slight pause] there."[26] We started in 1946, when the front page of the *New York Times* featured a computer called ENIAC built at the University of Pennsylvania. Paul's original book had also started in the 1940s, bucking the tradition of spending early chapters with Charles Babbage, with office machines, or with mechanical calculators.

Providing that longer history has become increasingly hard now as computers have supplanted so many other technologies. Mahoney's diagram encapsulates his insistence that the parallel stories of the communities of computing should be told separately, looking in each case at their practices before and after the introduction of computer technology. We cannot hope in one book to do justice to the stories of all these domains of use. To follow the path of each community before and after the horizontal bar that Mahoney labelled simply as "computers" would take 14 books, not 14 chapters. Why talk about adding machines but not movie projectors; filing cabinets but not pinball machines; or astrolabes but not telephones? No single book has room to do justice to all those things, and ours is already big enough.

The question "What is the history of computing the history of?" cannot be given a single answer. To fill in the arc of history needed to frame the computer, for example, as a business machine, is to forgo the opportunity to frame it in another way, for example, as a media device. The computer has become too many different things, been put to too many uses, and taken the place of too many other technologies.

When Paul wrote his book, dominant computing platforms were shifting rapidly and figuring out which developments really mattered was very difficult. He acknowledged this in the introduction to the second edition: "There is a finite time between sending a completed manuscript to the typesetter and the delivery of a book or journal article to the reader.

When the subject is computing, Zeno's paradox takes control: enough happens in that brief interval to render what was just written obsolete."[27] Computing was changing so fast that no history could ever get to the present day, however hard its author tried to close the distance.

By the time we began work on the new book the story of *the computer* had begun to seem like something with a recognizable end as well as a clear beginning. Computers were big boxes with lights and switches into which people plugged peripherals such as disk drives, monitors, and keyboards. Over time, the boxes got smaller and their capabilities changed, but *computer* was a fairly robust category from the 1940s all the way to the early 2000s. Even when laptops came along, people recognized them as miniaturized and portable versions of the computer that used to sit on or under their desks.

From the viewpoint of the computer scientist, there are more computers around than ever. Smartphones are computers, as are televisions, tablets, thermostats, GPS systems and dozens of other things in cars, Fitbits, smart watches, cameras, and Bluetooth speakers. But nobody ever pulled out their cellphone to send a text and said, "I need to do some computing" or invited a prospective sexual partner over to "compute and chill" with a streaming movie. The amount of time people spend using devices they conceptualize as computers is dropping. That's why the title of our final narrative chapter, which centers on the smartphone and other mobile devices, is "The Computer Is Everywhere and Nowhere."

Technology has not stopped changing, but the innovations are coming mostly in new applications of the Internet and machine learning technologies. Today's big technology news stories still have a lot of drama: Will Uber ever become profitable? Can crypto enthusiasts find a compelling and legal application for their technology? Is the metaverse anything more than a multibillion-dollar demonstration of the folly of letting founders exercise unfettered control over major public companies? Yet none of those questions need to be resolved for us to sketch the technology and platforms that underlie mobile apps, self-driving cars, and cloud-based social media platforms.

Scale and Scope

We began our project doing something authors often dread: reading Paul's reviews on Amazon. To judge from their comments, readers continued to enjoy the book but were eager above all to see its story continued into the current century. A positive (four-star) 2016 Amazon review that called the book "a delightful read" nevertheless notes that the book's "closing date . . . seems to be a long time ago," so that topics such as the commercial Internet, smartphones, and digital photography are not addressed. The reviewer observes that "'communicating,' 'doing work' and 'having fun' are at the core of today's computer applications, and these central uses do not emerge from Ceruzzi's history of manufacturers and model numbers."[28] The reviewer issued us a number of challenges and the new structure of the book addresses them by bringing the book's coverage of computing fully up to date, and by restructuring the work around the co-evolution of computing platforms with new applications, from scientific computation to video games and personal media consumption.

Our first step was to set up the new structure as a hierarchy of headings and subheadings. We made a first effort on a clean sheet of paper. We then cross-referenced the new outline with Paul's existing text to see which topics were already covered and which would need to be written from scratch. Next, we reassembled chunks of the existing text in the new structure. After this some chapters were already too long while others were largely empty. As we worked there was constant interplay between the "bottom up" development of the text and the "top down" refinement of the new outline.

A History of Modern Computing was written during the early and mid-1990s. Its narrative stopped in 1995 with Netscape's IPO, but other than a few pages on the Internet, it did not have much to say about events after the mid-1980s. Even if we take 1945 as the origin of the first edition (the story nominally started a little later with the first UNIVAC, but much of the first chapter consists of flashbacks) and 1995 as its end point, then its text covered fifty years in 400 pages. Our new history runs through 2020—a span of seventy-seven years. So, by this crudest of measures, we had at least 50 percent more history to cover. The second edition added a bonus chapter consisting of an essay on several important trends of the late 1990s: Linux, Java, the Microsoft antitrust trial, and the commercialization of the Web. This now seems much more dated than anything else in the book, precisely because it dealt with material from which Paul had no historical distance.

The second edition of *A History of Modern Computing* covered forty years in 400 pages, or approximately 150,000 words. Should our replacement cover sixty years in 225,000 words to retain all the existing text and cover new developments to a comparable level of detail? No. The existing text was already close to the limits of what readers and publishers could be expected to tolerate. And it was not just the last twenty years that needed attention. Even in the time periods covered by the earlier book we had a lot of new material to add. Only a few topics dealt with in that book have vanished completely from the new text, but many of them appear in new contexts. As in any area of history, from the vantage point of a different present we see new things in the past. The ARPANET was only briefly mentioned in Paul's original text. In the new one it is the centerpiece of a chapter called "The Computer Becomes a Communications Medium." However, we are also alert to the dangers of Whig history—the same chapter includes online services, bulletin boards, BITNET, USENET, Minitel, and the other forms of online communication that were more representative of computing practice in the 1970s and 1980s.

Excluding bibliography and index, our new book weighs in at around 187,000 words for a net gain of around 25 percent. That's not a small volume, but it does reflect several rounds of relentless cutting that eliminated many tens of thousands of words. Even portions based on material from Paul's earlier book were heavily trimmed and rewritten line-by-line.

New Topics

With respect to the history of computing technology, the book is as close to comprehensive as reasonably practical. There are hundreds of topics our readers would reasonably expect to turn up at some point in the narrative, each of which becomes an obligatory point of passage. Leave out Unix, OS/360, COBOL, or ENIAC, and people will notice. To write the big story of computing is like being handed a bunch of jigsaw pieces and asked to figure out a way to fit them together into a plausible picture.

This time around there were many pieces to add to the jigsaw. As the Amazon review quoted above made clear, today's readers would also expect to encounter digital media devices, video games, and smartphones. The Commodore 64 home computer, which sold more units than any other desktop computer, was not mentioned in *Computer* or *A History of Modern Computing*.

When Paul wrote his book in the mid-1990s, he faced the challenge of writing a synthetic history of a gigantic and amorphous topic with an underdeveloped literature. So did Martin Campbell-Kelly and Bill Aspray when they wrote the first edition of *Computer*. I read it after I had done most of my graduate coursework but before I really delved into my dissertation research. The specific assortment of topics seemed idiosyncratic, but a little later when I spent a few days in the library methodically flipping through the two-decade run of *Annals of the History of Computing* I realized that the contours of that book closely followed that of the literature from which it had been synthesized. Topics such as Charles Babbage, SAGE, SABRE, ERMA, and the IBM System/360 were there not just because they were historically important, but also because they had been written about.

Paul relied less on the secondary literature, though parts of his book stick closely with stories told, for example, in the monumental IBM histories produced by participants.[29] In other places he drew on original research carried out for the book, or on oral history interviews, to cover topics such as the history of minicomputers, UNIX workstations, or SDS timesharing computers, which had not then been adequately treated in the secondary literature. A lot of good history appeared over the next 20 years, letting us put coverage of topics like PLATO, SAGE, the Internet, and ENIAC on a stronger foundation by engaging with and summarizing the conclusions of that research.[30]

We made the decision to cover events as close to the present as possible, and to cover each decade in roughly the same detail. The last five narrative chapters of *A New History of Modern Computing* consist almost entirely of events since 1990. Their topics are less familiar, neglected not only in Paul's original book and its main competitors but in many cases by the scholarly history of computing as a whole.

"The PC Becomes a Minicomputer" looks at the maturing during the 1990s of personal computer hardware and software to replace other computing categories such as graphical workstations. It begins with the limitations of MS-DOS computing before following the development of the PC platform from Windows 3.0 and the 386 through to Windows 2000 running on a Pentium III laptop. The speed of that development was quite remarkable and set a foundation for personal computing in the twenty years since. Intertwined with that are the stories of the RISC challenge to Intel and the evolution of the PC itself from compatibility with specific IBM models to a cluster of open standards maintained by industry groups. This story has not previously been told by professional historians, but it's a crucial pivot point in the history of computer technology.

"The Computer Becomes a Universal Media Device," looks at the digitization of media consumption, with the addition of multimedia capabilities and 3D graphics to conventional personal computers, the rise of on-demand streaming video, and the integration of embedded computers in digital cameras and digital audio players. Media theorists have written about many of these topics, so our contribution here is more to broaden the scope of the core history of computing narrative by integrating some of their findings than to break new ground for media studies.

"The Computer Becomes a Publishing Platform" focuses on the commercialization of the Internet and the rise of the World Wide Web as a universal platform for electronic publishing and online commerce. At a time when both *Internet history* and *Web history* often signify scholarly communities with little connection to the broader history of computing, our challenge was not just to cover the development of the Internet but to do so in a way that makes clear the benefits of talking a longer perspective. One of the benefits of starting in 1946 rather than 1991 is to uncover all the ways in which the Web rested on existing hardware and software infrastructures. These were shaped in turn by the academic context of the original ARPANET. In particular, the lack of integrated security and the lack of a built-in billing mechanism, both unnecessary in the original context, have had profound implications for things such as spam, online fraud, and the devastation of the traditional news industry (and liberal democracy) when online publishing settled on an advertising-supported model.

"The Computer Becomes a Network" explores cloud computing and the remaking of the Web into a universal front end for online applications of all kinds rather than a platform for electronic publishing. Web browsers essentially replaced the terminals of earlier generations as a way of rendering the dynamically generated output of software running in remote data centers. There is a lot being written about the cloud, but we follow historians such as Martin Campbell-Kelly by positioning this in the broader history of interactive computing (going back to timesharing and teletypes).

"The Computer Is Everywhere and Nowhere" looks at the growth of new, mobile networked computing platforms such as smartphones and tablets. These have largely subsumed the brief proliferation of specialized gadgets such as GPS receivers and personal digital assistants. Earlier overview histories have had little or nothing to say about smartphones. Our aim was to contextualize these recent events in the longer history of computing rather than to attempt a definitive history of the smartphone.

These new chapters stretch the chronology of the core narrative of the history of computing. The Apple II, for example, appeared closer in time (1977) to the first UNIVAC (1951) than we are to the debut of Windows 95. Yet earlier histories tended to put the UNIVAC close to the beginning, look in comparative detail at the years between that and the Apple II, and then collapse everything from the 1980s to the present day into the last chapter or two. This is made apparent in Table 2.2, which documents the point in the narrative at which the authors of several overview histories of computing placed some of the most prominent of the jigsaw pieces we mentioned above.

Our decision came with a cost. Whereas earlier chapters each have pages of citations to scholarly works, the last few must have only a smattering of citations and those mostly to newspaper and magazine articles. The new chapters will again be the quickest to go out of date and will need the most attention in any future revision as historians finally get to grips with personal computing, smartphones, and the Web.

Table 2.2. Chapters in which key technologies are introduced in the most recent editions of several overview histories of computing. The technologies are in approximate chronological order.

	Computer (3rd edition)[i]	*A History of Modern Computing (2nd edition)*	*How the World Got into the Computer*	*A New History of Modern Computing*
ENIAC	4 of 12	1 of 10	2 of 6	1 of 14
Univac I	5 of 12	1 of 10	2 of 6	1 of 14
SAGE	7 of 12	2 of 10	4 of 6, footnote	4 of 14
IBM 360	6 of 12	5 of 10	5 of 6	3 of 14
DEC PDP-8	9 of 12	4 of 10	5 of 6	5 of 14
ARPANET	12 of 12	6 of 10	6 of 6	6 of 14
Unix	9 of 12	5 of 10	N/A	5 of 14
Xerox PARC	11 of 12	8 of 10	6 of 6	9 of 14
Apple II	10 of 12	8 of 10	6 of 6	7 of 14
IBM PC	10 of 12	8 of 10	6 of 6	8 of 14
Netscape	12 of 12	9 of 10	N/A	12 of 14
Windows 95	11 of 12	10 of 10	N/A	10 of 14
iPhone	12 of 12	N/A	N/A	14 of 14

[i] Late in the production of this chapter a fourth edition of Computer was published. It keeps the existing chapter structure largely as-is, but adds two new authors and a new section consisting of three chapters titled "Globalization," "The Interactive Web," and "Computing and Governance." These focus primarily on events since the 1990s and have a different character to the rest of the book.

Confronting Progress

Herbert Butterfield long ago warned against the dangers of the "Whig Interpretation of History," a caution against the construction of teleological narratives that has been taken to heart most strongly by historians of science. For example, if we believe that the big story of the history of computing is about explaining how we got to the iPhone, we might be tempted to pick out historical details and arrange them in a path that runs straight to Cupertino, leaving out all the messiness of actual history. Butterfield warned that the danger was most acute for projects of the kind that we are undertaking: "There is a tendency," he warned, "for all history to veer over into whig history," but history becomes "more whig in proportion as it becomes more abridged."[31]

The challenge of avoiding a triumphalist narrative of progress looms particularly large because we are dealing with a technology that has effectively come to define progress in the modern world. Moore's Law is held up as a challenge to other technologies and as an explanation for all kinds of cultural and economic phenomena. Paul has even argued for Moore's Law as a reason to take seriously the unfashionable idea of "technological determinism."[32] Institutions and industries are compared to an idealized version of Silicon Valley and are criticized for making too little progress or being insufficiently disruptive. The dramatic advance of computer and communication technology is used to justify claims that society is being transformed by technology at an unprecedented rate, even though technological change and productivity growth in recent decades were slower than at any time since the Industrial Revolution.

The new structure gives us some scope to address this challenge. Over time our protagonist, *the computer*, really does become spectacularly cheaper, smaller, and faster, even as it develops remarkable new capabilities. We cannot deny that computer technology has followed a unique path, and it would strain even the most determined skeptic to deny that this constitutes a kind of technical progress. Within chapters, however, our narrative has space for paths not taken in computer architecture, business models, and applications. Entire classes of machines, including graphics workstations, minicomputers, and single-processor supercomputers, come and go from the narrative. Technologies introduced for one purpose find unexpected applications, and predictions for the future are usually wrong.

Neither is it at all clear that the social and economic changes facilitated by these new technologies will make the world a better place. The ending of the book underscores the outsized importance of Silicon Valley (and the Seattle area) to our lives and to the global economy. Most of the companies we talked about in our final chapters were headquartered within a short drive of the original Tesla factory. In the book's epilogue, we look briefly at these changes and at role of the COVID pandemic in heightening our dependence on computer systems. As computers have expanded their roles so profoundly, they have come to matter in different ways. For most of the book, IBM was the only major global firm involved in the computer industry. The fate of firms such as Apple and Commodore in the 1980s mattered greatly to their fans, but less so to the world as a whole. In contrast, the last few years have seen a growing "techlash" against firms such as Facebook and the enormous power their leaders exert over our lives. The first five companies to reach market valuations of $1 trillion were Apple, Microsoft, Amazon, Alphabet, and Tesla.

Perhaps the style, indeed the genre, of historical writing must also change as the book progresses. It is almost a twist ending. While critics were of course warning about the perils of computer technology all along, our narrative mostly followed the development of computer technology without editorializing. Was our protagonist secretly evil all along? Some readers will wish we had been focused on oppression and inequality throughout the text, while others may wish we had stuck with bits and bytes to the very end. But the change of focus does highlight the sudden realization of many in the computing field that they are no longer plucky underdogs and that it is past time to take some responsibility for the state of the world.

Imagined Readers

A book like *A New History of Modern Computing* has multiple readerships with their own distinct needs. The interests of these different readers pull in different directions. No single book could fully satisfy all of them, and indeed to inch closer to making one of them happy is likely to annoy others. Trying to accommodate all of them, as we did, runs the risk of creating a book that is a jack-of-all-trades but master of none.

Paul's original book was not a textbook in the usual sense, with a high price tag, exercises, bullet points, and supporting materials. But as history of computing classes are neither plentiful enough nor standardized enough to justify creation of an actual textbook, it was often used as one. Indeed, the textbook market prompted MIT Press to request that Paul produce a new edition of this book. On the positive side, this made it easy for us to publish the new history as a larger and heavily illustrated book rather than a typical academic monograph. It also naturalized the otherwise unusual practice of incorporating chunks of an old book into a new one. On the other hand, thinking of it as a textbook made the press assume it would be best launched as an expensive paperback and limited the potential to promote it more broadly.

One imagined reader was therefore a student assigned the book in class, as a first and perhaps only exposure to the scholarly history of computing. To these readers we owed a clearly told story without too much academic jargon that can be digested, with supplemental readings, in a single semester. We could assume little background knowledge of computing, or indeed of anything else. I am currently teaching an undergraduate seminar with the book, which regularly reminds me of how much undergraduates do not know. To pick just from the topics discussed in chapter seven, most have not heard of Moore's Law, Microsoft BASIC, the Apple II, or the Velvet Underground, though the *Oregon Trail* game seems to have almost universal recognition among Americans regardless of age. It also seemed important to trace the story close enough to the present to reach the technologies this audience could remember using.

Another imagined academic reader is a graduate student looking to do research in the history of computing or a topic that touches on it. Books like this are useful early in the process as a grounding in the bigger story and as a gateway to relevant literatures. They help students identify research topics and shape their own work in response to stories that have already been told. As these more scholarly readers are more likely to follow up citations, they would expect us to identify, summarize, and cite important work. For these readers it was important for us to offer clear narratives within or against which new work could be situated. Our new book has already been cited in one dissertation just to point out that it does not mention a certain topic, continuing an important role served by Paul's book in legitimating dissertation research to fill its gaps.

A related group are scholars, perhaps already well established, working in areas such as science and technology studies or Internet studies, who find themselves in need of a quick grounding in the history of computing for a particular project. *A History of Modern Computing* has 1,946 citations tallied by Google Scholar, as of this writing. The citation pattern shows many citations from computer scientists, media scholars, economists, and business scholars.

The book might also be useful to historians focused on other areas of science or technology looking for a one-stop way to catch up with the history of computing literature, or for historians working in other fields entirely who need to engage in some way with information technology. These audiences might expect explicit historiographical discussion, narrative driven by people rather than machines, and a focus on the analytical categories foregrounded in their own subfields (knowledge, for example, or oppression). The demands of these historical traditions would be hard to square with each other, or with those of other readers.

On the less academic side, it's clear from their Amazon review pages that Paul's original book and our new one both found substantial non-academic audiences. From the reviews, many of those readers appear to be older people who worked with computers and lived though some or all the history described. These readers have the most patience with technical details, and indeed are likely to complain if topics close to their heart are left out or, as is inevitable, simplified. Input from some of them while the book was in draft form helped us fine-tune the book. I have been pleased with their generally enthusiastic response to the book, though of course being too close to the viewpoint of one's historical actors is not always a good thing. The most flattering responses have come from readers who say that we tell a story that is both startling and convincing, making them reassess the meaning of their own life experience.

Some of those veteran readers are computer scientists, though (as I discuss below) this is a history of computing practice, technology, and applications. It is only incidentally a history of computer science, in places where ideas and technologies nurtured within the discipline prove important in the broader world. As neither computer scientists nor computer users typically have much idea of how and when computer science has mattered to the broader world, looking for ways to make those connections seemed important.

How Else Could It Be Done?

Our goal, then, was to produce a comprehensive and scholarly narrative history of the computer that could work as an introduction for general readers, a textbook, or a one-stop solution for scholars looking to get a handle on how their specific topic fits into the broader story. Paul's original book worked in all three roles. It could be that today the time for such a book has passed, and that in future it will be necessary to aim squarely at one of the three audiences.

Public curiosity is today more likely to be satisfied by a trade press publication, such as Isaacson's *The Innovators* or by other books on allied topics such as the history of Silicon Valley or the history of Apple. Paul's original book was greeted by reviews in a wide range of venues including *Linux Journal, The Economist, New Scientist, PC Today, Nature, The Washington Post,* and *Publishers Weekly*, as well as a broad range of academic journals. This reflected unmet public curiosity in the topic. Our new book has sold respectably for an academic title but received no comparable attention. (Of course, in the intervening period the technologies we write about have obliterated the book-reviewing ecosystem.)

Neither does a book of this kind set the scholarly world alight. There are three journals focused on the history of information technology: *IEEE Annals of the History of Computing*, *Information and Culture*, and *Internet Histories*. None of them have reviewed it, though it did receive strongly positive evaluations for technical audiences in *SIAM News* and *Computing Reviews*. Two years after publication it has received only two reviews in historical journals. I was pleasantly surprised that *Isis*, a pillar of the history of science establishment, reviewed it. Victor Petrov gave it a very strong endorsement: "the standard by which other general histories are measured."[33] My own community, the Society for the History of Technology, was less impressed. Writing in its journal *Technology & Culture*, Martina Hessler complained that when it came to "conceptual issues" we "aren't tackling any at all."

A similar sentiment came up in one of the few tweets (technically a subtweet) that greeted the arrival of the book, posted by historian of science Patrick McCray. According to McCray, an unnamed "newly revised master narrative history of computing book . . . reverts in many places to—dare I say—internalist history of tech." The word "revert" is a revealing one, suggesting a faith in the unidirectional workings of progress that historians would be unlikely to sanction in any area except scholarship in their own field. Hessler similarly identified the question of "how much technical knowledge does the history of technology need" as one of three particularly pressing conceptual challenges we had failed to address. I had thought this was something we tackled showing rather than telling, but to clarify: I believe that the unique trajectory followed by computer technology versus other complex mechanisms challenges historians to get to grips with what is happening inside the box.

The ongoing shift of gravity among people gainfully employed in writing the history of computing towards media and communication programs will likely accelerate the shift to other modes of historical writing. While some media traditions are more open than present-day history of science and technology to engagement with materiality and affordances, writing in these areas tends towards narratives that jump around in time and connect topics in ways that differ from the causal explanations favored by traditional historical approaches. The shift seems likely to emphasize cultural histories, particularly those in which the underlying logic is exploring technology to explain intellectual trends, artistic movements, or cultural theories.

Even professional historians looking to tell big stories are more likely to do it by choosing a set of tightly focused chapters that, taken together, illustrate some kind of historical arc rather than by following our approach of attempting a comprehensive history that goes through dozens of required passage points and connects a lot of dots. One popular recent mode has defined projects along the lines of "A History of [Topic] in [Small integer] [Object Class]." Jacob Gaboury's recent *Image Objects* seems an exemplary application of this kind of approach to the history of computing.[34] Books that give snapshots over time in the evolution of programming practice, computing labor, user experience, and so on might offer detailed case studies from different decades rather than connecting all the intervening dots. Yet to select and contextualize their examples the authors of such narratives would need deep knowledge, so I hope our own book will be a useful resource in their creation.

The Missing Chapters

For the rest of this chapter, I focus mostly on what we did not do in the book and why we did not do it. This might seem eccentric, but from my perspective any book is an engineering project that involves a set of inescapable trade-offs. This is clearest in a big and broad book like ours. Anyone writing a review is likely to have two impulses: to complain about the length and to confront us with a list of important topics we left out. Something I took from my engineering training was the importance of including in the specification for a project an explicit statement of the goals it will not attempt to meet. In this case I think it is instructive to give a sense of how we approached those trade-offs. What would we have included if we had space, if we were better qualified or had a secondary literature to draw on, or if we had a little more historical distance? When we took one perspective over another, was it because we were blinkered or because we could not figure out how to integrate something important into the structure of the book?

We chose the stories told in the book to give us the materials we needed to make sense of the development of computer technology. Inevitably, we left out many of the communities and contexts of computer use. Hessler complained that the book shirked the "essential" task of including the history of robotics and AI, through which "computers have started to perform inherently physical activities . . . since the 1980s."[35] If we could squeeze in another chapter and had the secondary sources to base it on, a chapter titled "The Computer Becomes an Industrial Control Technology" would indeed cover an important application area. From the viewpoint of the development of the core features of *the computer*, however, a chapter on industrial automation and robotics would have been duplicative since chapter 4, "The Computer Becomes a Real-Time Control System," explores the processes by which computer technology was remade from the 1950s onward to control air defense systems, missiles, planes, and space probes. These are the same affordances required for industrial control.

Looking at the burgeoning literature on the history of computer security, I am beginning to imagine a chapter called "The Computer Becomes Mostly Secure and Highly Reliable." This would combine the history of computer security, an important field we barely mention, with the history of fault-tolerant systems, redundancy, transaction processing, and efforts to produce reliable, bug-free systems. This chapter would give another opportunity to show ways in which work by computer scientists on techniques such as model checking and the creation of software development methodologies has had a real-world impact. It would also look at the development of the hacker as a cultural role. In any revised version of the book, this chapter would fit well because of our focus on the creation of core capabilities that become part of the standard stacks of computer technology.

One could go on forever imagining new chapters, but crypto enthusiasts would no doubt bemoan the absence of a chapter called "The Computer Becomes Money." We could certainly tell the story of computerized banking infrastructures, from ERMA through ATMs, credit card networks, the push for a "cashless society" (which I have written about elsewhere), electronic payment systems, and so on.[36] But despite the huge wave of crypto hype, it is not yet clear that cryptocurrency will ever find real applications beyond crime and fraud. Bitcoin itself seems entirely and irredeemably useless; blockchain technologies more generally may one day find applications but so far have not. We just do not yet have enough historical distance on this story.

The contributions of the artificial intelligence community to computing practice are real but, until recently at least, consisted primarily of developments in computing infrastructure, algorithms, and so on rather than through the creation of thinking machines. One prominent AI researcher described the story of his field as "a history of failed ideas."[37] If we did add a chapter on twentieth-century AI, the most obvious title would be "The Computer Fails To Become a Thinking Machine." The recent proliferation of AI-branded automatic learning systems is already reshaping the capabilities of computer technology and any future revision of our book will have to deal with this technology, though at present the technologies and applications are changing so rapidly that no history could hope to properly document them. Artificial intelligence deserves its own overview history, and in fact I am currently writing a concise history of the topic, which I see more as an intellectual brand than a historically stable set of technologies.

This Is NOT the History of Computer Science

AI would have an important part in any broad history of computer science, but ours is a history of computer technology and use. Let me explore that distinction. What Paul and I tried to produce is a history of computer technology and practice that is engaged with understanding how computer systems work, meaning that we talk about computer scientists only when their work leads directly to changes in computing practice. Donald Knuth famously complained that the shift of "historians of computer science" away from technical material had reduced him to tears.[38] My response then was that very little work on the history of computing actually fell into that category of "history of computer science," because neither the history of the science community nor the computer science community could offer a plausible career path to somebody who identified thusly.[39] On the copy of the new history I sent to Knuth, I wrote, "This is not the history of computer science, but I hope you like it anyway."

There have been no broad histories of computer science or scientific computing. Princeton historian Michael S. Mahoney spent many years contemplating the writing of a history of theoretical computer science but produced only a set of densely suggestive essays.[40] Instead the history of computer science has been told only in fragments where it touches other stories such as the development of biomedical science, the Internet, the Association for Computing Machinery, programming work, bioinformatics, computer graphics, missile defense, computational support for natural science, logical inscriptions, automated proof, artificial intelligence, or timesharing systems.[41] Though the last decade has seen a move towards technically engaged history, most of this work has focused on the two decades immediately prior to the firm establishment of computer science as a discipline in the mid-1960s.[42] With the exception of several studies of funding agencies, little has been written about the era in which computer science did exist.[43] More scholarly attention has been paid to the meaning of Soviet efforts to conceptualize computer science as part of cybernetics than to the core story of the growth of computer science in the United States.[44]

A history of computer science that took an approach comparable to the one we took in the new history would look at the growth of the academic discipline, leading university programs, key areas of research, changes in career patterns and funding, relations between industry and academia, and above all at the rise and fall of subject areas including artificial intelligence, computer graphics, computer architecture, and numerical analysis within the loose federation of research communities that calls itself computer science. No such history exists. The closest thing we have, useful but schematic, comes from a computer scientist rather than a professional historian.[45]

Why are there no overview histories of computer science? Let me point to the *science* in computer science. In the United States, the histories of science, technology, and business are separate communities with their own journals and conferences (albeit with some overlap in membership). Of the three, the history of science is the highest-status field, with the clearest potential path to academic employment, the strongest institutional support, and the clearest connection to elite institutions. In contrast, authors who have produced broad scholarly histories of computing technology and industries, such as William Aspray, Martin Campbell-Kelly, JoAnne Yates, James Cortada, Paul Ceruzzi, and Jeffrey Yost, do not hold tenured faculty positions in elite history or history of science programs. Like me, they identified more with the history of technology and business than with the history of science. While historians of science often interpret their mandate to apply broadly across *technoscience*, authorizing them to write about technology and business, the reverse is rarely true. Historians of technology are likely to leave the core narratives of scientific disciplines to others.[46]

Even a historian of twentieth-century science with no interest in computing itself might be drawn to questions posed by the federated nature of computer science, its assemblage from a scattered array of disciplinary traditions, its relationship to computer technology and practice, and its rise to become one of the most popular fields of undergraduate study. I am not close enough to the history of science community to explain with confidence why its members have not told these stories. My impression is that the history of science long ago shifted its emphasis away from ploddingly comprehensive histories of disciplines, professional societies, or academic departments and towards more intellectually exciting studies of culture and epistemology. Historians of science are today unlikely to identify themselves as historians of particular scientific disciplines. Those transitions mirror a more general shift within the historical profession during the 1980s towards richly detailed, narrowly drawn studies that uncovered the agency of overlooked actors and argued against simplistic generalizations. For topics such as the history of evolution, quantum mechanics, or the scientific revolution, where well-trodden narratives were ripe for reinterpretation, the new approaches worked well. Yet because scholarly incentives no longer rewarded the production of comprehensive narrative histories of scientific disciplines there was never a grand narrative of the history of computer science, or any of its constituent communities, for ambitious young scholars to audaciously undermine on their way to tenured positions at great universities.[47]

I have also heard it said that the history of science is increasingly focused on contemporary history and on making explicit connections with present-day controversies. To the extent this is true, we might expect it to follow science studies in focusing disproportionately on hot topics, such as AI today or nanotechnology a decade ago, rather than on longer-term and bigger narratives—particularly given the difficulty in producing big narratives without a large mass of more specialized studies to synthesize.

Globalizing the History of Computing

The United States will, naturally, loom large in any overview history of computing. It installed far more computers in the 1950s than the rest of the world combined, and in later decades US firms such as IBM, DEC, Microsoft, Apple, and Google exerted an outsize impact in shaping the hardware and software platforms used worldwide. Any history concerned primarily with explaining where widely used technologies come from must grapple with this. Paul's existing text included important case studies of US-based user organizations such as NASA and the IRS, which we recycled for the new book. Some of these could have been replaced with non-US examples, as David Gugerli did in his recent book, though Gugerli's overall narrative appeared to be a universal one illustrated with US and German examples.[48]

Back in 2011 I was a little snarky about the conflation of global and American narratives in the history of computing:

> American scholars tend to view the history of information technology as a fundamentally American narrative. The US is the only country with a sufficiently central role in most areas of information technology that a coherent (if skewed) overall history of computing can, and often has, been written with minimal reference to the world outside its borders. There are a few Englishmen who force their way into the narrative: Charles Babbage, Alan Turing, the teams behind the Manchester Mark 1 and the EDSAC, the LEO group that pioneered administrative computing, and Tim Berners-Lee. Konrad Zuse flies the flag for Germany. Jacquard's automatic loom earns France a paragraph somewhere in an early chapter. But these can be dismissed in passing as brilliant figures whose seminal technical accomplishments quickly passed into the hands of Americans for practical exploitation. The history of personal computing, in particular, is told by Americans entirely without reference to the existence of a world beyond the oceans, or in most cases beyond the Valley.
>
> Historians based in the United States have therefore been much less likely than those working in other countries to attempt to isolate peculiarities of their own national experience or relate the development of information technology industries to the influence of government policy. . . . Scholars in other countries have tended to focus on national narratives, generally framed with perceived differences between local developments and those in America.[49]

Given the opportunity to create a new kind of history of computing narrative, what did we do? In all honesty, the US story remains our default, perpetuating an awkward hybrid of national and global history. We go overseas more often than previous overview histories, but we still make the journey only when necessary to illustrate the origins of some important new application of computer technology. For example, we visit the UK for LEO, the early computers at Cambridge and Manchester, the Sinclair and Acorn machines of the 1980s, the BBC's computer literacy project, ARM, and so on. Continental Europe shows up mostly for Cyclades, Minitel, and the World Wide Web; while the USSR, which contributed far more to computer science than computer technology, exists mostly as an off-stage bogeyman to justify US government spending. East Asia shows up later in the narrative, as the new home of microelectronic manufacturing, a source of challenges for text input and display and, eventually, as the world's most important e-commerce marketplace. Africa is mentioned only very briefly, as a pioneer of cellphone-based financial transactions. India appears only as a growing center for the outsourcing of software development. As the secondary literature matures, I hope that authors will document other unique forms of computer use in non-Western contexts.

Any future iteration of our book might most obviously benefit from the inclusion of a chapter called "The Computer Becomes Globally Inclusive" or perhaps "The Computer Becomes Non-Western." Early machines were hardcoded to Latin character sets, with minor national differences. Representing Greek, Cyrillic, Hebrew, and in particular Chinese and Japanese characters was a major challenge, central to the accessibility of computers and smartphones to most of the world's population (or, viewed from another perspective, the survival of those modes of writing in an Internet world).[50] Its resolution was tied to fundamental changes in user interface technology, graphics, hardware capabilities, and software infrastructures. Hints of that story show up occasionally in our book, but it would fit well with our focus on the interplay between users and core capabilities and should be told properly.

Centering our book on the emergence of core computing technologies and platforms compounded the issue. Several countries had significant national mainframe and minicomputer industries that go unremarked on in our book. Early computers were hand built in the USSR, Israel, and Australia, while Western-built machines were exported to Africa, Asia, and the Eastern Bloc from the 1960s onward. Yet the existing literature does not demonstrate that these machines introduced important new architectural features to computing, or that, for example,

Western and South American applications of mainframes and minicomputers created new capabilities. For example, two of the most acclaimed books on non-Western computing are framed around systems that never came into being: Eden Medina looked at Chile's doomed dream of building a management information system to facilitate cybernetic socialism, while Ben Peters asked why the USSR did not build an Internet. Neither absence offered an obvious hook for integration into our narrative.[51]

THE MISSING PERSPECTIVES

As well as missing chapters, we might think productively about missing perspectives. An obvious omission, and one I hope to remedy in any future edition, is discussion of the environmental costs of computing. Historians of computing have traditionally had little to say on computer manufacturing, or, following the material components of computer systems up the supply chain, about the creation of chips and other components or the mining, refining, and synthesis of the materials that go into those components. But work on the environmental cost of the materials that go into iPhones and batteries is starting to pile up, as is analysis of the environmental degradation of Silicon Valley back when silicon chips were made there. The materiality and geopolitics of semiconductor production have recently returned to the headlines, making me uncomfortably aware that we had little to say about those topics after the 1970s.

Aside from an apocalyptic epilogue set amidst the techlash and the pandemic, we generally avoided overt editorializing about the social consequences of the adoption of computer technology. Astute readers will notice the prevalence of military and aerospace applications in the early chapters, driving everything from the invention of the modern computer to the SAGE project, Minuteman missiles, and the earliest markets of silicon chips. Yet we did not, for example, include discussion of the role of statistics and big data in racist social and criminal justice policies or the political agendas of US computer companies. Neither do we get far into discussing the role of IT in economic transformations that have increased inequality. As these topics are explored in the historical literature, I hope to find ways to integrate them more fully into any future revision of the book.

Gender, race, and sexuality are becoming increasingly central to the history of computing. We acknowledged them more fully than previous overview histories, but that is a low bar. For example, we discussed gendered labor in data processing, around the ENIAC project, and in the computerization of office work. We included Grindr as an example of location-based smartphone applications built on cloud infrastructures. Race is even less prominent, though the social historian in me did feel obliged to note racial disparities in the adoption of personal computers and the Internet in American households. Given our organizing principles such asides inevitably work more as garnishes than structural elements.

This absence is a little wrenching for me. My most frequently taught class is an undergraduate survey on the history of race and health in the US. I regularly teach another on the history of capitalism. Race, class, and gender are structural to the narratives told in both classes, just as they have been to mainstream scholarly understandings of US history for many decades. And yet, here I am as the lead author of a big heavy book which is by no stretch of the imagination structured around class or race and is only intermittently concerned with gender. There is a trade-off though. While racial, ethnic, and caste-based discrimination is found around the world and throughout history, the specific history and construction of race in the United States is distinct even from that of other rich countries. Histories of gender and sexuality have likewise played out differently around the world. Aligning the structure of a book more closely with the American history of those categories would push against the impulse to globalize the history of computing. I do not see how both could be accomplished by the same book.

Even the communities of computing that we do explore may appear abstractly, as the sources of new requirements rather than fully depicted social spaces. We do have a chapter "The Computer Becomes Office Equipment," but Hessler faulted it for failing to offer a "well-rounded description of the transformation in office work" by looking at different appropriations of the computers by office workers, resistance by workers, and so on. To that I must plead guilty as charged. I had dealt with all those issues in an earlier paper but found it structurally impractical to engage with them in a synthetic book of this kind.[52]

Neither did we get to grips with the overwhelming whiteness of our cast of characters. There is no obvious template for dealing with this. We could comment frequently on the sexism, privilege, racism, heteronormativity, masculinity, and whiteness of the communities we are describing. Joy Rankin does this in her book on the history of educational timesharing usage, and the effect is bracing.[53] These asides situate her characters in the larger narratives of US history and undercut heroic narratives around computer technology. Yet emulating this might be hard to integrate with our structural focus on the development of computer technology. Rankin's asides function more as critical judgements on the dominant cultures of postwar America, which permeated computer-using communities, than as insights into why computer technology was shaped in one direction over another. Fully integrating the stories of technological and cultural development is a daunting challenge, and one that probably needs to be met by a range of specialized studies before it can realistically be attempted in an overview history.

Embodiment is an interesting category, and one that makes me realize how little concerned narratives around the history of computing have usually been with the biological materiality of human existence. This perspective has been central to the work of Elizabeth Petrick on the history of accessibility in computer interfaces, and I was struck by the response of my students to Laine Nooney's recent essay on ills wreaked on human bodies by the demands of computer technologies.[54] Embodiment has an obvious connection to the history of gender and sexuality. Yet we all have bodies, which differ of course but as variations on a common theme, so perhaps working them more systematically into the history of computing might offset both the persistent hostility to the realities of human existence found in geek culture and the difficulties of writing history across lines of class, race, gender, nationality, and sexual orientation.

Centering the Computer—Is That Bad?

The ongoing protagonist that ties together the many chapters of our book is *the computer* itself. If, as discussed above, *the computer* now seems to belong more to the past than the future, how stable is *computing*, the putative object of study for the *history of computing*? And should the history of computing community follow Eden Medina's influential suggestion that we "decenter the computer"?[55]

It is of course true that the history of computing should be far more than the study of computer hardware, and that examinations of the historical use of computer technology can and should be structured around people, cultures, institutions, practices, and the like. The *history of computing* identity was chosen to be broader than the *history of computers*, which already sounded dry and trivial in the 1970s. Most histories of computing explore specific, focused topics with the aim of better understanding aspects of human life rather than improving our understanding of computer technology.

As a verb, *computing* mandates attention to the use of technology, but it has no resonance with public or humanities audiences that might have some interest in history but do not think of what they do with their phones, games consoles, or cars as *computing*. In truth, the rhetoric of *computing* never resonated far beyond the professionalization discourse of computer science. For organizations such as the ACM and the long-defunct AFIPS computing worked as an inclusive description of computer use that spanned many otherwise distinct human domains. The involvement of these groups in the history of computing has ebbed, and even computer scientists are highly specialized these days and thus less likely to offer pronouncements about computing as a whole.

The history of computing never institutionalized very successfully. The few institutions founded to pursue it include the journal *IEEE Annals of the History of Computing*, SIGCIS, the Computer Museum (replaced by the Computer History Museum), the Charles Babbage Foundation (now the IT History Society), and the Charles Babbage Institute. In one way or another, each has broadened its scope beyond history, struggled to stay in business, or both. Given that many of the institutions most associated with the history of computing are decentering not just computers themselves but also computing and history, I suspect that *historian of computing* is not a scholarly identity that is ever going to loom particularly large.

Most historians of computing, broadly conceived, who have found gainful employment in recent years have done so in media or communication programs where foregrounding media theory or currently fashionable objects of study has far more professional utility than aligning oneself with the study of computing. The same is true for those who have won recognition from inside our professional community, for example, the recent winners of the book (Computer History Museum Prize) and article (Mahoney Prize) awards from SIGCIS. There must already be more people in the world who identify themselves primarily as historians of video games, algorithmic inequality, platforms, Silicon Valley, artificial intelligence, and big data than as historians of computing.

With respect to *A New History of Modern Computing*, I hope you have been convinced that the exceptional trajectory of computer technology from 1945 to 2020 justifies a book centered on the computer and the ways in which its affordances developed over time. But should the history of computing as a whole center the computer? In a way, it is the willingness to do so that might cause someone to identify as a historian of computing. The thing that defines the scope of the history of computing is, unavoidably, the movement of *the computer* into new social spaces over time.

Historians of all kinds who work on our current era will have to grapple with the role of information technology in the stories they tell. Given that computer technology today mediates so much human activity, the potential scope of the history of computing edges ever closer to being the history of everything. But not even I believe that a significant fraction of them will ever choose to identify as *historians of computing*. For example, Medina's own MIT home page identifies her as a scholar of the history of science and technology in Latin America. It includes the words "history of computing" only to describe the Computer History Museum Prize.[56]

The small group of people who do identify as historians of computing is, by definition, a community constituted by our shared connection to the computer even if we often decenter it in our individual studies. We have the collective responsibility of interpreting computer technology in ways that can inform the work of the much larger population of historians who need to understand its role in their own narratives. As for those broader historical communities, they cannot decenter something that they would never have contemplated centering in the first place.

Notes

1. Michael S. Mahoney, "The Roots of Software Engineering," *CWI Quarterly* 3, no. 4 (1990):325–334, quote p. 326

2. Karl E Weick, *Sensemaking in Organizations* (Thousand Oaks, CA: Sage, 1995).

3. Most notably in Michael R. Williams, *A History of Computing Technology* (Englewood Cliffs, NJ: Prentice-Hall, 1985).

4. N. Metropolis, J. Howlett, and Gian-Carlo Rota, eds., *A History of Computing in the Twentieth Century: A Collection of Papers* (New York: Academic Press, 1980).

5. Herman H. Goldstine, *The Computer from Pascal to von Neumann* (Princeton, NJ: Princeton University Press, 1972).

6. Edmund C. Berkeley, *Giant Brains or Machines That Think* (New York: John Wiley & Sons, 1949), p. 4.

7. Berkeley, pp. 10–13.

8. Berkeley, p. 5.

9. Martin Campbell-Kelly and William Aspray, *Computer: A History of the Information Machine* (New York, NY: Basic Books, 1996).

10. Paul E. Ceruzzi, *A History of Modern Computing* (Cambridge, MA: MIT Press, 1998).

11. Thomas Haigh, "The History of Information Technology," *Annual Review of Information Science and Technology* 45 (2011):431–487, quotation 339–340.

12. I have discussed some the limitations of Isaacson's view of history in Thomas Haigh and Mark Priestley, "Innovators Assemble: Ada Lovelace, Walter Isaacson, and the Superheroines of Computing," *Communications of the ACM* 58, no. 9 (Sep 2015):20–27.

13. Jeffrey R. Yost, *The Computer Industry* (Westport, CT: Greenwood Press, 2005). Jeffrey R. Yost, *Making IT Work: A History of the Computer Services Industry* (Cambridge, MA: MIT Press, 2017). Martin Campbell-Kelly, *From Airline Reservations to Sonic the Hedgehog: A History of the Software Industry* (Cambridge, MA: MIT Press, 2003).

14. Paul E. Ceruzzi, *Computing: A Concise History* (Cambridge, MA: MIT Press, 2012). Eric G. Swedin and David L. Ferro, *Computers: The Life Story of a Technology* (Westport, CT: Greenwood Press, 2005). Doron Swade, *The History of Computing: A Very Short Introduction* (New York: Oxford University Press, 2022) is an example of an overview history aimed at high school students.

15. David Gugerli, *How the World Got into the Computer: The Emergence of Digital Reality* (Zurich, Switzerland: Chronos Verlag, 2022).

16. If you want to know more on my opinion of Dyson's book, you will find it at https://sinews.siam.org/Details-Page/an-unconventional-history-of-the-early-ias-computer.

17. For example, the imagined users of Doug Engelbart's influential but never commercialized NLS system, as explored in Thierry Bardini, *Bootstrapping: Douglas Engelbart, Coevolution, and the Origins of Personal Computing* (Stanford, CA: Stanford University Press, 2000).

18. Thomas Haigh, "The Chromium-Plated Tabulator: Institutionalizing an Electronic Revolution, 1954–1958," *IEEE Annals of the History of Computing* 23, no. 4 (October–December 2001):75–104, page 95.

19. Haigh, "The History of Information Technology," quote p. 441.

20. Michael S. Mahoney, "The Histories of Computing(s)," *Interdisciplinary Science Review* 30, no. 2 (2005):119–135.

21. Ibid., here 127.

22. JoAnne Yates, *Structuring the Information Age* (Baltimore: Johns Hopkins University Press, 2005).

23. That is (almost) the question that Tom Misa challenged the history of computing community to explain. Thomas J. Misa, "Understanding 'How Computing Has Changed the World,'" *IEEE Annals of the History of Computing* 29, no. 4 (Oct–Dec 2007): 52–63.

24. Cyrus C. M. Mody, *The Long Arm of Moore's Law* (Cambridge, MA: MIT Press, 2017). Ethan R. Mollick, "Establishing Moore's Law," *IEEE Annals of the History of Computing* 28, no. 3 (July–Sept 2006):62–75.

25. Walter Isaacson, *The Innovators: How a Group of Hackers, Geniuses, and Geeks Created the Digital Revolution* (New York: Simon and Schuster, 2014).

26. Pulp, "Common People," 1995.

27. Paul E. Ceruzzi, *A History of Modern Computing*, second edition (Cambridge, MA: MIT Press, 2003), ix.

28. Review by Paul F. Ross, May 7, 2016. https://www.amazon.com/History-Modern-Computing/dp/0262532034/ref=mt_paperback?_encoding=UTF8&me=

29. Charles J. Bashe et al., *IBM's Early Computers* (Cambridge, MA: MIT Press, 1986) and Emerson W. Pugh, Lyle R. Johnson, and John H. Palmer, *IBM's 360 and Early 370 Systems* (Cambridge, MA: MIT Press, 1991).

30. Among the texts, many of them now classics, we leaned on heavily in improving coverage were Julien Mailland and Kevin Driscoll, *Minitel: Welcome to the Internet* (Cambridge, MA: MIT Press, 2017), Jonathan Sterne, *MP3: The Meaning of a Format* (Durham, NC: Duke University Press, 2012), Joy Lisi Rankin, *A People's History of Computing in the United States* (Cambridge, MA: Harvard University Press, 2018), Matthew G. Kirschenbaum, *Track Changes: A Literary History of Word Processing* (Cambridge, MA: Harvard University Press, 2016), Janet Abbate, *Inventing the Internet* (Cambridge, MA: MIT Press, 1999), Paul N. Edwards, *The Closed World: Computers and the Politics of Discourse in Cold War America* (Cambridge, MA: MIT Press, 1996), Yost, *Making IT Work: A History of the Computer Services Industry*, and Nick Montfort and Ian Bogost, *Racing the Beam: The Atari Video Computer System* (Cambridge, MA: MIT Press, 2009).

31. Quoted in William Cronon, "Two Cheers for the Whig Interpretation of History," *Perspectives on History* (September 2012), https://www.historians.org/publications-and-directories/perspectives-on-history/september-2012/two-cheers-for-the-whig-interpretation-of-history.

32. Paul E. Ceruzzi, "Moore's Law and Technological Determinism: Reflections on the History of Technology," *Technology and Culture* 46, no. 3 (2005):584–593.

33. Victor Petrov, "Review of A New History of Modern Computing," *Isis* 114, no. 1 (2023):219–220.

34. Jacob Gaboury, *Image Objects: An Archaeology of Computer Graphics* (Cambridge, MA: MIT Press, 2021).

35. Martina Hessler, "Review of A New History of Modern Computing, by Thomas Haigh and Paul E. Ceruzzi," *Technology and Culture* 63, no. 2 (April 2022):580–581.

36. Bernardo Batiz-Lazo, Thomas Haigh, and David Steans, "How the Future Shaped the Past: The Case of the Cashless Society," *Enterprise and Society* 36, no. 1 (March 2014):4–17.

37. Michael Wooldridge, *A Brief History of Artificial Intelligence: What It Is, Where We Are, and Where We Are Going* (New York: Flatiron Books, 2021).

38. Donald E Knuth and Leonard Shustek, "Let's Not Dumb Down the History of Computer Science," *Communications of the ACM* 64, no. 2 (February 2021):33–35. For Knuth's full talk, see https://www.youtube.com/watch?v=gAXdDEQveKw.

39. Thomas Haigh, "The Tears of Donald Knuth," *Communications of the ACM* 58, no. 1 (Jan 2015):40–44.

40. Michael S. Mahoney and Thomas Haigh (eds.), *Histories of Computing* (Cambridge, MA: Harvard University Press, 2011).

41. Joseph November, *Biomedical Computing: Digitizing Life in the United States* (Baltimore, MD: Johns Hopkins University Press, 2012), Abbate, *Inventing the Internet*, Nathan Ensmenger, *The Computer Boys Take Over: Computers, Programmers, and the Politics of Technical Expertise* (Cambridge, MA: MIT Press, 2010), Hallam Stevens, *Life Out of Sequence* (Chicago, IL: University of Chicago Press, 2013), Gaboury, *Image Objects: An Archaeology of Computer Graphics*, Rebecca Slayton, *Arguments that Count: Physics, Computing, and Missile Defense, 1949–2012* (Cambridge, MA: MIT Press, 2013), Atsushi Akera, *Calculating a Natural World: Scientists, Engineers, and Computers During the Rise of U.S. Cold War Research* (Cambridge, MA: MIT Press, 2007), Thomas J. Misa (ed.), *Communities of Computing: Computer Science and Society in the ACM* (San Rafael, CA: Morgan & Claypool [ACM Books], 2017), Donald MacKenzie, *Mechanizing Proof* (Cambridge, MA: MIT Press, 2001). AI has been a boom area recently, with completed dissertations from Stephanie Dick, Jonnie Penn, and Xiaochang Li awaiting publication as books and many more dissertations currently underway. Its literature seems destined to substantially outweigh work framed as the history of computer science, even though AI has intellectually and institutionally existed primarily as a subfield of computer science.

42. For example: Thomas Haigh, Mark Priestley, and Crispin Rope, *ENIAC In Action: Making and Remaking the Modern Computer* (Cambridge, MA: MIT Press, 2016), Liesbeth De Mol, Martin Carlé, and Maarten Bullynck, "Haskell before Haskell: An Alternative Lesson in Practical Logics of the ENIAC," *Journal of Logic and Computation* 25, no. 4 (August 2015):1011–1046, Mark Priestley, *Routines of Substitution: John von Neumann's Work on Software Development, 1945–1948* (Cham, Switzerland: Springer, 2018), David Nofre, Mark Priestley, and Gerard Alberts, "When Technology Became Language: The Origins of the Linguistic Conception of Computer Programming, 1950–1960," *Technology and Culture* 55, no. 1 (January 2014):40–75, David E. Dunning, "The Work of Writing Programs: Logic and Inscriptive Practice in the History of Computing," *IEEE Annals of the History of Computing* 43 (Oct 2021):27–42. Not to mention work focused on earlier forms of computation, such as Matthew L. Jones, *Reckoning with*

Matter: Calculating Machines, Innovation, and Thinking about Thinking from Pascal to Babbage (Chicago, IL: University of Chicago Press, 2016), and many books on Alan Turing and John von Neumann, whose work was foundational to computer science but were not themselves ever part of the discipline.

43. Funding-centric histories include Alex Roland and Philip Shiman, *Strategic Computing: DARPA and the Quest for Machine Intelligence* (Cambridge, MA: MIT Press, 2002), Arthur L. Norberg and Judy E. O'Neill, *Transforming Computer Technology: Information Processing for the Pentagon, 1962–1986* (Baltimore: Johns Hopkins University Press, 1996), and Peter A. Freeman, W. Richards Adrion, and William Aspray, *Computing and the National Science Foundation, 1950–2016* (New York: Association for Computing Machinery, 2019). Sam Franz, a graduate student at the University of Pennsylvania, is looking squarely at the disciplinary evolution of computer science but as I write has yet to publish.

44. On Soviet cybernetics, see Benjamin Peters, *How Not to Network a Nation: The Uneasy History of the Soviet Internet* (Cambridge, MA: MIT Press, 2016), Slava Gerovitch, *From Newspeak to Cyberspeak: A History of Soviet Cybernetics* (Cambridge, MA: MIT Press, 2002), Ksenia Tatarchenko, "'House With the Window to the West': The Akademgorodok Computer Center (1958–1993)," (PhD dissertation, Princeton University, 2013).

45. Matti Tedre, *The Science of Computing: Shaping a Discipline* (New York, NY: CRC Press, 2015).The history of computer science literature is reviewed in Stephanie Dick, "Computer Science," in *A Companion to the History of American Science*, ed. Georgia M. Montgomery and Mark A Largent (Malden, MA: Wiley-Blackwell, 2016):55–68.

46. But if historians of science also write about technology, what grounds do I have for saying that they do not by this very act become historians of technology? In a sense they do, but for the purpose of demarcating the two social communities I am crudely conceptualizing historians of science as people who attend the History of Science Society and publish in journals with "science" in the name and historians of technology as people who attend the Society for the History of Technology meetings and publish in journals with "technology" in the name. Some scholars are both, of course.

47. Something similar stymied Mahoney's efforts to write the history of theoretical computer science. In introducing his papers, I noted his debt to Thomas Kuhn, his dissertation advisor, and observed that Kuhn's approach was hard to apply to the history of computer science because of the lack of existing written historical narratives. Mahoney "wrote as if he had in mind an audience for whom the broad outline of early computer science, its theorists, ideas, and milestones, were already familiar. Much is alluded to rather than explained. Instead Mahoney worked to challenge our presumed assumptions about how all these things fit together, disrupting and rearranging a received narrative that has, alas, never been written in the first place." Thomas Haigh, "Unexpected Connections, Powerful Precedents, and Big Questions: The Work of Michael Sean Mahoney on the History of Computing," in *Histories of Computing*, ed. Thomas Haigh (Cambridge, MA: Harvard University Press, 2011):1–18.

48. Gugerli, *How the World Got into the Computer*.

49. Haigh, "The History of Information Technology."

50. Thomas S. Mullaney, "QWERTY in China: Chinese Computing and the Radical Alphabet," *Technology and Culture* 59, no. 4 (October 2018):S34–S65.

51. Peters, *How Not to Network a Nation: The Uneasy History of the Soviet Internet*; Eden Medina, *Cybernetic Revolutionaries Technology and Politics in Allende's Chile* (Cambridge, MA: MIT Press, 2011).

52. Thomas Haigh, "Remembering the Office of the Future: The Origins of Word Processing and Office Automation," *IEEE Annals of the History of Computing* 28, no. 4 (October–December 2006):6–31.

53. Rankin, *A People's History of Computing in the United States*.

54. Laine Nooney, "'Have Any Remedies for Tired Eyes?' Computer Pain as Computer History," in *Abstractions and Embodiments: New Histories of Computing and Society*, ed. Janet Abbate and Stephanie Dick (Baltimore, MD: Johns Hopkins University Press, 2022):416–433.

55. Eden Medina, "Forensic Identification in the Aftermath of Human Rights Crimes in Chile: A Decentered Computer History," *Technology and Culture* 59, no. 4S (2018):S100–S133. The concept was discussed in Janet Abbate and Stephanie Dick, "Introduction: Thinking with Computers," in *Abstractions and Embodiments: New Histories of Computing and Society*, ed. Janet Abbate and Stephanie Dick (Baltimore, MD: Johns Hopkins University Press, 2022):1–19.

56. https://edenmedina.mit.edu/.

Part II

Creating a New Story of Information History

Chapter 3

Writing Information History Using New Frameworks, Theories, and Approaches

James W. Cortada

This chapter is a small history of how one historian—me—confronted the question: how to study the history of information. The answer will come by describing an intellectual journey: Why study the topic in the first place, what were the research and writing efforts undertaken along the way, what issues came up that were worth investigating (or not), and what outputs of all that emerged from this effort in the form of articles and books? Like a journey into a relatively new land, ambiguity of definitions remained a constant challenge, where the lane was and where it went—since some of it appeared like a footpath and later as a highway cluttered with a combination of bicyclists, donkey carts, and muscle cars. Each discipline looking at information did it in their own human languages, used the jargon of their academic calling, and brought different paradigms and practices to the debate, such as the extensive use of theories and frameworks—which historians tend to eschew, but which I came to see as useful. In short, the intellectual structures one normally becomes familiar with in an academic discipline were not always there—often not at all—and so they had to be created. Put another way, to be an historian of information sometimes meant acting like a cultural anthropologist, economist, sociologist, project manager, or literature professor. Yet we were taught in graduate school to be loyal to our discipline's canons, anything less was seen as intellectual promiscuity. How did one person navigate all these issues? That is the task described here.

To accomplish this task, I have elected to take an autobiographical approach, which is an unusual approach in an academic chapter. It is a subjective journey of one individual, rather than a discussion of, say, methodologies that all historians can adopt. Others will have to draw what conclusions they find appealing from such an account.

Interlude: Who Am I to Fight Windmills? Confronting New Topics in History

Historians are trained in specific ways in how to explore the past, such as to understand broad contextual trends and events, then to dive deeply into archival materials to document what was actually said, thought, and happened—such as quoting from a letter written by, say, an American president in the 1850s—then how to pull all of that material together to tell a story. In recent decades historians have become increasingly facile in using statistics and even toying with theories of human behavior. But most of us, when confronting a new topic—such as information's history—know a great deal of some other history. I earned my Ph.D. in European and diplomatic history, which is about as classical as one can get in this discipline. My early books were about Spanish diplomatic and European political history. I could work in multiple European languages and spent the next half century continuing to fiddle with these subjects, while working for nearly four decades in a large multinational corporation (IBM) in sales, consulting, and management. In the course of my work, I trained myself in the history of computing, IBM, and business management—all subjects taught to very few historians when I was a graduate student (1969–1973). When I became interested in IBM's history—one of the largest multinational corporations in the early 1970s—I could only find a half dozen books relevant to the subject. In time, I was able to explain to other historians how many dozens had appeared and what they had to teach us about this firm and the world of computing.[1]

I came to learn that people did not acquire computers because they were fascinating technologies—they are not to me, instead, just refrigerator-sized complicated machines—but they had one redeeming feature: they spit out information (data, numbers, reports),[2] and that is why companies such as IBM, Microsoft, and Apple could thrive and I could explore myriad historical topics without being a professor teaching history. Mine was not the only definition of information, but one that linked closely to notions about the computer age and technology floating through various academic disciplines at the time. My business experience taught me how to educate myself on a new topic, actually twice! The first time was about the history of computing and business, the second about the history of information. It is to the second experience that we now turn. The history of information was as new to this historian as was the history of computers, technologies, and business four decades ago.

I had to travel down a path that increasingly transformed into a lane and now rapidly is becoming a multilane highway. Many students of information and its history have had to make the same journey, including some of the authors of chapters in this book. Required attributes to travel on such roads are curiosity; a willingness to put in the hours to read what was written about it before you; skill and resources to build a private library on the topic since academic libraries are too slow to build relevant collections; sufficient imagination to (a) transfer to the new subject questions you learned to ask in prior subfields of history and (b) others not asked before that became evident from multiple experiences (e.g., from working with computers and businesses or by other scholars in different disciplines looking at information), and (c) an ability to organize narratives that make sense to readers. Experts on how the human brain works have increasingly pointed out that organizing information is a human attribute and that insight facilitates and encourages historians subconsciously to do the same.[3] We care about these three points because they are part of the story about the history of information as used by humans.

Today my study of information's history is limited to *only* what humans did with it, ignoring largely the processing of information by trees, other mammals, and living creatures. That is a remarkable statement to make because so much work has been done by scientists in understanding other uses of information in the past three decades. This statement also points out what experienced historians do: constrain sufficiently their research and writing to make their findings digestible to their readers. Nobody wants to read a 1,500-page book or a 150-page scholarly article! And remember, the mind wants to organize information and develop shortcuts to only what is needed at the moment. Such cognitive organizational skills are crucial to apply to any subject and none more so than a relatively new one.

I have wrestled with two fundamental features of information's history: writing full-length histories about it and introducing theoretical frameworks into such work, the latter an action normally not welcomed by historians. While historians of information rightfully begin by applying methods canonical to their discipline—I do, too, to a large extent—I am also influenced by approaches more widely in evidence in other disciplines. I found them useful because they gave the subject of information a more specific identity, much as historians encounter when discussing specific individuals, events, institutions, and objects. Put in simpler terms, I wanted to make information appear more like a thing—an object—than as a conceptual, even abstract, notion, so I borrowed from those disciplines that deal with more sharply defined objects and tasks that otherwise would seem abstract, too. The reader will have to judge if the use of borrowed approaches works with our relatively new subfield of information history.

Writing Broad Histories of Information

Specifically, I discuss here how I put together two books that were broad in their conceptualization: *All the Facts: A History of Information in the United States since 1870* (2016) and *Birth of Modern Facts: How the Information Revolution Transformed Academic Research, Governments and Businesses* (2023).[4] What decisions needed to be made regarding what arguably are big topics requiring lengthy books? Both volumes focused on the period from the mid-nineteenth century to the near present, periods of American and European history with which I was already familiar, thanks to my academic training. For the same reasons, I focused on activities occurring on both sides of the Atlantic, with a tip of the hat to other regions when relevant. I recognized that both projects would result in big monographs, but I had learned through prior projects how to write long books and, just as importantly, how to persuade book publishers to invest in their publication.[5] So, I needed to shape the discussions of each volume in a way that readers would find them relevant and publishers would find them profitable. But both projects began out of my own selfish interests.

In the process of writing that first book I observed that information never stopped changing, so I sought out a volume that explained how all manner of information was evolving simultaneously as Americans were increasingly relying on organized bodies of information. At the most basic level, I always sought out books about specific topics; and when they did not exist, I chose to write them. I wanted a history of how Americans used information but could only find bits and pieces of the story, such as about the history of books on a few specific types of publishing; yet, I knew that children used Boy Scout manuals and textbooks, their mothers used cookbooks, their parents used operating manuals for many kinds of devices (including computers), and that entire corporations had thick manuals on various types of procedures.[6] But nobody had pulled together that story; and for me to do detailed studies of some aspect of American information history, I needed a book to provide context first, before more narrowly focused monographs.

Bits and pieces of that story existed, e.g., about the rise of scientific methods and about the histories of physics and biology. But no broad contextual history had been written, especially for the period in which I was interested, that situated its history into the communities in which I was interested, beginning with the United States, my home country. I had noticed that lacuna earlier when I had written a three-volume history of how computers were used in American industries. I had found that many industries discovered for themselves how best to use information technologies, with minimal insights from the experience of other industries.[7] That observation led to very human questions: "Hmm? Why was that?" "Did it matter?" "In particular, how were they affected by the transformation and increase in the amount of information in the past 150 years?"

Once curiosity is raised, an historian must answer the difficult question of "what am I going to do about it?" The orthodox answer that an experienced historian typically provides is, "I'll write a book and some articles about that." Experience breeds courage to do so, but in these two instances it was not so clear what these books should look like. What would they discuss, what would be the sources upon which a narrative could rely, how could one prevent themselves from spending decades researching just one volume, what would the table of contents look like?[8] In the end, answers and decisions were made that may seem arbitrary but that could be—indeed, had to be—defended as the rationale for why a reader should spend hours with the book and why a book editor would publish it.

Project management skills acquired working in a large corporation and experience writing a three-volume history of computing in business played key roles in shaping both books. In project management and consulting, one begins by asking what questions need answering, what information is needed for that and why, what the final "deliverable" (report, slide presentation, etc.) would be and by when. While historians may focus attention on all these questions, editors complain that authors are poor project managers as reflected by their notorious inability to complete a manuscript by a pre-agreed date. All these questions had to be answered for these books. Notice that one was published in 2016, the second in 2023—not too distant in time from one to another given that they were large books. However, in the interim years, I published several other books as well as three to four articles each year in scholarly journals. Some of these are discussed later, but *all* of them were part of the broader initiative to understand the role of information in the past two centuries. These efforts shaped the narratives presented in these two volumes.[9]

Guardrails were thus required. In the case of *All the Facts*, a historian at the Harvard Business School, Walter Friedman, had read a short business book I had written, *Information and the Modern Corporation*, which described what information existed and where in a modern enterprise.[10] He suggested that I write a similar book, but as a history, on that same subject from the 1870s forward—the period during which large modern enterprises came into being, which many historians had acknowledged were made possible by new bodies of information. A close friend of both of us, Alfred D. Chandler Jr., considered one of the most important business historians of the twentieth century, had encouraged me to undertake a similar project years earlier; indeed we had collaborated on a contributed volume that explained how the supply of information had increased in the United States.[11] So, for months I worked on an outline for a history of information in business, with Walter giving me feedback back and forth.

The chronology chosen was important. Historians are in general agreement that humankind underwent a massive transition, beginning with the use of steam and factories in the early 1700s, and again between the 1840s and 1870s with the creation of new corporations and government agencies, use of electricity, steel, and petroleum, and the birth of new fields of knowledge. Scholars most often select 1870 as a convenient date for distinguishing prior and post events in humankind's history, although they have not yet reached consensus on when that period ended (a debate I chose not to engage in with my history books). It was important for me to select chronologies, institutions, and activities familiar to scholars, readers, and publishers if I was to assert that information was integral to already-recognized features of modern societies. Not only was that intellectually a solid approach, but I also wanted to avoid distracting debates with historians who would more naturally gravitate toward what they knew—such as the start or end dates for an event—than to engage on the role of information.[12] My choice of dates conformed nicely with the many changes that occurred in business, the public sector, and private lives. Let future generations of scholars debate my choices; my dates and scope fit in nicely with today's historiographical practices.

The little book teed up obvious topics: new accounting practices, market surveys, advertisements, personnel manuals, manuals on how to do the work of manufacturing, new professional classes of workers such as managers and office workers, where the information was housed and used in the enterprise, and so forth. Questions faced by firms today, about how they used such information, needed to be answered; and prior histories of pieces of that answer proved helpful.[13] For this new project, I asked, should I concentrate just on corporations? Clearly not, because more than half the nation's workforce worked on farms, so farmers had to be included. What about their wives, who, too, worked on farms but also held ancillary responsibilities such as raising healthy children, cooking for everyone on the property, and taking care of the home? It turns out that they, too, had information essential to carrying out their duties.

Once armed with these questions about what information people had and used, a table of contents began to emerge. But I had noticed that much of the information collected and used in business was also reflected in the operations of government agencies at the local, state, and national levels, and was increasingly evident in the activities of individuals in their churches, community organizations, and private lives (e.g., hobbies and vacationing). So, what to do about these? Were they not part of the American experience? What would discussion of those aspects do to the table of contents? It would double the length of the book, but it would be far more comprehensive than my initial thinking.

Along the way structure—guardrails—proved helpful, indeed necessary. In writing the history of how computers were used in business I had settled on the notion of organizing that story by industry, e.g., computing used in banking, manufacturing, or transportation. This was the customary approach in traditional business historiography; and it squared with what I had learned working in corporations about the actions and identities of companies. Each industry was a community—think ecosystem[14]—and it worked for me. Research materials were already organized by industry, such as trade magazines, newspapers, and archival collections. Books were frequently organized by industry and by the needs of members of a specific industry or profession, such as being a banker, insurance executive, or automotive manufacturing executive. I chose to apply the same principle in the new book, identifying three major groups: people working for a living in the private sector, those doing the same across the public sector, and what people did in their private lives. Three large components of American society that existed throughout the entire period of consideration became the organizing principle for telling their collective story. To avoid discussions about the biology of the human mind processing information 24 hours a day, 7 days a week, which had created problems for other scholars,[15] I simply decided that I would only write a history of what people did with information during their waking hours.

The decision to include what people did in their private lives was innovative because historians writing about business, government, and education tended to apply guardrails that kept them more focused, say on business history or the history of higher education or government, for example. Yet I knew that how people use information flowed back and forth and through all three sectors of society in a simultaneous fashion, influencing its nature and use. While activities in private life were not yet fully integrated stories presented by historians, largely because these activities were far more diverse and fractured than those of such large institutions as corporations and government agencies, these needed to be included.

I could then answer several questions for each group: What did they do in their waking hours? What information did they have, create, and use? Where could I find copies of the materials and what archival records could I consult to explain within each of these three groupings of people and their activities what was happening? And how did those questions evolve over time? The study of how industries used computers helped here because I could identify, for example, students in scouting (that sector of society had had its own literature for over a century), farmers with their farm magazines dating to the seventeenth century (farmer almanacs), and cookbooks from the 1700s. Material for each aspect of human activity appeared everywhere in massive quantities, but—importantly—because we now had an organizing principle. Holding in my hands a travel guide for African Americans published in the 1930s, a cost accounting report from DuPont Corporation from the 1890s, and an issue of *Southern Living* from 2000 made sense, eliminating the terror of uncontrollable scope creep and disorganization.

Nevertheless, problems loomed—the biggest of which was what to do about defining information. Since the 1980s, it seemed that anyone writing about the nature and history of information had to engage in lengthy discussions about what it was. Was it wisdom, knowledge, facts, data, philosophy, and/or something else? That conversation continues today. I, too, had to wade into it; otherwise nobody would really know what I was discussing. So, after having written several chapters, I needed to stop and write a new first chapter entitled "Introducing American Information," in which I provided a definition, described types as they varied in American society, explained how it was used and where it came from, and indicated what made its use possible or not. Building from American economic and cultural history, I had to explain what facilitated its use: an economy that could afford to create, publish, and buy information; the rapid diffusion of literacy since most information was written and not just oral; and how it was organized (e.g., in books and by topics). In undertaking this process, it became evident that an information infrastructure did exist, made up of schools, colleges, publishers of books and newspapers, later a telegraph system, national telephone company (AT&T), roads and railroad networks, and so forth.

An important organizing principle that spoke to the issue of whether anybody was using organized information was that the nation treated information as physical objects—think in terms of *tools*—not as a kind of intellectual activity. American blue-collar workers and farmers were not intellectuals, but they were practical adopters of organized information that they could use to optimize work in a factory or to plant crops, yielding increased volumes of fruits and vegetables; while their spouses learned to curb diseases caused by harmful germs and to feed their children more nutritional meals. To all, information was like hammers and plows; screwdrivers and shovels; pens, pencils, paper, books, and pamphlets—namely, part of their tool kits. Answering that question led to the insight of information being treated as tools. Much of the insight historians gain from their research occurs to them during the act of attempting to write cogent chapters. That happened here. It is such a well-known phenomenon that experienced scholars in any discipline will admit to witnessing and partaking in that action. I knew not to continue conducting research forever on what was a never-ending project. The guardrails made it possible to decide frequently that I had enough to tell this story as I moved from one chapter to another.

The research, writing, and publishing activities proceeded as with any other history book. Reviewers recognized that the topic was expansive, but often concentrated on the guardrails. For example, one distinguished historian commented that "this encyclopedic compendium provides a brisk and accessible guide to a vast and sprawling theme." Another saw it more narrowly as "a pioneering way of thinking about American economic history" which "adroitly manages a torrent of information about information." A historian of libraries wrote, "Finally, a comprehensive highly readable history of practical information which locates its uses in the daily public, private, and professional lives of Americans who, Cortada ably shows, have always lived in an information age." I, and most reviewers, saw this book exactly as I had intended it to be in the beginning: an introduction, a contextual narrative that could help other scholars identify research agendas relevant to their niches of interests, "paths that he encourages others to explore."

These comments remind me that, at the beginning, I had to make one additional decision: to what extent should this book be a definitive magisterial magnum opus? Given that I thought the subject of information history was embryonic and that what I knew were far more definitive histories of diplomatic or national political histories, to even consider this nearly 700-page book as anything close to definitive would have been absurd. So, it was positioned and written to serve more humbly as an introduction to a very large topic in the hope that, in years to come, other equally long books covering aspects of America's information history would appear, in addition to many hundreds of smaller, more narrowly-focused monographs and thousands of history articles. In support of that goal, it was helpful to provide many detailed endnotes to sources (some 90 pages in small print). Just as importantly, I included a chapter-long bibliographic essay that reviewers lauded. It may have been more important than the main body of the book because it could point people to new research projects, while also identifying already published studies.

With publication of this book assured, and in the nearly yearlong period before one could hold it in their hands, came the question of what to do next. Experienced historians tend to learn from their prior projects what to explore next, influenced by extant sources, but more importantly by questions and issues raised but not fully answered in some prior project, all with the intent of building one on top of another. Chandler's three major books on business management, for example, are seen as one progressing over the next. You can even surmise that relationship by reading their titles.[16] Since Chandler influenced my thinking, I also wanted to build from my three-volume computer history, to the *All the Facts*, to something next. Pause to consider what you just wrote and you come to some useful conclusions.

Briefly, here, the two prior projects situated information into communities (e.g., industries, corporations, schools), defined in ways widely accepted by historians of societies and communities of the past two centuries. Situating information in such groups was well accepted, and I found these to be practical both as a construct and for how my sources were organized. I wondered: Could that strategy be applied to whatever book came next? Answering that question led to some reflection about what I had learned so far. I was particularly impressed with how individuals, communities, groups, and entire institutions interacted with each other. Those who collected, used, and were affected by information were obviously individuals, but they were also influenced by other people clustered within groups, e.g., members of their church, school, company, or profession. Each participating person or group affected everyone else around them. One example: an accountant working in a corporation using an adding machine and spreadsheet might also apply that same technology and methodology for tracking expenses on behalf of their church; or they might create a written budget for controlling their personal household expenses. In time, that accountant's wife would take over that mathematical function on behalf of other community organizations. Companies and government agencies followed the same pattern, borrowing information-handling activities from each other. So, looking at the discovery, creation, and use of information through some sociological or anthropological lens made good sense because the historical evidence demonstrated that reality. As this behavior was unfolding, people—individuals and then groups and employers—shaped the type, features, and appropriation of information to optimize its usefulness.

So, both projects led to the conclusion that information was being seen and used almost the way one thinks of tools, as objects. That conclusion proved to be of paramount importance because I already knew that one gap in our understanding about information's history was its massive increase in quantity and variety over the past two centuries. That realization became the heart of what next needed to be studied.

Now we had a problem. There are histories that explore knowledge, wisdom, and data in ways that focus on more conceptual and abstract themes such as truth and context, or more recently on falsehoods, lies, and misinformation—not centered on farmers, workers, or professors. Philosophers had dominated the conversation for centuries—not historians or sociologists. It did not help that computer scientists began in the 1950s to discuss data structures, files, and databases, or that even earlier librarians had tried to organize knowledge (in the form of books and journals) into large categories, such as History, Literature, Natural Sciences, and Law. What was a historian to do in the face of a substantially large literature accumulating for more than 2,000 years, most of it published during the years I had chosen to study? These previous approaches are not what historians of information were necessarily pursuing when, for example, they explored micro case studies or more focused research. So much was in flux, and thus, between that reality and the organization of my prior work, I decided that I needed to find a different path.

The prior projects had made clear that modern, organized (i.e., published) information was created and used primarily by people working in and identifying with organizations. That is to say, discussing information in the abstract as if it were an orb floating in space made little sense to a historian interested in the activities of humans in the past two centuries. That raised the questions of by whom, how, and where was new information discovered, created, and used during the past two centuries? Its discovery, creation, and use—three activities concerning information's history—was not really in dispute; it seemed that every historian wrestled with these issues, regardless of what era they examined.[17] The disparate but at least extant literature made it suddenly obvious that much modern information was largely the product of new professions, their associations, and the academics in rapidly increasing numbers of ever-larger colleges and universities. The table of contents for this new book almost popped off the screen when pondering how to organize the research and the subsequent narrative. The table of contents for *All the Facts* and *Birth of Modern Facts* (displayed in table 3.1) indicates the direction of this research.

Table 3.1. Table of Contents for *All the Facts* (2016) and *Birth of Modern Facts* (2019)

All the Facts
Introducing American Information
Short Biography of American Information, 1870–1945
Roots of Early Uses of Information, 1600s–1870
Big Business & Small Farms Rely on Information, 1870–1941
Big Government & Small Universities Take on Big Information, 1870–1941
How Citizens Became Dependent on Information, 1870–1945
From Pearl Harbor to 9/11: Government & Education in a World of Computers, 1941–2001
Informing the Business of America, 1945–1998
Information & the Modern Knowledge Worker, 1945–1998
Uses of Information Everyday Life, 1945–1995
The Internet & Modern Uses of Information
How Americans Used Information to Shape Their Society
Introducing American Information
Short Biography of American Information, 1870–1945
Roots of Early Uses of Information, 1600s–1870
Big Business & Small Farms Rely on Information, 1870–1941
Big Government & Small Universities Take on Big Information, 1870–1941
How Citizens Became Dependent on Information, 1870–1945
From Pearl Harbor to 9/11: Government & Education in a World of Computers, 1941–2001
Informing the Business of America, 1945–1998
Information & the Modern Knowledge Worker, 1945–1998
Uses of Information Everyday Life, 1945–1995
The Internet & Modern Uses of Information
Birth of Modern Facts
How Librarians, Scholars and the New Professions Defined Modern Information
Second Industrial Revolution Encounters Information
How Librarians Organized Information
Early Encounters by Computer Builders
Mathematicians & Statisticians Create New Tools
Scientists & Medical Experts Shape Information
New Business & Government Information Ecosystems
What Information Economists Created
Contributions of Political Scientists & Historians to Modern Information
How Information Evolved

Academic disciplines and their related professions became the logical choice to study; hence, the chapters discuss librarians, physicists, doctors, economists, biologists, historians, political scientists, computer scientists, accountants, business managers, political scientists, pollsters, and politicians. The challenge was to organize their activities both in terms of their importance (as measured by impact on modern societies) and the chronological influence of one group over another. That is why, in this new book, librarians were discussed before computer scientists, and mathematicians before economists. I knew from earlier research about computing in organizations that nothing was free, so a chapter was needed to discuss the role of the Second Industrial Revolution, which had made possible the discovery, creation, and use of information. There was also need for a chapter defining in broad terms how information changed in form and use over the past two centuries. That table of contents made it possible to define the research needed in support of the project.

In working on my industry computer project, *The Digital Hand*, I had realized that not all readers would know how an industry was structured or how it made its living, which was background information needed to understand how computers could, then did, assist in this task. So, as part of each new industry's discussion, I devoted a few pages to describing what that industry did and how they profited. That was less important in *All the Facts*, but was essential for the next project because not all readers might know what a mathematician, biologist, or economist thinks about or how they go about doing their work creating and using information. So, the same disciplined approach used with *The Digital Hand* proved a valuable way to introduce a reader to the role of information in modern times. As with the earlier book, the manuscript's referees and later reviewers found organizing activities involving information largely around recognizable academic disciplines and professions made sense. As in the prior book, it was acceptable not to be comprehensive in covering all professions and disciplines because, still, nobody wanted to read a 1,500-page book.

Thus, the questions I had to research proved similar: what information was needed by each discipline; how was this information discovered, created, and communicated; how was its veracity and relevance assured;

and what effects did this have on individuals and groups? That exercise led to discussions about the origins of academic societies and the proliferation of tens of thousands of academic journals and magazines aimed at trades, industries, and professions. These people and their associations and employers shaped the cultural norms needed for information to be accepted and used, again emphasizing the essential finding that information increased in quantity, variety, and use when integrated into the lives and cultural norms of a society. At the risk of possibly falling into hundreds of academic holes filled with intramural debates about the meaning of information or the role of theories, I chose to remain one level above the detail of these discussions, advising readers that, if interested in those issues, they should research specific histories of these more restricted topics.

The three major projects—computers in industry, facts available in the United States, and the nature of the new information—represented a long trek to what so many scholars had jumped to in earlier, narrower, shorter, more superficial studies that chatted about the nature of information, declaring its omnipresence in modern society (an historical inaccuracy inasmuch as organized information had been in use for thousands of years), when they had not yet done sufficient research to describe what actually happened. What the three projects did was to make it more obvious how little that modern users of information knew about the formation of, nature of, and use of information in recent times. But by my efforts, at least we could begin to realize how much more work would be needed.

Role of Theory, Frameworks, Ecosystems, and Infrastructures

It turned out that increasingly, over time, just situating information into organizations and professions was not enough. There were other activities that could be observed in both a historian's work and in that of other disciplines and professions. Because of the pervasiveness of information, historians could no longer simply rely on the rules of how to function drawn from within their discipline. There were many interesting new things and thinking going on in the new discipline not present two centuries earlier. For me, the problem was exacerbated by the fact that I had looked in detail at what was happening in many industries, professions, and academic disciplines. Where historical discourse ended and new ways of studying information appeared began to come into play well over thirty years ago. I read anthropology, economics, business management literature, and even biology specifically about neuroscience; and all that before the massive increase in studies prepared by journalists, library science professors, and the media industry and its observers. I routinely had to spend one to two hours each day, seven days a week to keep up with the general literature, all before engaging in specific research for a particular chapter or in writing.[18] The crunch of too much information was real. After leaving my corporate career at the end of 2012, which freed up more time to conduct research and writing, the opportunity presented itself to think more abstractly and broadly, and to contemplate new ways of exploring contemporary history. Methods of thinking as scholar/business professional began merging, resulting in new ways of viewing history.

So, we need to discuss how I managed this glut of information. Let me begin with the application of Bradford's Law, followed by the introduction of how at least one other historian dealing with similar issues addressed complexity and quantity of information. Then I discuss my developing model—if we can use that word as historians—that made it possible to discuss complex issues as part of social and information ecosystems with information flowing through these. It was in that process of sorting and organizing that I borrowed approaches from other disciplines.

Bradford's Law, even though a fabulous secret among librarians since the 1930s, seemed increasingly outdated.[19] His law of diminishing returns in research essentially held that you could keep up with changes in your discipline by reading a dozen articles published in the three key journals in your field and that to find an additional 12 useful articles you would have to consult an additional 6 journals, and so forth. The problem is more than what are the three main journals in one's discipline, since these are easy to identify. It is more an issue of how many different disciplines one now had to keep up with. I have to read about the history of technology, business, and information, and all of that before I even consider topics specific to whatever research I am conducting. I also still read about European history.

The historian soon learns that their subject is frightfully more complicated the more they study it. The distinguished British historian and world's leading expert on the Thirty Years War, Peter H. Wilson, addressed this issue of information glut: "The history of the Thirty Years War is rich in specialist studies, but poor in general accounts. Few authors provide more than short overviews intended for students. It is easy to see why. To cover all aspects would require knowledge of at least fourteen European languages, while there are sufficient archival records to occupy many lifetimes of research."[20] Concerning the large Treaty of Westphalia (1648), which ended the war and established most of the boundaries of European countries familiar to our readers today, he reported there were over 4,000 titles.[21] Over time he formulated points of view and identified patterns of behavior that enabled him to write behemoth books. Thus, the problem of how to contain within some scope of control is not limited to my work; many historians face the same challenge.

It is at this point that theories, frameworks, and my work on information ecosystems and infrastructures come into play. My earlier research led me to conclude that information and its use were tied up in communities and their cultures, what anthropologists have long called "ecosystems." Much like animals, bugs, and plants cohabit successfully in a Brazilian jungle, so too did people, often for hundreds of thousands of years. It was only the rapid and jarring emergence of new ways of living emerging at the beginning of the First Industrial Revolution in the early 1700s that humans diverged from this much longer form of cohabitation—one that today is being called to be renewed as part of the climate change crisis. Following a path that some sociologists were also following, I began seeing information ecosystems in American and European societies and communities, within and across institutions, professions, academic disciplines, and information.[22]

The historian in me wondered if there were some common features here. If so, what were they? Practicing a classic historical research process, it was time to test such questions through case studies. Knowing a great deal about Spanish diplomatic history, and its literature, I looked at what information diplomats used, collected, and reported exploring the period 1815 to 1936.[23] That exercise shed new light on the nature of their information and how it affected their lives. Although this case study provided promising results, one case study does not make a new framework or model true. Did the same behaviors occur elsewhere or were the diplomats outliers? I turned to another community I knew: farm families of Central Virginia. I examined the same issues concerning farm wives and their daughters from the 1950s to the 1990s. This was an extreme counter test because, while the diplomats had no choice but to work in certain ways for their employers, the ladies of Virginia worked only for themselves. But it turned out that their behaviors mirrored those of the Spanish diplomats.[24] They collected and used similar bodies of information, that is to say, the same information as each other used, such as recipes, how to grow kitchen gardens, sew, raise children, and so forth. They did this in a routine, even scheduled way, shared verbally what they learned and in a set of common physical places (churches instead of embassies, living rooms instead of offices), kept similar records, and built tiny home libraries. I then repeated the exercise with IBM's salesmen from the 1910s to the 1980s, and again identified similar patterns.[25] I even went so far as to test out these emerging patterns of behavior by exploring in greater detail the case of a single individual—a retired lady in her 70s—and again the pattern held.[26] Figures 3.1 and 3.2 illustrated her information ecosystem and infrastructure, using the exact same schematic as applied to the diplomats, farm wives, and salesmen.

Thus, with confidence I could begin to describe in general and also in case-specific terms the work of different groups and academic disciplines. I could also advocate for such an approach, by pulling together case studies and more formal descriptions of such ecosystems, thinking of them as *Building Blocks of Society*—the title I gave to a book on this subject.[27] This approach also reflected increasingly similar ways of viewing the workings of society—and its information—proffered by economists, political scientists, media experts, and sociologists.

This exercise required that we understand the wherewithal that made all that sharing and use of information, which meant more than being

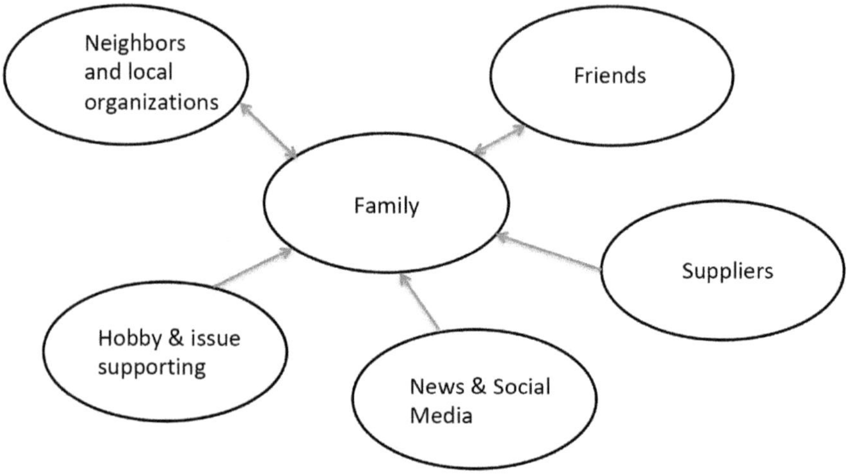

Figure 3.1.

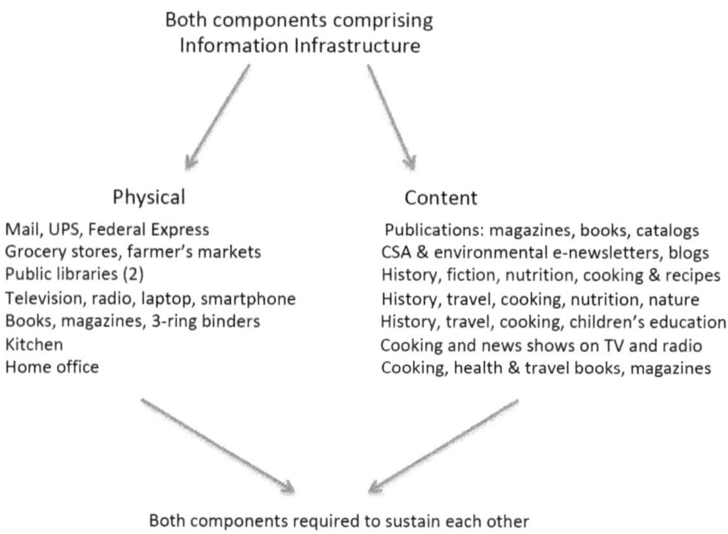

Figure 3.2.

able to buy publications or travel to meetings. Historians of libraries, cultural anthropologists, and communication scholars in general had already made it evident in their disciplines that just as there were patterns of behaviors in a metaphorical jungle, paths through the thickets were needed, watering holes, resupply of food through annual fruiting of essential plants and the birth each spring of more mammals. What did they look like in information ecosystems and were they also similar? This question led to a similar result: identification of ways in which information moved—the idea of *flow*, a conceptual contribution of information for scientists and librarians—analogous to what happened on barges, stagecoaches, and trains; through postal systems every country developed even before the industrial revolutions; telegraphy; newspapers; computers and other systems. They existed and were crucial to the development and use of information. They were also far more diverse than one might think. They also served as checklists to ask, "Did I account for all possible components of a specific ecosystem or infrastructure?"

How could I draw from other disciplines ideas that historians could either embrace as compatible or that could be jerry-rigged to assist in their work? We could offer up a book to answer that question, but several ideas will have to suffice to make the point that we can and should borrow from other disciplines. I began by recognizing the possibility that economists, sociologists, and cultural anthropologists sought to define, in very high-level terms, patterns of behavior much as economists do when they analyze large bodies of economic data to discern statistically and mathematically evident trends. That behavior melded nicely with the more widely accepted practices in applying the "scientific method" of research, which most disciplines had accepted as proper behavior since the seventeenth century. Where speculation entered the discussion, as when faced by cultural anthropologists of the unknown in applying meaning to, say, shards or fixation on stars and volcanoes, default to some interpretation that they were confronting a ritual or religious belief, the historian could brush that aside as speculation not rooted in evidence, facts to be precise. Similar to how cultural anthropologists treated the unknown, e.g., in applying meaning to shards by defaulting to some interpretation that they were confronting a ritual or religious belief, the historian could brush that aside as speculation not rooted in facts. As a lawyer once told me, "If you can't prove it with hard evidence in a court of law, you don't have a case." Historians tend to follow that advice, if not as strictly as a lawyer. Or consider business writers: they do an excellent job in describing repeatable activities and one-off events—they use the word *processes*—by naming each step (task) within its own box on a graphical representation of steps read left to right, start to completion, with narrative descriptions below each box. This is a simple artifact, but it clarifies for an audience what goes on either in a PowerPoint presentation or as a figure in a publication. I have found that tool to be as useful in writing history as I did in explaining to customers some concept in how computers could do work for their organization.

Using explanatory methods applied in business, social sciences, and hard sciences, I began to describe what they looked like—using, for example, graphical representations rather than just the older method of lengthy descriptive narratives. I continued, however, to embrace only slowly and selectively theories as I often found that they were being offered up with more enthusiasm on the part of a scholar than his or her evidence would have pleased my lawyer. The combination of the two approaches—graphical representations and narratives—proved effective in solving one key problem in the historiography of information: its ambiguous fuzziness. It turns out that scientists studying how the human mind works were discovering that such a combined approach did much to clarify one's thinking about fuzzy abstractions, because ideas were actually not physical, rather "fictions of the mind" that people are willing to act upon.[28] So, knowing that such models displayed in a figure clarifies thinking about an idea, I wrote about ecosystems and infrastructures, about graphical representations and frameworks, all to help sharpen my understanding of information's history.[29] That it resulted in various publications and recognition of these new approaches, was secondary. The prize was better history.

Much seemed to be coming together into some as yet ill-defined universal modern theory of the role of information in the past two centuries. For an economist or social scientist that would be considered wonderful, the basis for some prestigious academic recognition. But for the historian, realities of human behavior suggested perhaps it was too artificial, theoretical, elegant, and artful, since history teaches that humans never act exactly as predicted and what happens is always a messy story. We have only Professor Wilson's problems to keep in mind. As I knew would happen, while the merging potential theory and/or frameworks were emerging in my mind, along came William Aspray, who trained as an historian, identified with the information sciences disciplines, and had published numerous books and articles laced heavily with the methods of the historian, to call my attention to a major problem: fake or faulty facts, what politely is referred to as *misinformation* but that during the years of President Donald J. Trump, for example, walked through the American political landscape most journalists called *lies.*

How were these issues of the collection, use, and consequences of faulty information to be accounted for? Good question, and the way historians normally address something like a newly recognized anomaly is to formulate test cases of actual experiences. So, my drifting toward the use of models encountered information usages and definitions that might not be accurate. I had to be wary of a fundamental problem historians have with theories and models in that such intellectual scaffolding suggests that human behavior does not change, and yet every situation is always different than prior ones.[30] So, when should a historian generalize, and when should they put their foot down and say something is different? Was Trump, for example, behaving differently from what a political science model would suggest based on prior presidential actions, or was he typical?

As seasoned historians both familiar with much in the history of information, Aspray and I worked on two parallel problems: the history of misinformation in the United States and how urban myths and political information was studied and used coming from the Internet. The first one required looking at the role of lies and misinformation dating to the dawn of the eighteenth century: election politics, conspiracy theories of presidential assassinations, industry manipulations about the impact of various medicines, smoking, and environmental debates.[31] We found misinformation everywhere we turned. For the second book, focusing on political fact-checking during the years since arrival of the Internet up to the time Trump ran for president in 2015, we documented how fake facts were conjured and used.[32] But a pattern of rigorous declarations about the certainties of fake facts was not forthcoming, unless one wanted to argue that there were always fake facts, then declare intellectual victory and move on. We also discovered that there were genuine differences in what constituted truth, varieties motivated by other factors, such as alternative worldviews, for example, the divergence between American southerners and northerners regarding how best to interpret events of their civil war of the 1860s or their mythologies of the American Revolution of the 1770s.

We decided to explore the role of authenticity in how people viewed their history through a rigorous exploration of three historical sites as they understood these. One involved a community that rebranded itself as an ethnic heritage site in Kansas (Lindsborg), complete with imagined nineteenth-century costumes and holidays; a second, the reconstruction of an eighteenth-century community in Virginia (Williamsburg); the third, the largest and longest-running battle and its battlefield of the American Civil War (Gettysburg).[33] We approached the subject by (a) thinking of authenticity as a variation of misinformation and (b) by containing our cases within the guardrails of heritage tourism, since that subject—heritage tourism—was a rapidly developing academic subfield with its own rules of research, issues, and questions, which would allow us to compare their findings to ours as historians. We identified some patterns, but more interesting from an historian's perspective, it seemed "everyone" (or at least millions of people) had their own views of what authenticity meant, what they expected when confronting the histories of these sites, and what such findings could teach historians of information and misinformation, information studies experts, and tourism scholars. Information ecosystems existed here, too, such as Civil War military reenactment organizations and museum and parks, but also infrastructures such as magazines, books, and videos of all three sites produced by advocates of differing points of view—from chambers of commerce, to tourism promoters, to park rangers. Everyone used the same information infrastructures as everyone else: postal services, libraries, publishers, the Internet of course, and delivery services.

What were we to make of these findings? More broadly, what was I to make of my own research across multiple projects? Psychologists may have the answer and the reader knows it too. Recall that when you buy a new car, notice how many other people are driving the same model over the next few weeks. If you are pregnant, notice how it seems so many other women were also pregnant at the same time. A few weeks later you no longer notice either circumstance. Known as the Baader-Meinhof phenomenon, there are not more of the same cars or pregnant women, it is just that you did not notice these before, because your brain was too busy taking in information more relevant at this moment. Buy a new car or get pregnant and your mind shifts priorities, hence you notice these. Historians are human, too, and so are not immune to the same phenomenon: once they discover that there are information ecosystems, infrastructures, and misinformation, they will see these features all around them.

As historians we are constantly surprised by what we learn. Early in this chapter I described modern information as objects—tools—implying an almost physical form. In fact, there are surrogate physical forms (some scholars call them vessels) such as books, letters, and diaries for representing information. But when we look at the role of authenticity, which too relied on such components as books and magazines, we might ask if authenticity should still be seen as objects. So far, the answer is yes since information is used like tools, just more malleable than we might have thought as historians and users. Only philosophers believed before in the plasticity of information; today they are joined by neuroscientists, while we historians must remain open to new twists and turns in our journey to understand the history of information. What scholars across most, if not all, disciplines have come to believe in the past two generations is that information is far more central to the nature and activities of all living creatures than imagined before. It is one of the central revelations of humanity's understanding of reality uncovered over the past century. Like Wilson exploring the Thirty Years War, the work is not done; the questions to ask next are not yet so clear. The history of information is even more underdeveloped than Wilson's war, so the path before us is even less obvious. Is it a bike path, a four-lane highway, or a rocket trip to Mars? What is certain is that the way historians think and conduct research, what questions they ask, and how they tell the story will be—is—evolving far from what it was when I studied political history in the 1960s and 1970s.

Some Final Thoughts

How I have studied the history of information so far was conditioned by several factors. First, there were the methodological techniques and values I learned in completing a PhD in history, although not focused on the history of information. Those approaches and values served me well when studying the history of computing and business, and later information. Second, many of the steps I took were out of a genuine curiosity—call it fascination—with what was an ambiguous topic (information) and for which there lacked the methodological prescriptive actions one might have used prior to engaging with the subject. Some existed, but they came from other academic disciplines such as those that valued theories, hypotheses, frameworks, and graphical representations beyond those historians normally embraced. I found it useful to try out these methods, adapting them to my needs insofar as they furthered the agenda of a practicing historian. Third, because this topic was entirely new, I had to maintain a humble attitude toward what I might uncover; and so I was surprised as I connected different themes to each other. I purposefully, indeed ruthlessly, sought out patterns of discovery, creation, and use of information because with so many people using facts over such a long period of time, shared values and practices certainly had to exist. The evidence did not disappoint: my diplomats, farm wives, Boy Scouts, political liars, and scientists behaved in reasonably predictable ways.

But as my current and as yet unpublished research is suggesting, humans and their computers are beginning to change how they collect and use information *as a general behavior*. One simple teaser of findings hints at what is unfolding. It used to be that to translate phrases from one human language to another the translator had to be fluent in both. Today, a computer looks at billions of uses of a phrase in both languages and calculates the odds that one phrase is a humanly acceptable translation, and it does an increasingly good job at that. The computer does not need to know either language, only how to compare quickly vast quantities of texts and then predict the odds mathematically as to what the right answer is and serve it up as THE answer. Another: artificial intelligence and other software can identify patterns of behaviors and relationships among facts by (a) rapidly comparing trillions of pieces of information (you and I cannot do that) and (b) establishing relationships by using network analysis tools first developed in electronic communications in the 1920s to1940s and now very mainstream in the hard and social sciences, not just in telecommunications and computing. But fear not, historians still must look at paper ephemera in archives; books and journals are still core to what we read; and studying objects as if they were three-dimensional documents and footnotes is still essential. Finally, we still must ask many of the same questions we did decades ago; it is just that we have so much new information to sift through to arrive at acceptable answers.

Where does our discussion in this chapter leave me? The answer is more questions to explore. Are we dealing with more or less misinformation and if so what are its features that need further definition? More important, I have concluded that such questions may be masking a larger one born from an observation gleaned from earlier projects: that information is plastic, malleable, and thus not always misinformation, just a variation of it. Misinformation or lies requires someone to take an accepted truth—fact—and twist its shape. Our study of authenticity suggested that people may come to a piece of information with genuinely different definitions, or out of ignorance of its factual full nature. So, does that mean there is no absolute truth? Facts do change as new evidence emerges, as scientists are constantly reminding us, while clerics despair over relativeness in our religious beliefs. How fixed is some information? I know that my birthday is September 7 (I consider that fixed), but what about the origins of our solar system or the causes of COVID-19? How fungible is information, and how do individuals, organizations, and disciplines massage it? Are these questions even worth asking? I say yes, as do other historians, because information is not fixed, and certainly not our understanding of it yet. Therefore, all the questions presented here historians of information can explore, even *ad infinitum,* thereby contributing to the richness of one's understanding of the history of information. Certainly, these questions can be studied by historians using today's methodologies. Because of that reality, I remain as uncertain about my understanding of information as I did decades ago. Now what should I pursue?

Notes

1. James W. Cortada, "Change and Continuity at IBM: Key Themes in Histories of IBM," *Business History Review* 92, no. 1 (2018): 117–148.

2. There are multiple ways to conceptualize information, other than as data, numbers, or reports, but that this was my approach was influenced by my experience at IBM.

3. For an approachable discussion of the human brain, Thomas Suddendorf, Jonathan Redshaw, and Adam Bulley, *The Invention of Tomorrow: A Natural History of Foresight* (New York: Basic Books, 2022).

4. James W. Cortada, *All the Facts: A History of Information in the United States since 1870* (New York: Oxford University Press, 2016) and *Birth of Modern Facts: How the Information Revolution Transformed Academic Research, Governments and Businesses* (Lanham, Md.: Rowman & Littlefield, 2023).

5. James W. Cortada, *The Digital Hand*, 3 vols. (New York: Oxford University Press, 2004–2008) and *The Digital Flood: The Diffusion of Information Technology Across the U.S., Europe, and Asia* (New York: Oxford University Press, 2012).

6. The existence and use of such materials were not, *per se*, just data or numbers, but information seen as useful by their users.

7. Cortada, *The Digital Hand.*

8. The histories of computing had more clearly defined contours since the subject had been studied longer. In the early decades of computing's history (1970s–1980s), historians wrote about specific machines and software and how they were created, then used; later added studies about their economic and social impact (1980s–2000s); and after the arrival of the Internet, began identifying their use in many nooks and crannies of society, first in economically advanced countries then increasingly elsewhere around the world.

9. Cortada, *All the Facts* and *Birth of Modern Facts.*

10. James W. Cortada, *Information and the Modern Corporation* (Cambridge, Mass.: MIT Press, 2011).

11. Alfred D. Chandler Jr. and James W. Cortada, *A Nation Transformed by Information: How Information Has Shaped the United States from Colonial Times to the Present* (New York: Oxford University Press, 2000).

12. An example of that diverting behavior involves the two world wars. Some historians speak about them as World War I and World War II as two distinct events, while a growing number of them see these as one with a pause in the 1920s and 1930s. That discussion has generated a large body of literature.

13. JoAnne Yates, *Control through Communication: The Rise of System in American Management* (Baltimore, Md.: Johns Hopkins University Press, 1989); James R. Beniger, *The Control Revolution: Technological and Economic Origins of the Information Society* (Cambridge, Mass.: Harvard University Press, 1986); Alfred D. Chandler Jr., *The Visible*

Hand: The Managerial Revolution in American Business (Cambridge, Mass.: Belknap Press of Harvard University Press, 1977).

14. I discuss the concept of information ecosystems in James W. Cortada, *Building Blocks of Society: History, Information Ecosystems, and Infrastructures* (Lanham, Md.: Rowman & Littlefield, 2021): 13–18, 35–44.

15. For example, sociologist/historian James Beniger, *The Control Revolution.*

16. Alfred D. Chandler Jr., *Strategy and Structure: Chapters in the History of the Industrial Enterprise* (Cambridge, Mass.: MIT Press, 1962), *The Visible Hand* (1977), and *Scale and Scope: The Dynamics of Industrial Capitalism* (Cambridge, Mass.: Belknap Press of Harvard University Press, 1994). He wrote other business history books but spent the early years of his career as an historian of American political history too.

17. A recent example, Paul M. Dover, *The Information Revolution in Early Modern Europe* (Cambridge: Cambridge University Press, 2021); see also bibliographic essays, Cortada, *All the Facts*, 575–609, and *Birth of Modern Facts*, 395–421.

18. Impossible, you say? If you read while waiting to board an airplane, then while on a flight, many evenings for an hour or so, and getting up 45 minutes earlier each day than needed for decades does provide the time.

19. Samuel C. Bradford, "Sources of Information on Specific Subjects," *Engineering: An Illustrated Weekly Journal* 137 (January 26, 1934): 85–86, and his "Sources of Information on Specific Subjects," *Journal of Information Science* 10, no. 4 (October 1985): 173–180.

20. Peter H. Wilson, *The Thirty Years War: Europe's Tragedy* (Cambridge, Mass.: Belknap Press of Harvard University Press, 2009): xxi.

21. Ibid. He continued his deep dive, publishing *Iron and Blood: A Military History of the German-Speaking People since 1500* (Cambridge, Mass.: Belknap Press of Harvard University Press, 2023) which, like the earlier volume, is over 900 pages in length. To drive home further the point of complexity increasing, he also wrote *Heart of Europe: A History of the Holy Roman Empire* (Cambridge, Mass.: Belknap Press of Harvard University Press, 2016), which was over 1,000 pages long.

22. Unbeknown to me as I lived inside Corporate America, academic scholars were also borrowing from disciplines other than their own, with sociology a strong example, Harvey J. Graff, *Undisciplining Knowledge: Interdisciplinarity in the Twentieth Century* (Baltimore, Md.: Johns Hopkins University Press, 2015).

23. James W. Cortada, "The Information Ecosystems of National Diplomacy: The Case of Spain, 1815–1936," *Information and Culture: A Journal of History* 48, no. 2 (2013): 222–259.

24. James W. Cortada, *Building Blocks of Society: History, Information Ecosystems, and Infrastructures* (Lanham, Md.: Rowman & Littlefield, 2020): 131–180.

25. Ibid., 181–224.

26. James W. Cortada, "One Senior Citizen's Information Ecosystem and Infrastructure," in William Aspray (ed.), *Information Issues for Older Americans* (Lanham, Md.: Rowman & Littlefield, 2022): 63–90.

27. Ibid., *Building Blocks of Society.*

28. Suddendorf, Redshaw, and Bulley, *The Invention of Tomorrow*; but also useful, Daniel Kahneman, Oliver Sibony, and Cass R. Sunstein, *Noise: A Flaw in Human Judgment* (New York: Little, Brown Spark, 2021).

29. James W. Cortada, "Revisiting 'Shaping Information History as an Intellectual Discipline': Shaping Information History as an Intellectual Discipline," *Information & Culture* 54, no. 1 (2019): 95–126; "The Magic of Debunking: Interrogating Fake Facts in the United States since the Eighteenth Century," *Library & Information History* 35, no. 3 (2019): 133–150—co-authored with William Aspray; "Gaining Historical Perspective on Political Fact-Checking: The Experience of the United States," *Libres* 30, no. 1 (November 2020): 1–32—co-authored with William Aspray; "Viewing Corporations as Information Ecosystems: The Case of IBM, 1914–1980s," *Enterprise and Society* 23, no. 1 (March 2022): 99–132; and on methodology, "New Approaches to the History of Information: Ecosystems, Infrastructures, and Graphical Representations of Information," *Library & Information History* 32, no. 3 (August 2016): 179–202; and "A Framework for Understanding Information Ecosystems in Firms and Industries," *Information & Culture* 51, no. 2 (2016): 133–163.

30. For one of the clearest discussions about theories and how historians deal with the topic, see Peter Burke, *History and Social Theory* (Ithaca, N.Y.: Cornell University Press, 1992): 28–33; but also, James W. Cortada and William Aspray, *Fake News Nation: The Long History of Lies and Misinterpretations in America* (Lanham, Md.: Rowman & Littlefield, 2019).

31. Cortada and Aspray, *Fake News Nation*.

32. William Aspray and James W. Cortada, *From Urban Legends to Political Fact-Checking: Online Scrutiny in America, 1990–2015* (Cham, Switzerland: Springer, 2019).

33. William Aspray and James W. Cortada, *Authenticity: Understanding Misinformation Through the Study of Heritage Tourism* (Lanham, Md.: Rowman & Littlefield, 2022).

CHAPTER 4

Writing Information History from the Perspective of Rhetorical Genre Theory

Laura Skouvig

GEOFFREY WINTHROP-YOUNG ARGUES THAT BOTH FICTION WRITERS and intelligence operatives are engaged in the staging of information.[1] I agree, but with the addition that we are all engaged in the staging of information. My general assumption about information is that it never arrives naked, so to speak, but is always formed, shaped, and defined by the texts in which it appears. And those texts are the result of certain human actions. As an information historian or as a cultural historian, I am interested in information as a historical phenomenon. I am curious about how people acquired information, what they did with information, and how information was understood as a historical phenomenon, concept, and practice.

Being in an information studies department has obviously influenced my focus on information as an object, as a historical phenomenon, and as a concept. Being an embedded historian in an information studies department has also made me prone to expectations of being the historical conscience of information studies. I have explicitly tried to avoid that position. Of course, my institutional surroundings have influenced my approach to studying (information) history beyond the phenomenon itself. In this chapter, I will discuss how I do information history, with concrete examples from my existing and ongoing research. Information history is an exciting field because it enables me to combine three different but related and interdependent interests of mine. First, information history is historiographically interesting because it draws attention to how historians understand information, history, and information history. Second, I find it important to explore and develop information history theoretically. In this contribution, I will discuss how rhetorical genre theory inspires my work on information as a historical phenomenon. Rhetorical genre theory understands genres in a broader perspective of everyday life and not as fictional genres alone. Finally, information history provides an astonishing variety of topics for empirical studies, and I will here present my own research on the Danish information state at the turn of the nineteenth century.

The chapter is structured according to my overall interest in information history. First, I discuss how I understand information and information history. I am mainly inspired by Michel Foucault, but my understanding of information is particularly inspired by rhetorical genre theory, which I turn to in the second section of the chapter. In the final section, I discuss how a genre perspective informs my analyses of laws and legal texts as communicative actions.

Information + History = Information History

My understanding of information reflects how I understand history and vice versa. Information is a particularly tricky term to work with historically. "Information" as a word had not been in frequent use in Denmark until after the Second World War.[2] During the German occupation of Denmark, the main illegal news service was named Information, which became the name of the newspaper when it appeared legally on May 5, 1945, when Denmark was liberated.[3] Before this turning point, the main meaning of "information" was educational instruction. The commercial newspaper "Adresse-Avisen" (founded in 1759) contained advertisements for "information" in, for instance, languages.[4] In this context, "information" meant instruction, and this use emphasized the processual understanding prevalent also in the English use of the word.[5]

In Danish, "information" seems to be a particularly modern word, which has inspired my historiographical work in information history. I have elsewhere discussed and presented different approaches to information history. One main argument here is that approaches to information history are indebted to and rely on how the historian in question understands and approaches history.[6] I have been deeply inspired by Michel Foucault's ideas about history as a series of discourses, ruptures, and discontinuities, as opposed to history as constant progression and refinement. Equally, his genealogical approach of unraveling past conceptions by mapping and explaining power relations in the past as it is done in the present has been a main inspiration.[7] Foucault inspires the historian to understand discourses of information as contingent and interchangeable, as they also relate to, rely on, and expose specific power relations. Although Foucault argued that past practices were accessible through discourses, the mapping of such discourses, in my view, tends to underestimate the concrete practices of information.[8]

Foucault's understanding of technology is quite different from the way we think about information technology.[9] For instance, he understood discipline as a technology. He was acutely aware of the role of writing and bureaucracy as means of documentation, and of how writing and bureaucracy produced the individual.[10] His reference to a "network of writing" and the masses of documents emphasizes the materialities of information—for instance, how documents and genres from books to handwritten letters, notes, and records of different kinds influenced the information contained in such documents. What were they? What kinds of documents? What did they contain? How were they recorded? What did they record? And how did the records mold, shape, and form information? The relationship between documents, media, technology, and information, and the relationship between information, documents, and the historical context (in all its complexity), are major aspects that information history is focused on.

The inherent premise of information history is that information is a historical phenomenon and concept formed and shaped by a given historical period. However, the present technological connotations of information, which have stressed it as a non-semantic, quantitative phenomenon, have had a huge impact on doing historical research on information.[11] The impact of Claude Shannon's mathematical theory of communication on the conceptualization of information as a technological phenomenon is still contested. Toni Weller has explicitly argued for regarding information not as a technological phenomenon but as a socially and culturally embedded phenomenon.[12] Weller's argument resonates with recent calls for the humanities to reclaim information as a phenomenon embedded in social and cultural practices.[13]

To a cultural historian, this seems obvious; but within the Danish context, it has been particularly important to get a grip on how to do historical research on information without risking an anachronistic interpretation by forcing a modern conception of information upon the past.[14] My main approach to information as a historical phenomenon seems to be of the *etic* kind as argued by Ann Blair. I apply information as a concept to a historical context in which there was no equivalent concept. But, as Blair continues, doing this is followed by a focus on what terms were then in use and on the contextual understandings and practices in the past.[15]

One of the main issues that I have worked on has thus concentrated on how to connect the premise that information must always be understood in its historical context, the practices of information, and its material instantiations in texts.[16] In the following section, I discuss how rhetorical genre theory enables me to combine the materiality of documents with information practices.

Information History and Rhetorical Genre Theory

Understanding information merely as a technological phenomenon means that information has been removed from the humanistic tradition.[17] However, if we want to understand the impact of the present information society, a technologically grounded understanding of information is insufficient. In this work, I have, besides the inspiration from Foucault, gained much inspiration from the field of rhetorical genre theory.[18] Rhetorical genre theory provides a lens that also includes the materiality of documents and has a sharp focus on the text, what a text is, and what a text does. Texts contain information, but they are not neutral containers of information but instead frame, sort, and shape information in certain ways.

As historians, we are trained to work with texts and to understand them in their historical contexts and as products of their historical contexts. This is stressed by the approach of rhetorical genre theory, which stresses texts as human constructs and as part of everyday life. In rhetorical genre theory, a text is understood as a concrete historical event that also has an impact on history or takes part in history.[19] Although genre theory is fundamentally communication-theoretical in its approach, I have found it inspirational because of its implied and distinct view on information.[20] In the following paragraphs, I will briefly present the main aspects of rhetorical genre theory on which I rely in my general work and specifically in this contribution before turning to how rhetorical genre theory affects my understanding of information.

Genre is a much-used concept within the humanities and is used as a classification of different text types in, for instance, literary studies, film studies, and music.[21] Rhetorical genre theory is different from literary genre theory in that it is pragmatic and focuses on intention.[22] It understands a genre not so much as a text type or as a text form but as typified social actions, which means that genres such as texts are in constant dialogue with their contexts.[23] An underlying assumption is, first of all, that texts (of any kind) are social and communicative actions and that genre contributes to establishing or understanding the relationship between text, its contexts, its writers, its readers, and other texts.[24] Genre is thus not a fixed classification but an open-ended and continuously evolving process.[25] Another particular feature of rhetorical genre theory is that it is interested in everyday genres such as user manuals, tax forms, and memos—or in my case, decrees, rescripts, memoranda, police bills, and other genres stemming from a state bureaucracy.[26]

In her seminal essay in 1984, Carolyn Miller defined genres as representing recurring typified rhetorical actions.[27] This means that genres emerge in social contexts and as a consequence of a specific rhetorical situation. Both context and rhetorical situation provide the genre with meaning. The rhetorical situation provides an exigency that requires generic responses. Such exigencies are socially constructed and are not to be understood as external, deterministic occurrences.[28] Hence, genres are shaped by ideologies, values, practices, and norms of a given social organization or community. The analysis of genres gives an analytical opportunity to explore the structures and values of the given communities and organizations.[29] In short, genres are historically embedded and reflect textual cultures of a given historical period.

JoAnne Yates emphasizes that genres do not operate individually.[30] Genres are also not merely an expression of an individual's need for writing but tie together the individual and the social level of an organization or community.[31] Genres are related to other genres of a given community, which is reflected in the set of genres—or the *genre repertoire*—of the community.[32] But such a genre repertoire is also connected to other genres of other communities. In 1994, Charles Bazerman introduced the idea of *genre systems* not only to stress how genres structure activities and processes but also to include genres of other parties or communities that participate in the "set of social relations."[33] Genre systems connect different genre repertoires.

When Yates and Orlikowski, and also Bazerman, discuss genre repertoires or genre systems, they focus on very specific communities, organizations, or rhetorical situations. I have previously discussed how the authorities in Copenhagen relied on a vast repertoire or system of genres in a situation in which the inhabitants of Copenhagen needed to be prepared for a coming siege. This situation and the different kinds of genres contained information about securing defense lines and supplies for the citizens and the military forces. But they also disseminated information about water supplies for the fire brigade and gave citizens instructions about how to react in case of bombs or where to seek refuge.[34]

The close relationship between genres and activity spheres means that genres not only help readers navigate the concrete texts but also direct users toward the activity sphere of the text. In this way, texts "help enlist and align people to larger social institutions and practices and text genres provide means of recognizing social relations, obligations, and interactions embodied within communications."[35] Genres do this because they also shape the content, the information, of the text because genre "creates a space that prompts the production of certain kinds of information."[36] Genres form and shape information according to different informational landscapes, which help the readers (as well as the writers) navigate the information at hand. The informational landscape, Bazerman et al. argue, consists of objects, actors, events, and a time-space setting.[37] From this point of view, it becomes interesting to study how decrees and instructions orient readers to a specific activity sphere and which actions define the concrete activity sphere—for instance, when it comes to navigating censorship rules.

Rhetorical genre theory emphasizes information as "a human creation for human purposes."[38] The idea of information as shaped by a given purpose is also stressed by Neil Postman in an analysis of the French encyclopedia and the 18th-century information culture.[39] Positing that information is a human construction created with a specific purpose in mind means that information is not value-free or neutral.[40]

To summarize, I do not have any fixed understanding of information other than that it is always about something and that it is defined, created, and formed by a certain socially perceived need and hence created with a distinct purpose. Information is formed and shaped by the communicative actions in which it emerges, which affect the production of genres. I find such an understanding helpful when inquiring and analyzing past cultures, in which "information" as a word was hardly used and in which perceptions, ideologies, technologies, and social relations were fundamentally different from those in our present society. In the following section, I illustrate how this approach works in the context of the absolutist Danish-Norwegian state in the 18th century.

Decrees as Communicative Actions

My work concentrates on conceptualizing the Danish absolutist state (1660–1849) as an information state and how information became a trading zone between population and state.[41] From 1660 until 1849, the Danish system of government was absolutist, which resulted in a strong centralization of power and bureaucracy in Copenhagen. At this time, the Danish state was a conglomerate consisting of the double monarchy Denmark-Norway, the duchies Schleswig and Holstein, the dependencies Iceland, Faroe Islands, and Greenland, and several colonies in the West Indies, Africa, and India—all governed from Copenhagen. My examples in this chapter deal only with the Danish part of the state.

The absolutist ideal was embodied in the constitutional Royal Law from 1665 and prescribed that the king was accountable only to God, and that he alone had the power in his hands. Idealistically, the king governed alone, but in practical everyday life, the role of civil servants and the rising bureaucracy increasingly gained power in a pragmatic realization that an absolute sovereign could not manage a state on his own. During the 18th century, the legitimization of power changed from the divine to the ideal of opinion-led rule. This ideal expressed the need for the state to include knowledge from its subjects.

As a result of an agreement between the three estates and Frederik III in 1661, the Act of Succession, the powers bestowed on the absolute sovereign meant that he and his government in the form of the state council had the prerogative of legislation without the participation of the population. This agreement meant that the three estates disposed of their (political) influence.[42] There was no representation of the population in the form of a Diet or parliament in Denmark until the Assembly of the Estates in 1834, which was only advisory. Theoretically, the population had no political interest and no part in lawmaking. In practice, the population had possibilities for making their voices heard.[43]

Laws are crucial for regulating social life and for the contract between the sovereign and the population. This made laws important as part of "state communication."[44] Several decrees had a short preface or contained short motivations for the actual orders of the text. In as early as the 1880s, the historian Edvard Holm argued that decrees were a means for the sovereign to publicly communicate with the population.[45] Holm indicated that this communication might reflect an oral tradition formed by traditional arenas, such as parliaments, for political participation.[46] Holm further identified different shades in the presentation depending on the content. For instance, the tone in decrees about religion was often rebuking, whereas new taxes were leniently explained. Regardless of the topic, the decrees always signaled a royal interest in the welfare of the subjects. Arthur der Weduwen has recently argued that the republican regents in the 17th-century Dutch republic used the laws to justify to the population their reasons for issuing legislation. Through the laws, the republican leaders presented the ideal of the state and persuaded the citizens of its superiority.[47] Der Weduwen is interested in the publication of the laws and how the authorities communicated with the population.

As part of state communication, laws were communicative actions that formed an informational landscape that enabled the king, the government, and the subjects to navigate the world of information. The concrete decrees did more than prescribe the correct behavior of people in economic, social, and religious life.

Recently, research has focused on how the Danish absolutist state operated different communication strategies with the population: a horizontal and a vertical strategy. The emerging public sphere defined the horizontal strategy, whereas the vertical communication channels concentrated on how the state used, for instance, petitions as a means of communication with the population.[48] These strategies also affected how information was understood: the petitions (and the contained information) belonged to the private sphere of the state, and only through complex management of censorship rules could the private and secret information access the horizontal communication means of books, journals, and newspapers in the public sphere.[49]

Law texts belong to a very specific genre repertoire. The legal world constitutes a large and heterogeneous genre system that contains legal texts such as codes, decrees, and instructions. It also contains legal treatises by scholars or men of learning, petitions from subjects to the king, and concrete files in police archives, as well as court rulings. When decrees are used as source material, they are usually interpreted as a relic of past legal worldviews and practices. The need for issuing a new decree, instruction, police bill, or legal document reflects a recurring rhetorical situation of a state that—in the Foucauldian sense—acted to control and discipline the population or, on a more positive note, acted to protect and secure the welfare of its citizens.[50] But laws also relied on the cooperation of the population in implementing and reinforcing their content.[51]

As legal documents, they initiated a diverse genre system, in the sense that the articles in the Danish Code initiated further legislation but also generated cases in the legal system. Decrees are particularly interesting since they are stabilized as genres. Yet, as becomes clear from the analysis, they also display change. The analysis concentrates on the rhetorical situation and the informational landscape of the genres. How was the necessity of the different legal initiatives on censorship framed? Or more precisely, which understandings of information can be identified in these different genres? How did they stage information, and how did they create informational needs that prompted new genres?

I focus on censorship rules because they also express ideologies and values about information. Scholars have been interested in the conditions for expressions and how they control not only targeted expressions in writing or speech but also the formation of associations.[52] Censorship rules were not new to the absolute state but dated back to the Reformation in 1536 and the Church Ordinances in 1537 and 1539. Following the changes in the media landscape in the late 17th and early 18th centuries, the state had to issue new decrees and direct instructions to keep pace.[53] The state handled the difficulties of controlling the media landscape in two different ways: (1) through laws that banned certain content and certain genres, such as lampoons and pasquinades, and (2) through instructions to civil servants, such as the instructions to the two newly appointed newspaper censors in Copenhagen.

In the next paragraphs, I analyze four different examples by using three different genres: the code, the decree, and the instruction. All three genres are part of a wider activity system defined by state bureaucracy, communication, and law enforcement, but they originate in different rhetorical situations. I begin with a short introduction to the overall activity system of legal texts before turning to the concrete genres.

The informational landscape of decrees

The genre repertoire of legal texts oriented readers and users of these texts toward an activity sphere characterized by definitions of legal or criminal behavior. Typically, these texts were born in state bureaucracy—more specifically, either in the state council or in the Danish chancellery. They were expressions of power regulation and exercise of power. The language would need to be precise and specific to ensure a correct understanding by the people, judges, and other civil servants. An impersonal tone, detached from individual and subjective judgment, would seem natural to concentrate on the issue at hand.[54] Yet the language was also bombastic, with the purpose of impressing the population and legitimizing the rule by demonstrating the glorious power of the absolute king.

As briefly mentioned, the genre repertoire consisted of different kinds of genres beyond the legislative texts themselves. Legislation prompts further genres in the judicial system, in which, for example, disputes resulted in documents from the court proceedings. My sample material consists only of texts issued by the chancellery, which include different genres such as decrees, ordinances, instructions, and police bills. How are these genres different? It is not only a question of who is speaking or by whose authority they are issued. Police bills were issued by the commissioner of the Copenhagen police but by request from the Danish chancellery. In the end, all legislation originated from the king's authority.

The differences between the genres find their origin in the different recipients and the different degrees of publicity. Decrees and police bills were aimed at the public, whereas ordinances and instructions addressed concrete persons, often clarifying or elaborating on specific issues or requesting these persons to take certain actions. The latter often began with promises of royal favor, whereas the former began with the full title and possessions of the king as well as his intention of making the specific content known and ended with the phrase "under our hand and seal." Yet the mere expression of making the information known was not enough to guarantee that the population got to know the content. Traditionally, decrees were read by vicars after Sunday sermons, which also included explaining the content of the decrees. Some decrees end with a command to members of the nobility and other local authorities, such as prefects, mayors, and aldermen, to make the decrees known to the public.[55] In the instruction to two newspaper censors in Copenhagen in 1701, the civil servants of the Danish chancellery requested that decrees were brought in the newspaper "the following mail day after its [the decree's] publication."[56] Legislation was embedded in the communication strategies of the state.

Following the initial formula phrases, decrees typically addressed the purpose of the given legislation. This part of the text is particularly interesting since the absolute sovereign here argued for the urgency of the legislation. The arguments were based on and revealed the information that was considered necessary for the population to obey. Here the vicar could also find the reasons if he needed to explain the content to his parishioners. Prior to the specific paragraphs, a short summary digested the main content of the decree. Finally, the decrees ended with a description of the penalties.

In the next paragraphs, I analyze how laws as a genre defined information in a specific way and oriented the readers toward a specific informational landscape. I begin with the Danish Code from 1683, which stipulated the overall rules for censorship.

"Nobody can let anything print"[57]

In this section, I analyze the Danish Code from 1683, because it compiled existing legislation and because it covered Denmark as a unified entity.[58] The Code was one of the first examples of how the absolutist state acted to modernize and standardize the existing legal frameworks. The ideal was that the Danish Code was a durable, general, and universal law book.[59] Divided into six books, the Code regulated all aspects of society from procedures of the judicial system to the penal code and maritime laws to the obligations of the clergy.

Rules regarding expressions are located in two different books of the Code. Censorship was the responsibility of the church and university, and concerned the relationship between subject, state, and (mainly) religion. But other expressions were regulated too—those between people—and they were regulated in the book concerning the penal code. The articles about censorship in the Code began by describing the process of censorship and introducing the professors at the University of Copenhagen as censors. Their approval was a cornerstone without which no books could be printed. The subsequent articles identify the content to be censored as the information that was not approvable: criticism of the government, the king's person, or the Lutheran confession, but they also identified certain genres as almanacs (if not the university-calculated version) and calendars with prognostications or prophecies.[60]

The rules were preoccupied with printing as action: to print—or to let others print—was forbidden. Besides printing, the import of books with banned content was prohibited. Besides the professors, the printers were important actors, as was the executioner, who was commissioned with the task of burning books that encouraged insurrection against the king. One actor, however, was not addressed: the author of the texts.

The informational landscape was defined by certain (prohibited) actions, a main actor who was a "nobody," and a very concrete procedure that defined a specific censorship time. Censorship time depended on the approval of the professors at the University of Copenhagen (or the local bishop if outside Copenhagen). In the form of pre-censorship, it defined a time span before which printing was possible, but it was a time span without any specific duration. The pre-censorship time span also meant that printing was illegal without the permission of a censor. Only the approval of the censor opened a new time span for printing and distribution.

Without this approval, printers were committing an offense for which they were punished. However, the Danish Code was far from explicit or specific in stipulating penalties for the offenses. The unclear definitions made the punishment arbitrary. Nobody reading the Danish Code could doubt that punishment was hard. Punishment regarding lese-majesty was included in the penal code. The penal code was structured in accordance with the Ten Commandments. The rules of lese-majesty reflected the fourth commandment (according to the Lutheran classification): to honor your father and mother.[61] In the Lutheran doctrine of the three estates, all individuals were either authority or subordinate, and all authority stemmed from parental authority.[62] As the king was, by divine power, the main authority in the state, lese-majesty was met with capital punishment, including the wheel.[63]

The informational landscape of censorship did not include the relations between the subjects. During absolutism, society was infused with honor, which made defamations unwanted information. As a very specific form of information, defamations were rumors, lies, or slander such as "thief, son of a whore, traitor."[64]

The spatial settings for conducting offending behavior in matters of honor were the inn, the mill's house, the smithy's house, the moot, or other places of "such audience."[65] These spaces reflected the traditional medieval structure of an oral society. However, defamations also existed in writing. Genres such as lampoons and pasquinades were specifically addressed as written (not printed), which reflected a social structure unaware of mass communication by print.[66]

The Danish Code presented an informational landscape defined by printed books and expressed a strong desire to control information about specific topics such as state, religion, and the king's person. Besides these topics, the landscape identified prognostications, prophecies, and other speculations about war, periods of scarcity, or coming plagues as unwanted information. This kind of information, which we today would probably identify as fake news, misinformation, or simply bogus, was feared because it risked creating a sense of insecurity in the population.

Information in the Danish Code expressed strong religious values but reflected also social norms of honor. In the next section, I explore how these different values merged and formed a rhetorical situation underlying a decree in 1689 about pasquinades.

"To be burned at the stake"

This section begins with a brief characterization of the media landscape in the late 17th century and how it changed. This serves as a background for the concrete analysis of the decree from 1689 about pasquinades. The informational landscape of the Danish Code identified a media environment of printed books, orality, and handwritten documents (ephemera such as lampoons and pasquinades). The perception of the media landscape reflected the intention of the king and his government to control especially the communication of news, which they also managed to use for their own propaganda purposes.[67] The young absolutist state was intensely preoccupied with critical voices against the new line of government, and a few cases of lese-majesty were conducted in the 1660s. Libels against the authorities were continuously banned as the same authorities tried to gain control of the new media of communication: the newspapers.[68] However, the late 17th century witnessed a change in the media landscape, driven by the communication of news in newspapers, pamphlets, broadsides, and street ballads, which existed to convey information in print.

As genres, pasquinades were characterized by defaming and conveying libels, but not only against the royal family or the government. They were often anonymous, which was seen as an aggravating circumstance.[69] It seems reasonable that concrete events led to renewed releases of decrees against pasquinades, such as the decree of June 8, 1689.[70] Christina Holst Færch argues that this specific decree emanated from a theological and personal controversy between a Court chaplain in Copenhagen and a philosopher, Christian Thomasius, in Leipzig. Some formulations in the decree clearly signaled a close relationship with the concrete situation, but it also illustrated a preoccupation by the state with pasquinades on a more general level.[71]

However specific the event leading to the 1689 decree was, the decree itself aimed at the broader issues of defamations. It opened with a generically stabilized formula presenting the king, Christian V, as the main actor regarding legislation in the geographical area identified in the formula. After this introduction, the 1689 decree continued with a brief explanation of the rhetorical situation leading to the decree: "it has come to the knowledge of the king that." In this way, the decree presented a short argument for its necessity, but it also defined the main event: lampoons against the state, the king's person, the religion, and several members of the clergy had "slipped into" the country and had been distributed by unknown actors.

The formulation in the passive voice indicates the concealed identity of the main actors who illegally circulated the pasquinades. In contrast to the unspecified reference (to "nobody") in the Danish Code, the 1689 decree explicitly addressed all subjects of the king. The king warned his subjects generally and the booksellers specifically, and the decree instructed those in possession of pasquinades to hand them over to the police commissioner for the executioner to burn them at the stake. Besides the general address to the subjects, the king explicitly prohibited booksellers from importing, selling, or hiding infamous and scandalous lampoons. If either booksellers or subjects did not obey, they would incur a penalty either in the form of a large fine or by "forfeiting the body" if they could not pay the fine. Hence, the decree recapitulated that possession, and dissemination of lampoons was a criminal act even if they were imported. The king did apparently not consider that his own subjects could write such libelous texts. As in the Danish Code, there is no explicit reference to a writing individual or an author.

The reference to the police commissioner in Copenhagen, Claus Rasch, indicated that the spatial setting was restricted to the capital. However, it becomes clear from the final part of the decree that the entire population was the intended audience. All counts, barons, prefects, mayors, and city councils were ordered to proclaim and read aloud the decree at appropriate places to ensure that everybody was informed about the content.

The concise and somewhat impersonal account of the illicitness of possessing pasquinades was balanced by the personal tone of the king's voice, which presented him as an individual legislator with an interest in his subjects. The decree initially stated how the troubles had come to his knowledge, and the decree expressed what he expected from his subjects. The decree is situated exactly between the events of the past (the lampoons that had slipped into the country) and the expectations of future behavior (that the subjects immediately would hand over the lampoons to the executioner). The informational landscape of the decree included three main actions: the warning of the subjects by the king, the handing over of the problematic pasquinades, and the burning of these pasquinades at the stake. Information such as defamation was scandalous and libelous in character. Finally, the decree emphasized that the state was able to control but also eliminate unwanted information at the stake.

The idea that information could be burned at the stake indicates a reified understanding of information. The pasquinade as a genre still reflected a traditional media landscape based on orality and writing. This landscape was, however, changing due to new media such as newspapers. How they influenced the attempts of the Danish state to control information can be seen in an instruction from 1701 that strengthened pre-censorship.

"What really happened"

Before turning to the instruction from 1701, I begin by situating the instruction in a brief sketch of the media landscape. Newspapers were not specifically addressed in the early legislative work of the absolutist state.[72] Presumably, the increasing complaints from foreign envoys against stories in newspapers from Hamburg led the new king, Frederik IV, to instruct two civil servants to revise the newspapers before they were printed. News from and about foreign states was particularly troubling to the absolutist state because this kind of news was accused of circulating rumors. The policy of the state regarding newspaper censorship seemed to have been rather volatile and indecisive. Another means of controlling the news sector was through licenses to the printers and editors of the newspapers.[73] It seems that the instruction of 1701 led to a firmer practice of pre-censorship, but after the last censor stepped down from the post in 1719, there was no pre-censorship of newspapers until 1738.[74]

In the instruction, the king appointed two civil servants as censors.[75] As a genre, the instruction individually addressed specific persons, and it became operational through its concrete descriptions of procedures in specific cases and its suggestions of what actions the civil servant in question had to undertake.[76] In 15 sections, the 1701 instruction elaborated on how the two censors should conduct the censorship.[77] The instruction summarized existing practices, stipulated new guidelines for enforcement of the censorship rules, and detailed severe restrictions regarding the content of foreign news and news in general. The instruction systematized control and gave the state the means to intervene at any moment, which internalized the self-censorship of the newspaper editors.[78]

The instruction summarized the tasks of the two censors in a short preamble. In the first section, the instruction outlined a process of control as it provided the two censors with the task of revising all newspapers before they were printed. The following sections elaborated on the permitted content, ranging from how to report political occurrences in Denmark to how to use news from foreign newspapers. The relationship with foreign powers and the stability of foreign policy were crucial topics in the instruction because they depended on the "conditions of time or . . . any great change in Europe."[79] This understanding of the position of Denmark in international politics infused the instruction not only by restricting what could be digested from foreign newspapers (section 8) but also by stipulating how the newspapers could write about envoys and view foreign powers (sections 5–7). It seems that the chancellery here suggests that the instructions might change if new power constellations emerged. The religious values of the Danish Code were replaced by international relations.

The two censors and the printer were the main actors in the instruction. Yet foreign envoys, allies, or foes played a significant role. It was important that the reported information in the newspapers did not offend foreign powers, and it was consequently emphasized that the newspapers could describe the events as they took place (for instance, the reception of a new envoy at the Royal Court). The idea of representing the events as they took place also affected the understanding of information from foreign newspapers. Section 8 specified how the newspapers should excerpt the "most important, extraordinary, and remarkable [information] and to present [it] without verbosity or partiality but in a useful and decorous manner."[80] This understanding of information as a factual representation of reality was stressed in the following section, which explicitly forbade futile speculations and reasoning. Newspapers should deal only with that "which is reported as having really happened."[81] Reality—or more specifically, reality according to the absolute state, political alliances, or enemies—defined the informational landscape in which the work of the two censors took place, which again affected printers and newspaper editors.

The instruction spoke to the bureaucracy and was not aimed at the public. It introduced an entirely new, political conceptualization of information. The state still intended to control information by using pre-censorship but also by providing a definition of information as reflecting reality. The opposite of this definition was information as reasoning, a speculative form of information. The distinction influenced the Danish media landscape during the 18th century, when newspapers merely accounted events and reasoning became the hallmark of the emerging periodical press.[82]

Controlling information through censorship was still the main strategy of the Danish state during the 18th century. However, there was an intermezzo from 1770 to 1772 with unlimited freedom of expression. In the next section, I begin by sketching out the years of freedom of expression before I turn to the analysis of the final example, the decree about the freedom of the press from 1799.

"Grumbling with acrimony"

During the 18th century, the state experimented with using public debate as a means of acquiring information about the economy as a well-defined area. Yet it was not a repeal of censorship rules. The repeal came, all of a sudden in 1770, when an ordinance issued on behalf of King Christian VII (1769–1808) by his private physician, Johan Friedrich Struensee, repealed all censorship legislation.[83] Ordinances were addressed to one or several specific persons or authorities containing an administrative command; and the 1770 ordinance was specifically addressed to printers, professors, and bishops, all of whom were actors in the activity sphere of censorship. The 1770 ordinance provided the public with unlimited freedom of speech and emphasized in this way a commitment of the absolute state to letting debaters express themselves.[84]

The repeal resulted in a lively debate. The writers, however, soon also vigorously debated Struensee's affair with the queen, which resulted in a retraction in 1771 regarding lese-majesty and criticism of religion.[85] Struensee fell from power in 1772, but the new rulers were irresolute concerning a restoration of censorship. Finally, the new rulers decided not to reintroduce censorship, but legislation in 1773 restricted what information newspapers were allowed to include. In effect, this ended the years of unrestricted freedom of expression.[86]

In the following years, especially after the coup d'état led by the crown prince (later King Frederik VI) in 1784, the execution of censorship was rather liberal. The more liberal interpretation did not imply a legally sanctioned freedom of speech but only that no pre-censorship was reintroduced as long as the press wrote in favor of the regime.[87] However, the 1790s also witnessed an increasing concern of the crown prince regarding what he considered improper use of the freedom of expression. Consequently, a new decree was issued in 1799 entitled "Decree regarding freedom of the press." To 21st-century readers, the title suggests the reinstatement of the full freedom of expression. However, using an almost Orwellian newspeak, the decree did not reintroduce freedom of expression but explicitly aimed at defining the "borders" of the freedom of expression.[88]

The 1799 decree opened with a preamble in the voice of the king stating certain principles.[89] These principles situated the concrete paragraphs and the regulated content and suggested penalties in a distinct informational landscape. The landscape was characterized by the benevolent king who considered that the freedom of the press was valuable for popularizing education and knowledge among all classes in society but that ill-wishers constantly attacked the very foundation of society. Within this continuum between proper and improper use, information took on different forms, such as rumor, untruthful accounts, and slanderous accounts if not factual.

The 1799 decree was positioned within the existing genre repertoire of the Danish Code and former ordinances in a space characterized by print (including engravings), writing, and orality. In the preamble, the voice of the king presented past grievances of improper use and evoked earlier decrees and ordinances as ignored warnings. Together, they formed the rhetorical situation of the new law. The preamble not only evoked past grievances but argued that the law was still violated. Hence, the preamble foreshadowed a future in which such violations could be avoided if the borders of the freedom of speech were properly defined. The rhetorical situation established in the preamble constructed these continued violations as a need for clarifying the limits of free speech, which was sketched out in the specific paragraphs on possible violations against the law.

The improper use of free speech would result in the erosion of the cornerstones of the state and the disturbance of civic society. To encircle the boundaries of free speech, it is useful to examine how specific communicative actions identified concrete kinds of information as questionable and unwanted. Besides more neutral verbs such as *to translate, to announce, to disseminate, to publish,* or *to present*, the paragraphs used verbs such as *to incite* or *to advise.* In combination with changes in the constitution or insurrection against the king, information would be violating the law. To disseminate untruthful accounts of any important aspect of the state's conditions, the government's decisions, or the government's measures would result in imprisonment in the House of Correction for between two months and two years, depending on the damage or inconvenience caused by the untruthful information.

Verbs such as *to blame* or *to insult* implied that information was offensive when it dealt with the constitution, the government, the monarchy as a form of state, or the royal family (paragraphs 1–4). Insults against members of the royal family were specified as rumors. Since the verb *to blame* (*at laste* in Danish) was crucial to the decree, it was specified how it should be interpreted in a postscript to the law itself. In this postscript, it was emphasized that the meaning was conceived as generic and that it originated from traditional legal language. Further, the meaning of the verb was clarified to be an attribution of defectiveness to persons or things, such as "the state's form of government and the religion, protected by the laws of the country, are blamed when they are attributed such deficiencies that the purpose of civil society (freedom, security, and the common good) is unachievable or cannot exist."[90]

The dilemma of the law was to prohibit unwanted criticism against the form of government but at the same time to permit honorable and educated men to publicly share their thoughts on how to improve the conditions of the common good, the laws of the country, or public arrangements. The author could share such insights when expressed with modesty and showing "the deference that the citizen and subject owes the government and the legislator."[91] If, however, the author "grumbled with acrimony" or expressed his considerations in unjustifiable and indecent words, he risked imprisonment.

The 1799 decree reflected a firmer understanding of the media environment of the late 18th century compared to the understandings reflected in the Danish Code and the 1689 decree. The government clearly addressed the (educated) citizens, specifically those who were writing. The author emerged during the years of press freedom and became the target of the 1799 decree, together with the publisher and the printer. The firmer understanding of the media landscape is also reflected in the attention to orality, writing (pasquinades), and print (including engravings) as media forms with different impacts. Defamations, for instance, were considered much more serious if they were publicly announced in print than if they were presented orally. The reason for this was not that the audience was wider but that the mere act of having libels printed was considered deliberate.

The 1799 decree was the last example in this chapter of how the absolutist state staged information in legal texts. As readers, we read legal texts with specific expectations. In a very concrete way, we read them to figure out what they forbid. As historians, we add to this that we read such texts for what they can tell us about past societies. The different genres "enlist[ed] and align[ed]"[92] the population in general and civil servants in particular to navigate in a world of information and to articulate information in accordance with the norms and values expressed in the rules. Laws inform people how to act in accordance with the law and what the state considers proper behavior and good conduct. In this way, the analysis illustrates how the absolutist state identified the good citizen.

The three examples of laws (the Danish Code from 1683, the 1689 decree, and the 1799 decree) illustrate how genres molded the content—the information—within the text. The initial opening of the two decrees immediately made the text recognizable as a law and oriented its readers (the population) toward a content of which they might not be specifically aware in advance but with which they would be familiar as a typified content that regulated yet another aspect of social life. The laws did so by explicating actors, actions, and consequences of infringements of the law. The 1799 decree at the same time challenged the generic stabilization through the elaborate preamble and by including an explanation of the word *to blame*. The instruction from 1701 illustrates how the censorship rules created a space of bureaucratic knowledge and information practices and how the ideological level of the laws was followed by a practical implementation in an expanding state bureaucracy.

Laws were meant to be publicly known, whereas the 1701 instruction—when not necessarily veiled in secrecy—was addressed to specific persons. This text informed concrete readers how to engage with censorship as a specific practice of controlling information. It is further an example of how the legal texts resulted in other genres and together formed a specific genre repertoire that also merged with other repertoires. The practices of information control and their activity spheres were intertwined with the academic world of the professors at the university in Copenhagen. They read and evaluated concrete manuscripts, leaving approvals or rejections as genres. The practices also intersected with the court system in dossiers and verdicts. As intended, the laws also intersected with the very activity of writing—especially critiques of the governing system.

Conclusion

What do these examples tell us about the relationship between information, documents, and the historical context? First, even stabilized genres such as laws (in the form of decrees) can be twisted when it comes to evoking proper conduct in their audience. Second, laws also address the level of context. Absolutist information policy is often associated with the divine power of the sovereign to regulate what the subjects need to know. It is, however, well known in Danish research that an absolute ideology in an everyday political culture was prone to input from the population—an important aspect that is not necessarily reflected in the ideologically infused legal texts.

My analysis of the three laws and the instruction about censorship from the perspective of rhetorical genre theory illustrates the intertwinement of concrete rhetorical situations and the formulation of an information policy, which had a profound impact on the organization of media during the years of absolutist rule in Denmark. It is not merely the changing media landscape and how the laws responded to such changes as the Danish Code and the 1689 decree, which both pointed toward an obsolete media environment. Likewise, the decree in 1799 reflected the complex news arena with actors such as printers, publishers, and authors. Despite the changes, the state still attempted to regulate and control information in public. The medieval tradition of declaring utterances "null and void" might be seen as a relic in the reified understanding of information in 1683 and 1689, when unwanted information could be disposed of at the stake. The 1799 decree expressed a different understanding of information as formed by different (communicative) actions.

Documents such as laws and instructions are socially, culturally, and ideologically embedded in their concrete historical contexts, and in concrete cases they also reflected complex and changing conceptions of how information was at the same time understood as utterances and as reified items.

Notes

1. Geoffrey Winthrop-Young, "Intel," in *Information: Keywords*, eds. Samuel Frederick, Michele Kennerly, and Jonathan E. Abel (New York: Columbia University Press, 2021).

2. http://labs.statsbiblioteket.dk/smurf?q=information (created and accessed May 30, 2023), 119.

3. https://www.information.dk/indland/2020/05/4-maj-1945-information-foedt-frygt-fare (accessed May 30, 2023).

4. See, e.g., Københavns Kongelig alene priviligerede Adresse-Contoirs Efterretninger, August 1, 1791, no. 165 [upag.], https://www2.statsbiblioteket.dk/mediestream/avis (accessed May 30, 2023).

5. John Durham Peters, "Information: Notes Toward a Critical History," *The Journal of Communication Inquiry* 12, no. 2 (1988), 10.

6. Laura Skouvig, "Present and Past: The Relevance of Information History," *Information & Culture* 58, no. 1 (2023).

7. Michel Foucault, "Nietzsche, Genealogy, History," in *The Foucault Reader*, ed. Paul Rabinow (New York: Pantheon Books, 1984); Michel Foucault, *The Archaeology of Knowledge*, trans. A. M. Sheridan Smith (London: Routledge, 2001), 76–100.

8. Paul Veyne, *Foucault: Die Revolutionierung der Geschichte* (Frankfurt am Main: Surhkamp, 1992), 7, note 1. See also p. 33ff.

9. Foucault's lack of interest in information technology has been heavily criticized within the field of surveillance studies. See, e.g., Kevin D. Haggerty and Richard V. Ericson, "The Surveillant Assemblage," *British Journal of Sociology* 51, no. 4 (2000), 605–622.

10. Andreas Marklund and Laura Skouvig, "Introduction," in *Histories of Surveillance from Antiquity to the Digital Era: The Eyes and Ears of Power*, Routledge Studies in Modern History, eds. Andreas Marklund and Laura Skouvig (Milton: Taylor and Francis, 2021), 7. See also Michel Foucault, *Überwachen und Strafen. Die Geburt des Gefängnisses*, trans. Walter Seitter (Frankfurt am Main: Suhrkamp, 1994), 243.

11. Ronald R. Kline, *The Cybernetics Moment: Or Why We Call Our Age the Information Age*, New Studies in American Intellectual and Cultural History (Baltimore: Johns Hopkins University Press, 2015), 202–228; Colin Koopman, *How We Became Our Data: A Genealogy of the Informational Person*, Chicago Scholarship Online (Chicago: University of Chicago Press, 2019), 16–19.

12. Toni Weller, *The Victorians and Information. A Social and Cultural History* (Saarbrücken: VDM Verlag, 2009), 9–15.

13. Eric Hayot, Lea Pao, and Anatoly Detwyler, *Information: A Reader* (New York: Columbia University Press, 2021); Michele Kennerly, Samuel Frederick, and Jonathan E. Abel, *Information: Keywords* (New York: Columbia University Press, 2021).

14. I have discussed the risks of anachronism and presentism in Skouvig, "Present and Past," 2–3.

15. Ann Blair, "Foreword," in *Literary Information in China: A History*, eds. Jack W. Chen et al. (New York: Columbia University Press, 2021), xvi.

16. Jack W. Chen et al., "Introduction for a History of Literary Information in China," in *Literary Information in China: A History*, eds. Jack W. Chen et al. (New York: Columbia University Press, 2021), xxiv.

17. Eric Hayot, "An Introduction," in *Information: A Reader*, eds. Anatoly Detwyler, Lea Pao, and Eric Hayot (New York: Columbia University Press, 2022), 1–14.

18. For an introduction, see *Genre Theory in Information Studies*, Studies in Information, ed. Jack Andersen (Bingley, UK: Emerald, 2015).

19. Kristin Asdal et al., *Tekst og historie: å lese tekster historisk* (Oslo: Universitetsforlaget, 2008), 25.

20. Asdal et al., *Tekst og historie,* 193.

21. Carolyn R. Miller and Amy J. Devitt, "Introduction to Rhetorical Genre Studies," in *Landmark Essays on Rhetorical Genre Studies*, eds. Carolyn R. Miller and Amy J. Devitt (New York; London: Routledge, 2019), 1–7.

22. Asdal et al., *Tekst og historie*, 193.

23. Carolyn R. Miller and Amy J. Devitt, "Introduction to Rhetorical Genre Studies," in *Landmark Essays on Rhetorical Genre Studies*, eds. Carolyn R. Miller and Amy J. Devitt) (New York; London: Routledge, 2019), 1–7.

24. Charles Bazerman, "Systems of Genres and the Enactment of Social Intentions," in *Genre and the New Rhetoric*, Critical Perspectives on Literacy and Education, eds. Aviva Freedman and Peter Medway (London: Taylor & Francis, 1994), 79. Asdal et al., *Tekst og historie*, 189.

25. Ralph Cohen, "History and Genre," *New Literary History* 17, no. 2 (1986): 204–205.

26. For genres in tax accounting, see Amy Devitt, "Intertextuality in Tax Accounting: Generic, Referential, and Functional," in *Textual Dynamics of the Professions: Historical and Contemporary Studies of Writing in Professional Communities,* eds. J. Paradis and C. Bazerman (Madison: University of Wisconsin Press, 1991), 336–357.

27. Carolyn R. Miller, "Genre as Social Action," *The Quarterly Journal of Speech* 70, no. 2 (1984), 151–167.

28. Miller, "Genre as Social Action," 155–156.

29. JoAnne Yates, "Values, Media, and Genres for Standardization," in *Historical Studies in Computing, Information, and Society*, Insights from the Flatiron Lectures, ed. William Aspray (Cham: Springer International Publishing, 2019), 52. See also Cohen, "History and Genre, " 214–215.

30. Yates, "Values, Media, and Genres for Standardization," 53.

31. Laura Skouvig, "Genres of War: Informing a City," in *Genre Theory in Information Studies*, ed. Jack Andersen (Bingley: Emerald Group Publishing Limited, 2015), 137.

32. Devitt, "Intertextuality in Tax Accounting: Generic, Referential, and Functional," 340–341; Wanda J. Orlikowski and JoAnne Yates, "Genre Repertoire: The Structuring of Communicative Practices in Organizations," *Administrative Science Quarterly* 39, no. 4 (1994), 541–574.

33. Yates, "Values, Media, and Genres for Standardization," 53. Bazerman, "Systems of Genres and the Enactment of Social Intentions," 98–99.

34. Skouvig, "Genres of War," 133–154.

35. Charles Bazerman, Joseph Little, and Teri Chavkin, "The Production of Information for Genred Activity Spaces: Informational Motives and Consequences of the Environmental Impact Statement," *Written Communication* 20, no. 4 (2003): 456.

36. Bazerman, Little, and Chavkin, "The Production of Information," 457.

37. Bazerman, Little, and Chavkin, "The Production of Information," 457–458.

38. Charles Bazerman, "Nuclear Information: One Rhetorical Moment in the Construction of the Information Age," *Written Communication* 18, no. 3 (2001): 260.

39. Neil Postman, *Building a Bridge to the 18th Century: How the Past Can Improve Our Future* (New York: Vintage Books, 1999).

40. In early-21st-century discourse, information is often placed in a hierarchy, with data and knowledge both having histories of their own. A main difference between information and knowledge is that the latter always implies a subject. Information seems to exist independently of a knowing subject. See Geoffrey Nunberg, "Information, Disinformation, Misinformation," in *Information: A Historical Companion*, eds. Ann Blair et al. (Princeton, New Jersey: Princeton University Press, 2021), 496–502. Data seems to have the same corpuscular qualities as information. Daniel Rosenberg argues that data today is understood as "information in numerical form" (p. 33). Daniel Rosenberg, "Data before the Fact," in *"Raw Data" Is an Oxymoron*, ed. Lisa Gitelman (Infrastructures Series; Cambridge, Massachusetts: The MIT Press, 2013), 15–41.

41. Laura Skouvig, "Rethinking a Danish information State: Information in the Trading Zone," *Library and Information History* 39, no. 2 (2023), 82. Pamela O. Long, "Trading Zones in Early Modern Europe," *Isis* 106, no. 4 (2015), 840–847.

42. https://danmarkshistorien.dk/vis/materiale/enevaelden (accessed June 22, 2023).

43. Michael Bregnsbo, "Struensee and the Political Culture of Absolutism," in *Scandinavia in the Age of Revolution: Nordic Political Cultures, 1740–1820*, eds. Pasi Ihalainen et al. (Farnham: Routledge, 2012), 55–65.

44. Arthur der Weduwen, *State Communication and Public Politics in the Dutch Golden Age* (Oxford: Oxford University Press, 2023).

45. Edvard Holm, *Danmark-Norges indre Historie under Enevælden fra 1660 til 1720: Indledning til den dansk-norske Stats Historie fra 1720–1814*, vol. 1 (København: Gad, 1885).

46. With the transition to an absolutist rule in 1660, the possibility for participating in the governing of the country was out of the hands of the estates. In general, until 1660, there had not been much opportunity for the burgher estate to participate in the governing of the country, with the power in the hands of the king and the council of the state (Rigsråd) consisting mainly of members of the nobility.

47. der Weduwen, *State Communication*, 4–5.

48. Jakob Maliks, "Vilkår for offentlighet: sensur, økonomi og transformasjonen av det offentlige rom i Danmark-Norge 1730–1770" (Norges teknisk-naturvitenskapelige universitet, Det humanistiske fakultet, Institutt for historie og klassiske fag, 2011), 171–176. Ellen Krefting, Aina Nøding, and Mona Renate Ringvej, *En pokkers skrivesyge: 1700-tallets dansk-norske tidsskrifter mellom sensur og ytringsfrihet* (Oslo: Scandinavian Academic Press, 2014), 60.

49. Maliks, "Vilkår for offentlighet," 331–333.

50. Toni Weller, "The Information State: A Historical Perspective on Surveillance," in *Routledge Handbook on Surveillance*, eds. Kirstie Ball, Kevin D. Haggerty, and David Lyon (New York: Routledge, 2012), 59.

51. der Weduwen, *State Communication*, 4.

52. See, for instance, Øystein Rian, *Sensuren i Danmark-Norge: vilkårene for offentlige ytringer 1536–1814* (Oslo: Universitetsforlaget, 2014). Rian focuses on the dual monarchy of Denmark-Norway, which ended in 1814 when Norway was ceded to Sweden as a result of the peace conference following the Napoleonic Wars.

53. Krefting, Nøding, and Ringvej, *En pokkers skrivesyge*.

54. Leif Becker Jensen, "Under henvisning til den af Dem fremsendte skrivelse," in *På given foranledning. En antologi om dansk forvaltningskultur*, eds. Karl Peder Pedersen, Grethe Ilsøe, and Ditlev Tamm (København: Dansk Jurist- og Økonomforbund, 1995), 112–115.

55. See, for instance, "Forordning om Politiens Administration," in *Kong Friedrich den Fierdes allernaadigste Forordninger og aabne Breve Aar 1701 udgangen* (Copenhagen: Johan Frederik Schultz), October 22, 1701, 96.

56. "Instruction Regarding Censorship of Newspapers," Copenhagen, March 19, 1701, in *Dagspressen i Danmark: dens Vilkaar og Personer indtil Midten af det attende Aarhundrede*, vol. 2, ed. P. M. Stolpe (Kbh: Rosenkilde og Bagger, 1977), 11.

57. I use the following version of the Danish Code: V. A. Secher, *Kong Christian den Femtis danske Lov* (København: Gad, 1929), 2-21-1. I use the traditional form of reference to the Danish Code in book (2), chapter (21), article (1).

58. Four years later, it was followed by the Norwegian Code governing Norway.

59. Ditlev Tamm, "Danske og Norske Lov i 300 år. En introduktion," in *Danske og Norske Lov i 300 år*, ed. Ditlev Tamm (Jurist- og Økonomforbundets Forlag, 1983), xxx.

60. Secher, *Kong Christian den Femtis danske Lov*, 2-21-2 and 2-21-3.

61. Jørgen Stenbæk, "En kirkeretslig vurdering af Danske Lovs 6. bog—strafferetten," in *Kirkehistoriske Samlinger* (1972), 58–91.

62. Nina Javette Koefoed, "I Trust You with My Child: Parental Attitudes to Local Authorities in Cases of Disobedient Children in 18th Century Denmark," *Journal of Historical Sociology* 33, no. 4 (2020), 491.

63. Secher, *Kong Christian den Femtis danske Lov*, 6-4-1.

64. Secher, *Kong Christian den Femtis danske Lov*, 6-21-6.

65. Secher, *Kong Christian den Femtis danske Lov*, 6-21-1.

66. Secher, *Kong Christian den Femtis danske Lov*, 6-21-8.

67. Rian, *Sensuren i Danmark-Norge*, 172.

68. Christina Holst Færch, *Smædeskrifter, sladder og erotiske vers i 1700-tallet: Hans Nordrups forfatterskab—med et udvalg af hans digte* (København: Museum Tusculanum, 2019), 113. Rian, *Sensuren i Danmark-Norge*, 172.

69. Rian, *Sensuren i Danmark-Norge*, 157.

70. *Kong Christian den Femtes Forordninger og aabne Breve, som paa Prent ere udgangne fra Lovens Publications Dato indtil Høystbemeldte Hans Majests. dødelige Afgang den 25. Augusti 1699* (1737), 532.

71. Færch, *Smædeskrifter, sladder og erotiske vers i 1700-tallet*, 114.

72. Stolpe, *Dagspressen i Danmark*, 2, 293–294.

73. Rian, *Sensuren i Danmark-Norge*, 245–246.

74. Stolpe, *Dagspressen i Danmark*, 2, 302.

75. The instruction is transcribed in Stolpe, *Dagspressen i Danmark*, 2, 350–355. In Stolpe (1977), the correspondence leading to the instruction is reproduced, as is one example of a censor report to the chancellery.

76. See the definition here: https://ordnet.dk/ods/ordbog?query=instruks (accessed August 3, 2023).

77. Stolpe, *Dagspressen i Danmark*, 2, 309.

78. Stolpe, *Dagspressen i Danmark*, 2, 298.

79. Stolpe, *Dagspressen i Danmark*, 2, 355.

80. Stolpe, *Dagspressen i Danmark*, 2, 353.

81. Stolpe, *Dagspressen i Danmark*, 2, 353.

82. Ulrik Langen, "Reportage, Rumours, and Conversation: Curating News During Political Upheaval in Eighteenth Century Copenhagen," *Media History,* ahead-of-print/ahead-of-print (2022), 1–14.

83. See a more elaborate discussion of the repeal of censorship in Laura Skouvig, "Rethinking a Danish Information State: Information in the Trading Zone." Christian VII was mentally ill and incapable of governing. Struensee seized power and ruled on behalf of the king until he himself was dethroned in 1772.

84. Thomas Lyngby, Søren Mentz, and Sebastian Olden-Jørgensen, *Magt og pragt: enevælde 1660–1848*, 1. udgave. ed. (Kbh: Gad, 2010), 169.

85. Henrik Horstbøll, Frederik Stjernfelt, and Ulrik Langen, *Grov Konfækt: tre vilde år med trykkefrihed 1770–73*, 1. udgave. ed. (Kbh: Gyldendal, 2020), 32.

86. Horstbøll, Stjernfelt, and Langen, *Grov Konfækt*, 35.

87. Rian, *Sensuren i Danmark-Norge*, 183. See also Jens Engberg, *Den standhaftige tinsoldat: en biografi om Frederik 6.* (København: Politiken, 2009), 282.

88. Rian, *Sensuren i Danmark-Norge*, 187.

89. I have used the following digitized copy: *Forordning, som nærmere forklarer og bestemmer Trykkefriehedens Grændser . . . den 27. September 1799.*

90. *Forordning, som nærmere forklarer og bestemmer Trykkefriehedens Grændser*, 14.

91. *Forordning, som nærmere forklarer og bestemmer Trykkefriehedens Grændser*, 6.

92. Bazerman, Little, and Chavkin, "The Production of Information," 456.

Chapter 5

An i for an I

Call and Response for the iSchools

Geoffrey C. Bowker and John Leslie King

The iSchools: Dream and Survival

Our title is a play on *lex talionis*, the principle of exact retribution ("an eye for an eye" from Exodus 21:23–2).[1] This has been abandoned by modern jurisprudence in favor of the sophisticated principle that punishment should be condign, "fitting" the crime, not the same, but equivalent.[2] Without exact reciprocity, justice requires judgment. It becomes more abstract. This notion is woven into the iSchools discussion that follows; we explore the tension between the i (at its simplest, the role of the schools in developing information professionals) and the I (a dream that information scholars could underwrite a revolution in the academy).

A caveat is warranted. Those seeking a conventional history of the iSchools will not find it in this chapter. As one reviewer put it succinctly, this is a gossipy story told over drinks at a conference. The chapter does not generally name names or discuss particular schools. In fact, it goes out of its way to avoid that. The simplest conclusion is that the authors are lazy; both are retired. The objective is to interpret the mystery of the iSchools. The more the iSchools emerged historically, the more approximation—between the i and the I—was needed. The notion of equivalence entailed in *lex talionis* became central. The iSchools seemed at once highly unlikely and inevitable.

iSchools emerged around what we call "Dream." At one extreme of belief, the thing being "informed-about" is embroiled in the hope that representation, if worked on enough, will represent precisely all of the features of the thing being represented.[3] We think that view is waning in favor of a probabilistic view in which information is "close enough" to *allude* to something, but cannot go farther than allusion without encroaching on point-of-view, judgment, wisdom, the contribution of the knower to the known. In this view, the transitive verb "to inform," substitutes for the reassuring noun "information." Information is not "good enough" to stop and use the results. It is "good enough" for the knower to take over and figure out how to best characterize what is being informed about. Persistent ambiguity does not signal failure in the "information system." There is no apparent way to disambiguate completely and objectively.

We started using the definite article "The," capitalized, no less, to construct The Dream. A reviewer kindly pointed out that this implied there was a shared dream. No, there was Dream. People adopted or rejected parts of it. Dream was the starting point of the iSchools movement. It is intellectually captivating, and has been for a long time, while it is fundamentally a reverie. This chapter argues that, after a flurry of hope and, frankly, surprise, the iSchool movement could not finish Dream so the iSchools did not draw direction from it. Most iSchools built on traditions and opportunities related to professionalism, teaching, and providing accreditation to students desiring professional careers in information-related fields. This strategy did well with enrollments, internships, and placements; and iSchools that took the professional route did well. They are not dependent on Dream. Nevertheless, Dream is alive and will continue to draw attention and the belief that it might, someday, achieve the breakthroughs hoped for.

One path of this chapter considers Dream as it gave rise to and arose from the iSchools. Another is the creation of the iSchool movement as an official enterprise. A third is professionalism. A fourth is odds and ends—residual categories—that resist premature categorization. Elements of the iSchool history can be recounted, but *the history* cannot. Odds and ends are specific to universities and their iSchools, central to their story.

> ***Call: What's the right way to think about the iSchool phenomenon?***
>
> ***Response: There's no one right way; it is still emerging. But there are important points to consider.***

The first point to consider is that the iSchools came together during the 1990s to form a new social movement within universities. But this social movement did not have a common cause. In many cases the iSchool phenomenon presented an ungainly amalgam of existing programs with little more than names in common, if that. No linear history tells this story, and attempting to emphasize iSchool unity threatens to occlude the local situations on campuses and with industry players that had *already* created the schools that would subsequently make up the iSchools. We do not emphasize the cobbling together of unlikely alliances, but beneath the surface and among some there was motivation to make "information" core to our society. We avoid the general problem of reversion to an early dream of the academy, revealed in the 16th century, when the notion of wisdom was core to the academic enterprise. Knowledge became a focus of the Enlightenment. The focus on information arose during the late eighteenth and early nineteenth centuries.[4] Now, it seems, the focus on data is eclipsing all. (We hereby recite the elements of the DIKW pyramid: data, information, knowledge, wisdom.) It is a work in progress.

Historiographically, this chapter examines three questions. One is how did the iSchools emerge in local political milieus like phoenixes rising from the ashes of closures of professional programs such as those of Schools of Library and Information Science (LIS) that were shut down in the 1970s and 1980s (some say over 80 were closed; some say fewer). Second, what did iSchools claim to do in a world where information was suddenly a magic word that opened doors? And third, how did the iSchools navigate the rise of data, the ups and downs of computer science (and other fields such as communication) that sometimes exhibited similar threads and that some people consider important today?

A relevant comparison here—and represented in this book—is the history of Computer Science from its early days, focused on how to build machines for optimizing performance of algorithms and data structures, to producing compilers, to writing code, to managing code libraries, and all points in between as well as "out there" (e.g., artificial intelligence). As times changed, both LIS and CS, to highlight two examples, had to reinvent themselves from within while maintaining institutional presence. Unlike most disciplinary histories such as, say, physics, always about doing more of the same according to a set of somewhat accepted and evolving paradigms (string theory complicates the physics story), the iSchools were always about working out what they were about as the world changed around them.

The questions raised here deserve one or more weighty tomes. We take a few shortcuts. This chapter concentrates on North America, especially the United States, not because of any US hegemony, but because it reflects our experience. We do not try to create a single through story, but follow several stories that are intertwined. Two in particular draw our attention. One is universalizing disciplines of information, close to what we call Dream. It is a key intellectual part of the iSchools. These universalizing stories are hard to write because their visions have seldom been realized. They are incomplete utopian projects—always just beyond the current temporal horizon. The quest is ongoing and will not end soon; it is subordinated to the iSchools as institutional endeavors. The second thing that draws our attention is institutional endeavors that first brought people together to form iSchools. Both are needed to create the basis of a coherent history, while neither is possible without the other.

We provide no overarching narrative. This chapter was drafted as a series of vignettes, reflecting the personal views of the authors. We consider it a participant-observer exercise, or autoethnography. Our biases surface from time to time; how could they not?

Call: Did anything pull people together intellectually to form the iSchools?

Response: Yes, Dream.

One reviewer pointed out that Dream is the protagonist in this chapter, a main character who influences the plot, drives the story forward, and is closely followed by reader or audience. To create the right setting for Dream we turn to Australian aboriginal culture. Certain ancient Australian people said dreamtime was *everywhen*, not in vassalage to conventions of time. It was dream *time*, but not time as usually thought of. Eternal ancestral figures of heroic proportion and supernatural abilities were to be found, but uncreated and uncontrolled by the material world.[5] The dreamtime suspended both time and materiality. A dreamtime for information is the realm of universalizing disciplines.[6] Repetition is central—as in Thomas Mann's *Joseph and His Brothers* or Yerushalmi's *Zakhor*, which discuss eternal repetition.[7] This is always about universal truths of this happening and then that to tell the same story—in our case the triumph of information. Much of the history of information is predicated on making representations as "real" as the things represented. Twisting time and materiality sparks confusion, but offers insight. Universalizing disciplines are both disciplines to be developed intellectually in their own right, and meta-disciplines to be applied in other disciplines.

One universalizing discipline was statistics as it evolved in the nineteenth century. Statistics was contemporary with grand national censuses of people to serve as information about the state.[8] It played a role in the perceived need for grand censuses of nature (e.g., natural history surveys).[9] Along with names of academic disciplines (physics, chemistry, biology, history, philosophy, etc.) that were inscribed in university layouts, and sometimes carved into building walls,[10] counts summarized in statistics were taken as a representation of universal knowledge. Academic disciplines settled into specialties, and statistics sat astride them. In the 1850s the first International Congress of Statistics was held.[11]

Statistics started to invade every scientific discipline, to become an obligatory passage point:[12] one couldn't be a well-rounded sociologist without statistics (cf. Durkheim on rates of suicide as a universal cultural constant).[13] Nor could one be a well-rounded physicist (cf. especially Gibbs' statistical mechanics of atoms in a gaseous state as lauded by Wiener).[14] Statistics itself was disorganized, congealed into multiple departments, many vestigial within the academy until recently when "data" appeared and scattered the field over biostatistics, social statistics, and so forth. Subdisciplines were created within disciplines, and "statistics" people sometimes could not or did not want to speak with others unless from their own domain.[15] The statistics dream of universality lived on, but it was often sequestered locally. Statistics presaged the story of iSchools.

A second universalizing endeavor was cybernetics, which tried to capture the spirit of the industrial age with the concept of feedback.[16] A 19th-century material symbol of feedback was the steam engine governor.[17] It is difficult in the early 21st century to recall the effect of steam power on popular imagination. It was the first exploitation of the thermal gradient that came to power agriculture, transportation, and manufacturing. It was behind most generation of electricity. It led to motor transportation (automobiles, trucks, etc.). It touched everyday life. It caused L. J. Henderson to say in the early 20th century that science owed more to the steam engine than the steam engine owed to science.[18] In cybernetics in the mid-20th century, science asserted itself. Cybernetics embodied the economy of industrial times, presaging the "information economy." As with statistics, in surveys touching on society and nature cybernetics was a universalizing vision. For example, it was hoped that deterministic Markov chains would unify mathematics,[19] sociology, and physics into a single discipline that sat atop others. A few cybernetics departments (for example, the one at the University of Brunel in the UK under Gordon Pask) were created. Still, cybernetics disappeared into the woodwork of different academic disciplines. In fact, it found favor mostly in the military where command and control were central. If cybernetics is resurgent, it is the "second order" form that includes the views of the observer so that now there is a balance between universalizing and localizing.

Statistics and cybernetics reveal two issues relevant to Dream and the iSchools. One is that inspiration can come from sources at once intellectual (e.g., statistics' base in mathematics) and practical (e.g., cybernetics in the steam engine governor that was an engineering feat long before the science behind it was explained). This is relevant to Dream and the iSchools given the manifest power of information technology, especially after the advent of the digital age in the mid-20th century. But much of that power came from intellectual insights about computation, data, information, etc. The iSchools have sought to marry the practical and intellectual. In the process, the second issue emerges: a tension between universal and local. Universal covers everything. Much local is tied to politics on the ground. Tension is core to the iSchools. Any claim of universality might have a flush of enthusiasm and then settle into the local. Individual iSchools are today continuations of universalizing efforts, with information as driver of the economy and the state. Through Dream is an attempt to make information a universalizing discipline, but universalization defaults to localization and information is distributed over existing disciplines; and Dream's survivors must respond to local institutional politics.

We produce a cartoon exaggerated for effect: consideration of the traditions of Library and Information Science (LIS) and of Computer Science (CS). Many iSchools are tied to both. We acknowledge the exceptions, noting that some iSchools follow the traditions of one or the other, and some focus on allied fields such as archives, information economics, informatics in areas like health, communication, organizations, and so forth. In the past, LIS was seldom consulted for fundamental insights about information. It was for professional training. While some in LIS pursued fundamentals (for example, the documentalists)[20] as an institutional entity, LIS usually helped with the flow of information for any possible use, and with training people to do such work. When the iSchools emerged in the late 20th and early 21st centuries, there was a supervenient vision of information central to every discipline and human enterprise (including universities).

A blend of LIS and CS made sense. The iSchools would educate people to use information sources, akin to instruction in how to use libraries, while leveraging technology (e.g., Babbage's concepts, Hollerith's card equipment, and subsequent exploitations of the electromechanical, vacuum tubes, and solid-state). This created a kind of universalism, but one in which the LIS and CS traditions seldom collaborated. In fact, there were deep disagreements. Some claimed computing devices would eventually subsume what LIS did, and perhaps exceed human general intelligence. Others argued that computing *was* a model of a clockwork universe where everything would be part of a massive computer.[21] This view grew in strength and alarmed those who felt it encroached on other disciplines, making them open to colonization.[22] As with statistics and cybernetics, this view was criticized for lacking an intellectual core, but it could be sown widely. Still, it could not provide an unassailable intellectual base for the iSchools. In many institutions there was already established one or more CS programs and few needed to create an iSchool for the purpose of incorporating computing.

The LIS tradition prepared intermediaries to work between those who need library services (librarians call them "patrons") and the information they seek. Patrons who do not know how to frame their questions work with an LIS specialist such as a reference librarian.[23] Technology applied to writing, books, catalogs, etc. that are amenable to computerization. Some iSchools extended their efforts in the human use of computers through, for example, computer-supported cooperative work (CSCW), part of a larger tradition seeking to understand how humans work with computers.[24] Some iSchools provide tools by which people do their work. The service ethic seen in LIS tradition was not as strong in CS. LIS was concerned with how people and information interacted. CS was concerned with how machines and information interacted It is not that CS cares nothing for people, but the LIS tradition of "patrons" contrasts with the CS tradition of "users."

When more than one tradition claims universalism, a struggle for dominance can emerge. This proved to be the case in some iSchools. Struggle can be exacerbated as the focus shifts from "information" to "data." What is dominant: "information science" or "data science"? It can become a conundrum. An iSchool universal approach to information for the benefit of all human enterprise requires the iSchools to be a "peaceable kingdom" where thought leaders solve the problems of information. The problem is the inherent tension, as in H. G. Wells' fascist and socialist notion of a universal brain, or the cybernetics-inspired vision of the socialist paradise of Chile's Allende.[25]

Alas, Dream carried the seeds of limitation. To be universal required bringing in of everything related to information, which increasingly means everything. Other fields cannot permit this if they are to build *their* fields rather than the "information" field. Institutional mechanisms governing the academy protect disciplinary prerogatives. The academy is good at *adding* disciplines (see the enormous proliferation after the Second World War),[26] but not so good at consolidation. There have been occasional and somewhat successful local attempts to integrate iSchools with other disciplines, but the difficulties are legion. An information program would have to be huge or full of polymaths to cover everything. The academy is ill-equipped to resolve such problems. Dream faced this challenge. Those working on Dream sometimes found themselves marginalized and admonished to pursue more practical concerns.

Nevertheless, iSchools are inherently dreamy and the idea of the iSchool is interesting irrespective of practicality. Dream does not go away simply because there is a need to be practical. The iSchool *movement* of North America, especially that of the United States and Canada, was "interdisciplinary," containing people from many perspectives. While in principle something that could be designed, it often emerged as the result of convenience. Information as a concept became nebulous, and Dream failed to produce a grand synthesis. Moreover, Dream might be moving away from the goal of a grand synthesis as more effort is placed on addressing local concerns. *The* vision of a future synthesis, always ten years away, around information might be, to use Judith Gregory's term[27] an "incomplete utopian project."

Dream has failed to provide a solid and sufficient intellectual rationale for the iSchools. Universalizing ambitions failed to persuade factions to agree, so Dream never materialized. *Claiming* universality became difficult in iSchools, where different traditions prompted different views on what the iSchool is and should do. In some cases, Dream led to conflict, the opposite of the "peaceable kingdom." To keep the peace, many reverted to parallel play within established traditions. In some cases, traditions are siloed by formal separations (departmentalization or such). The iSchools face the relative youth of Dream's intellectual movement, which does not convince all those who must be convinced to come together. Dream might be moving counter to the universal, paying more attention to the local. Dream does not yet supply the intellectual ground necessary to draw the iSchools together and probably will not for the foreseeable future.

Dream should treat iSchool idiosyncrasies as *features*, not *exceptions* or *bugs*. The iSchools are academic programs that embrace information, but show variance beyond that, including focus on particular topics others ignore. There has never been a successful effort to normalize structural or other characteristics of iSchools. Even when accreditation is central, it is unevenly pursued. Some iSchools are not even schools, but departments in larger schools. Some people in iSchools do not acknowledge iSchool identity. Some iSchool programs have unique concerns not relevant to other iSchools. Some iSchools seldom if ever mention iSchool affinity and do not signal that their program is an iSchool. The term "iSchool" usually refers to programs that are members of the iSchools *movement* described below. There is considerable ambiguity about what a school *must* be to be an iSchool. This ambiguity can be *constructive* as part of Dream because intellectual difference is a basis of the iSchools movement, an inherent feature of the iSchools, not something to be fixed. It is a problem one wants to have, but a problem nonetheless.

A purpose of this chapter is to discuss the *problematic* of the iSchools. The chapter focuses on iSchools as phemomena. It does not attempt to concoct a coherent iSchool history. The authors have been on faculties of several iSchools, on external reviews of iSchools, in historical analysis relevant to information, in administrative and technical roles, and involved in creating the contemporary iSchools movement. They and their precursors of the iSchools were around for decades before the movement began in the 1990s. The iSchools are stuck in an "eternal now" where things stay the same but everything is changing. Dream recognizes that information is fundamental to the contemporary academy.[28] Information encompasses the university. Some argue it can encompass the universe. However, university admonitions to "Stay in your lane," despite the name "university," suggests being universal is harder than it looks. Dream does not answer the questions about information; it has not settled anything. The iSchools, with no clear lane to stay in, stay in multiple lanes to avoid falling down the information rabbit hole.

It is not clear where Dream goes from here. In the 19th century, British prime minister Benjamin Disraeli said that, as a general rule, the most successful person was the person with the best information.[29] During the 20th century, information became pervasive to the point that few doubted its importance.[30] Our "genetic code" was said to contain all the information needed to understand humans. Decrypting the code would reveal who would be gay, extroverted, a dangerous criminal, and so forth. This story was blown to hell by many things: genetics proved to not be up for the job, humans and their illnesses and behavior remained a mystery even after the human genome was sequenced, the human microbiome materialized to show that a large fraction of cells in a human "body" have non-human genetic codes. This is typical of Dream. Saying a problem is all in the genes does not make the problem easier to solve. One might as well say it is all in the human brain, or that human consciousness can be reduced to 0s and 1s of neuron firing.

Stories get muddied. Who anticipated the microbiome? Who anticipated that gut bacteria would affect how humans think and behave. (Do humans have a "second brain" in their guts?) The increasingly obvious symbiotic nature of humans and other organisms upends established notions. Is the immune system a sixth sense?[31] If so, it decenters "intelligence" from the brain. Or what about *distributed cognition*[32] in which intelligence is a distributed, social phenomenon that includes social ties and technology? To claim "all is information" is followed quickly by "no it isn't, quite." Unfirm territory may not be the preferred hunting ground of the iSchools, but it is what Dream has provided, akin to Thomas Aquinas' simultaneous theological benefits of faith and doubt.[33]

There are many possible paths for Dream. We provide two examples. In one, Dream is represented by quantum computing. Why compute over one universe when you can compute over 10^{500}?[34] Proponents say quantum computing will move humans beyond the linear, binary arithmetic common to current technology. (Will it instead escape the binary arithmetic from China,[35] that Leibniz felt was the greatest invention in mathematics?) In any case, what comes back from quantum computing is still information, an answer to a query. It will be muddied. Quantum computing is emergent and will provide emergent technical answers, truths remain approximate. The hope for pure and precise information remains, but the best probability to hope for is that people will check quantum computing against more quantum computing. There is no *deus ex machina* to correct the problems of *hominem ex machina*. This will be messy.

A second vision is exemplified by artifacts such as ChatGPT. The same messy process seen with quantum will accompany "complete AI," hoping for *better* than neuron by neuron mapping. Perhaps humans will delegate morality and sociality to machines. This triumph of the binary might be the end of binary. How will the iSchools deal with this muddy picture?

Other things might come along that move forward Dream. In their short history, iSchools have revealed themselves to be capable of this, but there are no guarantees.

Call: How did the iSchools get going as a social movement?

Response: Dream proved problematic in giving a base for the iSchools, so they formed a community in which all could defend any.

Formal iSchools started in the late 20th century, but the iSchools movement was not formalized until well into the 21st. Other papers on the iSchools might discuss which schools were pioneers, who was involved, and so on.[36] This chapter does not. All wrestled with Dream to some extent. We repeat our earlier observation that Dream gave rise to the idea that became the iSchools as creatures of the institutions they came from. In many cases these institutions profoundly influenced the formalization of each iSchool. The iSchools collectively became a social and structural movement of schools that had already been created on each campus. Most iSchools were established entities before the iSchool movement was created. The iSchool movement was fashioned by established schools seeking a defendable identity.

The operative word is "school," although in some places the term is "college." By the late 20th century North American higher education recognized schoolhood as independence of thought. Schoolhood conveyed academic and programmatic authority denied to departments in a school, which were guided by a school. In some cases, academic departments have considerable autonomy and are large in their own right. The relative independence of a school was particularly important to professions that had to meet the expectations of external professional constituents. Many iSchools formed around pre-existing schools, often in LIS and other "information-related" topics. Many LIS programs were schools. Most CS programs were departments of divisions of departments in schools of engineering or sciences.

Pre-existing, stand-alone schools were occasionally "takeover targets" for creation of new schools that might or might not incorporate the original school's vision. Nearly all those that became iSchools embraced information. Faculty in an area such as LIS might feel invaded and colonized if powerful institutional leaders used "their" school to build something the leadership desired but did not have. For the overwhelming majority of iSchools, the plan was to do something "big" involving information. Within "big" the intention of the iSchool could be idiosyncratic, local to the institution that created the iSchool. The iSchools movement provided institutional legitimation for any given iSchool. It was assumed that each institution would have only one iSchool. If that iSchool lacked academic standing, it could leverage what has been called institutional isomorphism to compete for political power and legitimacy within an institution.[37] Individual iSchools were already legitimate *as units* in their own institutional context, having been locally recognized. With the iSchool movement, they could broaden their legitimacy, be seen as compatriots of iSchools at other respected institutions. The search for aspirational exemplars was a driving force behind the iSchools movement.

The iSchools movement established low barriers to participation. Any school that *saw itself* as an iSchool was welcome to join as long as it was in good standing at its parent institution. This was more or less guaranteed by the fact that the school had been approved within its institution. It was generally a stand-alone school, usually having a dean and committed to information (though this need not be spelled out in detail). Together these qualified a school to be an iSchool; and even then the demands were not high, and different local configurations were accommodated. For example, schools that featured Communications, or Media Studies, or CS were accepted. Members committed to stand with each if questioned about academic legitimacy or called on to explain themselves. This arguably went back to an earlier trauma. LIS experienced program closures, while CS had seen major enrollment declines. Solidarity lessened the need for intellectual rationale. Pointing to an iSchools community of strong institutions offered legitimation. The iSchools at high-ranking institutions were among similar high-ranking institutions. The iSchools at lower-ranking institutions could associate with high-ranking institutions.

The initial iSchools enterprise was informal, mainly a membership list created by a handful of iSchool deans who called themselves the Gang of X, where X was the number of members at any given point. This was a tongue-in-cheek reference to the Gang of Four, a faction of the Chinese Communist Party. The "gang" designation was a joke, abandoned before the membership exceeded a dozen. Informality gave way to a formal name, the Information Schools (iSchools), and an organization of iSchools was incorporated as a non-profit organization. Bylaws were established and a web page created. Membership expanded to over 120 iSchools, all exploring information, without favoring any particular methodologies in the social and behavioral sciences, computing, artificial intelligence, linguistics, and so forth. iSchool people work in data science, human-computer interaction, information organization and access, bibliometrics, information integrity, and so on.[38] The iSchools began informally but grew more formal over time, while retaining separate identities.

Some computer science (CS) schools attained school or college status, and banded together under the Computing Research Association (CRA), an existing association of CS programs. The CRA Deans, as they were known, contained a few schools who were members of both the iSchools and the CRA Deans. Most CS schools went with the CRA Deans that remained part of CRA. Increasing recognition of the importance of information technology affected both iSchools and CRA Deans. The growth of the Internet, personal computers, graphical user interfaces, and the commercial potential of information technology had been made manifest in the dot.com phenomenon of the 1990s. Before long many organizations established and sustained online presence, even following the dot.com crash. Search engines and smartphones made it clear to the public at large that "information" was central to everything.

Little slowed down the recognition of information. The dot.com crash and the Great Recession proved to be blips. Information and information technology recovered and proliferated. The iSchool movement became fixated on growth (the default path to progress in the academy). If academic legitimacy is conferred by a mix of financial and intellectual justification, and Dream was proving a difficult project for providing intellectual legitimacy, programs without intellectual justification found they could survive if they had large enrollments. Most iSchools have been successful via enrollment, a virtuous cycle of prospective students seeing that graduates of iSchool programs are seen by prospective students to get good jobs. The prospective students pursue iSchool programs and get good jobs to perpetuate the cycle. Any given iSchool could show purpose by enrollments that are large enough.

In the history of education, Terry Shinn[39] and others have argued an institutional dynamic here. Practical education becomes valued (see the Prussian model in the nineteenth century for Shinn). This is then followed by emphasis on blue sky research. US research funding reflects such cycles. Research programs focus on the practical while also being attracted to blue sky. This is happening in AI and other fields. The practical can create internal turbulence: for example, research projects often now need a data plan to permit later use of data. Many research funding applications require a statement of social relevance along with a statement of intellectual merit. This cycle has been going on for centuries, but it can be hard to spot because institutional memory focuses on decades. The iSchools, despite their name, found it hard to grasp information intellectually. Many learn to leave the harder parts alone. This is true of other topics than information. Mary Douglas[40] says "social memory" was invented and forgotten at least three times during the nineteenth century. Most iSchools defaulted to professional artifacts and artifact clusters in the face of Dream challenges. Examples of iSchool focus on computing artifacts are seen in human-computer interaction (HCI), computer-supported cooperative work (CSCW), software engineering (SE), or dedicated application of information to a field like health care (health informatics). Abandonment of Dream's broad "information" focus is common now, not because it is uninteresting (it is interesting) but because it is hard.

Some argue that processes turn data into information, information into knowledge, and knowledge into wisdom in the data-information-knowledge-wisdom (DIKW) pyramid.[41] Each is part of a hierarchy, with each level "above" subsuming those below. The DIKW pyramid is controversial but fertile. The iSchools rediscovered it (it had been discovered before), but simply pointing to something that is, itself, poorly understood does not make the iSchools understood. A saving grace is that sub-systems operating through computing platforms facilitate cooperation and generate innovation among individuals and groups. Computing helped create the Internet. Innovation creates value. Information professionals handle whatever arises. The question of information entailed in Dream is large and complicated. Everyone knows that information is important. A key to iSchool success in the long run might be understanding information, at least in the public imagination and responsive to local demands. It is not clear that a universalizing understanding is needed.

The iSchools perhaps bit off more than they could chew in claiming all of information. It is easier to follow established academic traditions and focus on *parts* of the whole. The whole will not be covered by Donald Campbell's "fish-scale model of omniscience."[42] The iSchool movement is relegated to a constrained reality that *size* matters. Information is too big for small programs, and no iSchool can be big enough to capture all of information, even if other disciplines agreed to let that happen.[43] As iSchools were forming, so too were for example geoinformatics and bioinformatics. It might be practically impossible to universalize.

iSchools that enroll and graduate students do not need Dream if they amass good student course evaluations, recruit good scholars, maintain research productivity, garner philanthropy, and are cited in the popular press. Any Dream to tie together the major elements of information has yet to be done. Perhaps this is the wrong epoch. Also elusive is a close relationship between information and innovation, though the two are probably connected. Templates for iSchool evolution are missing, but the iSchools were never an idea whose time had come. Rather, the iSchools are part of Dream that will be evolving for a long time.

Call: What do you do if Dream proves elusive?

Response: Professionalism. Many iSchools are successful with professional programs.

Most iSchools grew from professional roots. Some were accredited by professional accreditors. Many LIS programs were accredited by the American Library Association (ALA), and professionalism in librarianship was enforced by employers requiring that professional librarians complete an accredited master's program in librarianship. Other professional work is less explicit. It does not depend on employers requiring hires to complete accredited professional programs. For example, U.S. computer science, once accredited by the CSAB (Computer Science Accreditation Board), and later by the accrediting structure for engineering, ABET (Accreditation Board for Engineering and Technology), allowed graduates to take professional jobs in computer science without having graduated from accredited programs. Fields related to information—communications, management, health care—might or might not enforce accreditation. Some iSchools took graduates of non-professional programs (for example, in the liberal arts) and provided professionalism in their programs.

The professional orientation is favored by some graduate students who seek professional work, especially if their undergraduate degree is from a non-professional program. Historically, LIS drew many students from undergraduate programs in the humanities who wanted professional jobs. iSchool instructional programs were often built to meet student expectations for professional jobs and identity. The iSchool movement did not adopt an official position on professionalism or accreditation, and each iSchool handled professionalism in its own way. Some iSchools continued accrediting programs as in the past (e.g., LIS programs seeking accreditation from ALA). Many had or built up placement services to help students find professional jobs. Some implemented engagement programs focused on practical experience (for example, internships) to help students get a start on a professional trajectory. Engagement attracted some students to otherwise fuzzy concepts.

Dream had limited direct effect on professionalism. It might help attract good faculty and instructors, or build the iSchool's reputation, or inform the curriculum; but professionally-oriented students seldom considered Dream a criterion for attending a program. They sought professional success. The link between Dream and the professional strength of an iSchool is unclear. A reputation for adding practical professional value might be tied to the reputation of the larger institution. Given the difficulty of Dream, the professional curriculum seldom reflects it. A noteworthy point given that professionally oriented thinking dominates many iSchool curricula.

As in other situations, the local forced attention away from the universal. Over the years higher education has become increasingly tuition dependent, and iSchools have created instructional programs to draw students. The popularity among students of good professional jobs reinforces the focus on professionalism, helped by the growth in professional information-related employment. Beyond professional jobs in libraries (K–12, academic, public, special), jobs have expanded in the "information-oriented professions" (programming, software development, coding, and others) often pursued by computer science majors. Data-oriented professional occupations (data scientists, for example) have been on the rise. Catering to such demand has been good for iSchool welfare if it depends on student matriculation and graduation. For many faculty, the choice is not between research and no research, but in research focus with aligned teaching and research programs. If Dream does not appeal, and information is too large an area to embrace entirely, an aligned professional focus for research and teaching makes sense.

Most iSchools have less than one hundred faculty, and cannot cover all of information. Growth creates pressure to departmentalize. Competition among departments seldom broadens the topics embraced, either within or across departments. There is a tendency to strengthen what is already strong and reinforce the *status quo.* This fits the professional objectives of students seeking professional jobs. They do not want oddball identities, but to be recognizable and in demand. Information-related fields have their own dreams and face similar pressures. Growth brings a tendency for individuals to identify with groups and conform to them. This can produce balkanization. The easiest course for the research-oriented is to work on problems that qualify them for promotion on research grounds, while making them professionally relevant to students.

Tuition-dependent iSchools must keep professionally oriented students satisfied. Thus the proliferation of professional master's degrees was often sold internally as supporting research. Professionally oriented programs increasingly require "engagement" that can take considerable time of instructors and/or professional staff and add to program costs. Similarly placement, which is sometimes needed for engagement but usually for launching or redirecting a professional trajectory. Professional career building is not at odds with Dream, but is not on Dream's critical path and can further separate research and instruction. Engagement and/or placement can initially be a lower energy-cost path than pursuing Dream, but is a one-way road; once begun it must continue. An iSchool that is, in the eyes of students, a professional school is orthogonal to Dream. Professional education is costly and resources for professional programs are "lumpy." It is difficult to shed 10 percent of someone (supply), given a 10 percent reduction in workload (demand). Cutting or adding can be complicated. A hypothetical "optimum" is seldom realized as steady-state. People cut when resources are tight. At the same time, they know that growth is a problem one wants to have, but it is still a problem. If Dream proves slow, the answer is "information-related" professionalism.

Call: What else is important?

Response: iSchools are emergent, with Odds and Ends, "residual categories" that do not fit readily into established categories.

Dream is old but despite venerable old age, it will not reach sufficient closure to serve as the conceptual foundation for the iSchools (or their successors) for decades more. Still, iSchools ratified by their respective institutions were able to create identities. Many developed professional instructional programs and have established or kept professional identities. The mutual aid aspect of the iSchools movement might have been good insurance, but it is not clear it was needed. Nevertheless, iSchools remained emergent, with a significant number of Odds and Ends that continue to have significant impact on iSchool development.

The deflationary terms "odds and ends" have specific referents. Derrida's concept of deconstruction shows that even smooth intellectual edifices can be defined and undermined by what is excluded (Derrida's canonical case is laughter in Philosophy *contra* Searle).[44] Applied to iSchools, a technocratic university arguably wants an ideal, smooth flow, such as scientific discovery leading to engineering of new products that produce economic growth. iSchools in this model deal with information, which has emerged as crucial to discovery, product creation, and management of growth. But at the same time, iSchools are unwanted. In an "ideal" world, iSchools would not be needed. Odds and Ends are residual categories, things "not elsewhere classified" as noted in the International Classification of Diseases.[45] iSchools can themselves be residual to the university, not really believed in but professionally successful, a convenient place to park programs that the university needs for local reasons but does not really want.

These "odds and ends" involve a large part of iSchool history. They are the residual of the difficult task of assembling something from a bucket of parts from various sources. They are not a tractable historical line or part of a linear history. Some notions even challenge fundamental notions of information. As residual categories, they allude to the failure of categorization in which meaning is not ascertained nor conferred. Established categories do not work. Thus far, no replacement categories work. The presence of too many residual categories challenges any categorization scheme. Odds and Ends raise questions about knowing. There is a long list of them. We focus on one: identity politics.

The 19th and 20th centuries were watersheds for identity politics relevant to the evolution of iSchools. We consider three kinds of identity: racial, gender, and disciplinary, using the cartoon of LIS, CS, and organizational information systems (IS). LIS was "born" in the late nineteenth century, and in North America it came of age slightly later in the 1920s and 1930s, due to problems in rapid expansion of public libraries.

The Carnegie Library Commission paid to establish about 2,500 libraries, without librarians to run them.[46] University administrators realized that a new core of intermediaries (librarians) could be created between people and information (somewhat ironic given subsequent calls for disintermediation and the claim of no further need for librarians).[47] A little later, starting in the 1950s, programs were created to train people to build and apply computers in engineering and science. Most computers were applied in organizations, and programs in information systems (IS) were created in business schools already training business professionals. Thus did LIS, CS, and IS establish professional programs. This foreshadows disciplinary identity, as discussed below.

All information-related professions have confronted issues of gender and racial diversity, and all have made progress. Professions and professional programs were dominated by whites (nearly everything was), but unlike CS and IS that were dominated by men, LIS was dominated by women. Even now, CS STEM education has a high proportion of male students.[48] Disparities were inherited by the iSchool movement. Female-dominated professions adopted social justice early, with public libraries tolerant or even welcoming of indigent people, often people of color, off the streets. Some LIS faculty worked on community freenets to bring the Internet to the poor. Many LIS programs saw community engagement as core to the field, asking whether the profession was more about books or community. Science, engineering, and business did not have this discussion as often. A residual category involving gender identity is an issue in many iSchools. It is a sensitive subject. LIS, a feminized profession from early in its creation, was a "service" field, like nursing was. Positions of power and authority (Director, Dean) might be held by men, but women made up the professional ranks. Librarianship was *seen*, at least in U.S. culture, to be feminized. (Compare Donna Reed as the timid "old maid" locking up the library in the film *It's A Wonderful Life*.) Most jobs originating in science or engineering or business went to males. This has begun to change, but the histories behind gender identity politics linger.

Gendered professional work in LIS tended to be in public, governmental, or non-profit education organizations, with training and hiring tied to accreditation. Professional positions tied to science, engineering, or business were in the military or for-profit firms. An ethos of LIS was to provide service to others: helping patrons find the information they sought. Professional deportment often transcended employers. The ethos of science, engineering, and business was often to assist in provision of computing capability as part of larger systems. Professional deportment in business or science and engineering jobs was shaped by employers.

Disciplinary identity politics can be a challenge for the iSchool professions. For example, within CS there is an important distinction between computer engineering (focused largely on electronic computational and data storage and management devices) and computer science. Some programs have remained divisions of larger EECS (electrical engineering and computer science) departments to retain control of computer engineering. One IS is information systems (earlier with an M in front, for Management Information Systems), located mostly in business schools and concerned with application of computers to organizational work. This IS field has tended to stay in business and management schools and typically does not connect with iSchools. Some engineering and science schools include a second IS, or "information science," concerned with digital communications of the kind explored by Claude Shannon.[49] A third IS is in LIS and tends to focus on the documentalist tradition. This tradition, though often tied to library science through the acronym LIS, sees itself as *different* from libraries. Fields such as archives, communication, information management, and others think of themselves as holding unique disciplinary identity. And this does not touch on the differences among librarians who find disciplinary distinction in public libraries, school libraries, academic libraries, and the vague category of "special libraries."

"Science" modifies the nouns including library, information, computer, and other academic fields. So, too, does "engineering," as with computer and knowledge engineering. These identity claims bolster appeals to intellectual solidity for power projection and are in flux. "How we got here" histories do not predict how the iSchools will develop. The current trend seems to be a weakening of LIS traditions and a strengthening of other traditions. Some LIS programs never fully embraced the iSchool movement or went back to LIS roots. Other LIS groups in iSchools made LIS into departments or divisions within the larger iSchool. People still ask what iSchools are, what they do, why the university has them, etc. Could this be because iSchools lack disciplinary identity, and if they do, is this in part because there are so many possible identities? The iSchools have not yet been able to harness the intellectual revolution of Dream. Still, iSchools are secure when they offer professional education programs that enroll sufficient students and place graduates in good jobs. This makes identity politics issues recede to "background" levels.

Call: What is the default to artifact?

Response: An effort at concreteness, which most people prefer to abstraction.

The difficulties of Dream default to the concreteness of artifacts related to information. These might be printed or inscribed (books or manuscripts and other documents) that are part of collections, or artifacts that store and provide access to collections. Artifacts include "finding aids" to facilitate patrons finding what they seek among many possibilities. These artifacts extend to collections generally in galleries, museums, zoos, and aquaria. A common organizing artifact is the catalog that lists what is in the collection and where it can be found. Sometimes the catalog provides for cross-indexing the collection to universalized information. Cataloging emerged early as a central plank of the library tradition.

For some the key artifact is the computer or its smaller chunks: architectures, devices, data structures, algorithms, protocols, computational geometry, standards, applications, systems to do specific tasks (such as operating systems), programming languages, repositories ("libraries") of written code, etc. Taken together these *define* computing. Few academic programs are big enough to be strong in all of them. Most try to cover a strong subset. Attention concentrates on what is "bound," meaning costly at the time. In the 1950s and 1960s, when computing was "processor-bound," improvements in artifacts such as algorithms and data structures could improve processor performance and were of paramount importance. People who worked on such artifacts became powerful. Power waxes and wanes as different artifacts become the focus, fade, and resurge.

The iSchools draw strength from the world-changing impact of computing. Libraries, galleries, archives, museums, and collections of many kinds bound by a "master narrative" (for example, DIKW) could help create or clarify concrete artifacts that are easier to grasp than some ambiguous concepts found in Dream. A "form" (for instance, the library form) can be applied to archives, museums, and so on, and is relevant to interface design. This artifact focus brought iSchool attention to Human-Computer Interaction (HCI), user experience (UX), "X Informatics" (where the X is some other focus), and digital humanities. iSchools cover a large waterfront, making it hard to establish disciplinary focus. "Interdisciplinary" can make it difficult to become a discipline. The question, "What are iSchools about, *really?*" is not easy to answer, even with respect to "natural" things in "natural history" museums. Once collections are cataloged (this is a star, cataloged under celestial bodies; that is a coelacanth, cataloged under fish) they became portable artifacts with a history. Academic fields emerge in consonance with artifacts in collections. In one way or another, the iSchools are following this path.

The artifact story is an important residual category in the larger iSchool story. Artifacts can mis-signal what is intended, be deeply embedded and be hard to abandon. In the late 20th century, two major information-related organizations tried to get rid of "machine" in their names—the Association for Computing Machinery, and International Business Machines Corporation. After trying to find substitutes that would allow them to use their acronyms, ACM and IBM, both simply used their acronyms. Renaming suggests a desire to evolve: Man-Machine Interface evolved into Human-Computer Interaction (HCI). "Machine" implied "mechanical" artifacts, acceptable at one time, but later deemed too narrow. Getting information "under control" by forcing it into a narrow construct was an animating spark of the iSchool movement, but a fool's errand. Machine was sufficient until people eventually wanted to move beyond it.[50]

Pharma and chemistry tinker, machine-like, to discover new things. Resources matter, and connecting science and engineering to the military is probably more than coincidental. The field of Social Studies of Science arguably does not handle standards very well, although standards are manifestly important to both science and social studies. Pornography, religion, and other forces probably play prominent roles in histories, and are core to innovation in Information or other fields—but are often not well-covered by iSchools. Perhaps Social Studies of Technology under-attend "obsolescence," given important path dependencies. How to account for the fact that much technology in daily use is obsolete? Fields engaged in organization can be weak on organization study and at odds with other concerns (for example, information retrieval). The input/output model has traveled well between information-related communities, but few other models have. Libraries might be seen as community-building resources *inside* LIS, LIS people also depend on how libraries are seen *outside* LIS. "Cyberinfrastructure" includes far more than the computer. But it sometimes fixates on the computer (typically "big iron," meaning powerful computing machines) running programs written in ancient computer languages like FORTRAN. Efforts to share data often fail to consider institutional reasons against sharing data, and wonder why sharing data is so difficult.

Call: How to do the history of iSchools?
Response: It's harder than it looks.

The history of iSchools is recursive, so "doing" the history is difficult. For one thing, the iSchools are probably but one episode in a protracted effort, Dream, that has seen many epochs and will see more. The iSchools might turn out to be relevant only to this particular epoch. Any focus on "information" contains vestiges of prior efforts at Dream, representing information-speaking peoples. iSchools focus on areas that are often broken down into artifacts (say, books) or abstractions (say, algorithms or data structures). They incorporate the concept of "systems" (say, library systems, database systems, operating systems). When someone asks, "What are iSchools about, *really?*" everyone goes silent until something concrete is mentioned, usually an artifact. The iSchools collectively show great heterogeneity and richness. This probably facilitates adaptability, waiting for Dream to deliver. When it fails to deliver, what constitutes an iSchool is hard to say. It entails finding a common understanding of "information" when there is, thus far, no real possibility of a common understanding.

The iSchool movement was undertaken by people motivated to create a community that would nod sympathetically when anyone said, "Information is a field, you know." The iSchools offered solidarity; as with Article 5 of NATO, an attack on any is an attack on all. It appears to have been unneeded; no iSchools have been attacked. With iSchools higher education institutions made new programs, swept up loose ends, created *Lebensraum*, or built apparatus to put butts in seats. First-tier schools claimed peerage with other first-tier schools. Second-tier schools could claim co-membership with first-tier schools. It had something for everybody.

For the historian, there is the thorny question of the unit of analysis. One can describe history of information since the Enlightenment—marked by a sine wave wending between universality and practicality. One can start with an institution (say the library) and look at its growth and attendant needs for a professional cadre. One can start with an artifact (say the computer) and trace its genesis and spread. Each trajectory described will be different and cannot be inscribed into a single narrative.

Conclusory call: Can you sum it up in one word?

Response: Brigadoon.

The intellectual history of the information-speaking peoples—what we call Dream—is the iSchool effort to deal with the problematic of "information." This is a long-standing effort, dating back at least to the Enlightenment. The iSchools movement emerged at the end of the 20th and the beginning of the 21st centuries. People who take it seriously engage Dream to provide solidarity and facilitate iSchool development within academic institutions. Dream is information high ground. People can pursue it without running afoul of bigger schools or colleges. Sure, it is difficult and easily diverted to professionalism. But the story of Dream is it recognizes the complexity of intertwining narratives and the heterogeneity of skills within iSchools. The response is a story of Dream and professionalization. The iSchools require both. An i for an I. Perhaps a truth for a Truth.

We sum this chapter up with the word *Brigadoon,* the title of a 1947 Lerner and Loewe Broadway musical about a Scottish village of that name that appears for a day every century. It is based on legends of disappearing towns in folklore, from various countries, including the mythical village of Germelshausen in Germany. The word applies to the iSchools in that intellectual engagement around information as in Dream happens episodically, usually at the intersection of developments in technology and/or technique that cause a flurry of attention to information issues. During the 20th century there were huge strides in information technology and technique, in computing, communication, information organization and search, applications to enterprise and governance, and so on. The salience of the potent if ambiguous "digital transformation" is testimony to this. Dream never goes away, but it does sleep . . .

Acknowledgements

We thank the reviewers for their constructive suggestions and Bill Aspray for his generous leadership of this project. Also, thanks to Roderick Crooks, Thomas Haigh, Cory Knobel, and Howard Rosenbaum for their helpful comments.

Notes

1. Call and response starts with a speaker's call (often a question) followed by a response (often an answer) from an audience. Each major section below follows this format. Author order is alphabetical; a convenience which benefits the first author on this occasion.

2. https://en.wikipedia.org/wiki/Eye_for_an_eye, accessed 10/19/23.

3. Compare here Ross Ashby's Law of Requisite Variety from early cybernetics—which stated that every representational system needed at least as many variables as the system it represented—there's just too many variables nowadays. In Ashby, W. Ross (William Ross), *An Introduction to Cybernetics* (New York, NY: J. Wiley, 1956).

4. Wellmon, Chad. *Organizing Enlightenment: Information Overload and the Invention of the Modern Research University* (Baltimore, MD: Johns Hopkins University Press, 2016).

5. Chatwin, Bruce. *The Songlines* (New York: Penguin Books, 1988). Cf. Mann, Thomas, and John E. (John Edwin) Woods, *Joseph and His Brothers: The Stories of Jacob,*

Young Joseph, Joseph in Egypt, Joseph the Provider. Translated from the German by John E. Woods (New York: Everyman's Library, 2005).

6. Bowker, Geof. "How to Be Universal: Some Cybernetic Strategies, 1943–70." *Social Studies of Science* 23.1 (1993): 107–127.

7. Yerushalmi, Yosef Hayim. *Zakhor, Jewish History and Jewish Memory*. Foreword by Harold Bloom; with a new preface and postscript by the author (New York: Schocken Books, 1989).

8. Burchell, Graham, Colin Gordon, and Peter M. (Peter Michael) Miller. *The Foucault Effect: Studies in Governmental Nationality: With Two Lectures by and an Interview with Michel Foucault*. Edited by Graham Burchell, Colin Gordon, and Peter Miller (London: Harvester Wheatsheaf, 1991).

9. Napoleon brought a group of scientists to catalogue Egypt. See https://www.napoleon.org/en/young-historians/napodoc/bonaparte-in- egypt-2-the-scientific-expedition/, accessed 10/19/23. Cf. Winchester, Simon. *The Map That Changed the World: William Smith and the Birth of Modern Geology*; illustrations by Soun Vannithone (New York: HarperCollins, 2001).

10. Wellmon (2016). Compare Blair, Ann, *Too Much to Know: Managing Scholarly Information before the Modern Age* (New Haven, CT: Yale University Press, 2010).

11. Goldman, Lawrence. "The International Statistical Congress, 1851–1878." In *Victorians and Numbers* (Oxford: Oxford University Press, 2022).

12. Latour, Bruno. *Science in Action—How to Follow Scientists and Engineers through Society* (n.p., 1987).

13. Lester, David, ed. *Emile Durkheim: Le Suicide One Hundred Years Later* (Philadelphia: Charles Press, 1994).

14. Wiener, Norbert. *Cybernetics; or, Control and Communication in the Animal and the* Machine. Forewords by Doug Hill and Sanjoy Mitter, 2nd ed. (Cambridge, MA: The MIT Press, 2019), chapter 1.

15. Hall, Rogers, Ken Wright, and Kären Wieckert. "Interactive and Historical Processes of Distributing Statistical Concepts Through Work Organization," *Mind, Culture, and Activity*, 14, 1–2 (2007): 103–127.

16. Or fourth, depending on how you count—the water mills of the 11th century being the first for many.

17. Beniger, James R. (James Ralph). *The Control Revolution: Technological and Economic Origins of the Information* Society (Cambridge, MA: Harvard University Press, 1986).

18. Attributed, to Henderson, 1917: Gillispie, Charles C. *The Edge of Objectivity: An Essay in the History of Scientific* Ideas (Princeton, NJ: Princeton University Press, 1960), p. 357.

19. Wiener (2019).

20. Buckland, Michael K. "The Reception of Suzanne Briet in the United States." *Bulletin of the American Society for Information Science* 39, 4 (2013): 40–41.

21. Zenil, Hector, ed., and foreword by Roger Penrose. *A Computable Universe: Understanding and Exploring Nature as Computation* (Singapore: World Scientific, 2013).

22. Cf. Ribes, David et al. "The Logic of Domains." *Social Studies of Science* 49, 3 (2019): 281–309.

23. See, for example, Dervin, Brenda, Lois Foreman-Wernet, and Eric Lauterbach, eds. *Sense-Making Methodology Reader: Selected Writings of Brenda Dervin* (Cresskill, NJ: Hampton Press, 2003).

24. Kling, Rob, and Walt Scacchi. "The Web of Computing." https://doi.org/10.1016/S0065-2458(08)60567-7. Compare the chapter by Howard Rosenbaum in this volume.

25. Medina, Eden. *Cybernetic Revolutionaries: Technology and Politics in Allende's* Chile (Cambridge, MA: MIT Press, 2011). Cf. Wells, H. G. "The Idea of a World Encyclopædia." *Nature* (London) 138.3500 (1936): 917–924.

26. Bush,Vannevar Bush. "As We May Think," 1945. https://www.theatlantic.com/magazine/archive/1945/07/as-we-may-think/303881/

27. Gregory, Judith. "Sorcerer's Apprentice: Creating the Electronic Health Record, Re-Inventing Medical Records and Patient Care," University of California, San Diego, 2000.

28. See Wellmon (2016) and Blair (2010) for a longer history.

29. Disraeli, Benjamin. *Endymion* (New York: John W. Lovell Company, 1899), p. 156.

30. Kay, Lily E. *Who Wrote the Book of Life?: A History of the Genetic Code* (Stanford, CA: Stanford University Press, 2000).

31. Gilbert, Scott F., Jan Sapp, and Alfred I. Tauber. "A Symbiotic View of Life: We Have Never Been Individuals." *The Quarterly Review of Biology* 87.4 (2012): 325–341.

32. Hutchins, Edwin. *Cognition in the Wild* (Cambridge, MA: MIT Press, 1995).

33. Discussed by many, including Thomas Aquinas. See the *Internet Encyclopedia of Philosophy* (https://iep.utm.edu/faith-re/#SH4e).

34. Deutsch, David. *The Fabric of Reality: The Science of Parallel Universes—and Its Implications* (New York: The Penguin Press, 1997).

35. Leibniz, Gottfried Wilhelm. *Writings on China.* Translated, with an introduction, notes, and commentaries by Daniel J. Cook and Henry Rosemont Jr. (Chicago and LaSalle, IL: Open Court, 1994).

36. There are few histories of the iSchools beyond the one at the iSchool web page https://www.ischools.org/ and the histories of particular schools that one can find on iSchool web sites. One exception is Ron Larsen's history that can be found at http://d-scholarship.pitt.edu/5852/1/The_iSchools_final.pdf.

37. DiMaggio, Paul J., and Walter W. Powell. "The Iron Cage Revisited: Institutional Isomorphism and Collective Rationality in Organizational Fields." *Economics Meets Sociology in Strategic Management*, Vol. 17 (Emerald Group Publishing Limited, 2000), 143–166.

38. https://www.ischools.org/, accessed 1//19/23.

39. Shinn, Terry. *L'École Polytechnique: 1794–1914*, préf. de François Furet (Paris: Presses de la Fondation nationale des sciences politiques, 1980).

40. Douglas, Mary. *How Institutions Think* (New York: Syracuse University Press, 1986).

41. The origin of the DIKW hierarchy is disputed. A good general introduction is in Wikipedia, https://en.wikipedia.org/wiki/DIKW_pyramid.

42. Campbell, Donald. "Ethnocentrism of Disciplines and the Fish-Scale Model of Omniscience," in M. Sherif & C. W. Sherif (Eds.), *Interdisciplinary Relationships in the Social Sciences* (Boston: Aldine, 1969), pp. 328–348.

43. Arney, William Ray. *Experts in the Age of* Systems. Foreword by Ivan Illich. (Albuquerque: University of New Mexico Press, 1991).

44. Navarro Reyes, Jesús. *How to Do Philosophy with Words: Reflections on the Searle-Derrida Debate* (Philadelphia: John Benjamins Publishing Company, 2017).

45. Bowker, Geoffrey C., and Susan Leigh Star. *Sorting Things Out: Classification and Its Consequences* (Cambridge, MA: MIT Press, 1999).

46. For an interesting, albeit corporate, view of the Carnegie Libraries, see Anderson, Florence, *Carnegie Corporation Library Program, 1911–1961* (New York: Carnegie Corporation of New York, 1963).

47. Shirky, C. (2005). "Ontology is Overrated: Categories, Links, and Tags." From the shirkey.com blog. http://shirky.com/writings/ontology_overrated.html, accessed 1//19/23.

48. One example is https://medium.com/@jcueto/race- and- gender- among- computer -science-majors-at-stanford-3824c4062e3a, accessed 10/19/23.

49. Shannon, Claude Elwood, and Warren Weaver, *The Mathematical Theory of Communication*. Foreword by Richard E. Blahut and Bruce Hajek (Urbana: University of Illinois Press, 1998).

50. Mumford, Lewis, *The Myth of the Machine* (New York: Harcourt Brace Jovanovich, 1967).

Part III

Reshaping the Mature Story of Computing History

Chapter 6

Writing the History of Computing

The First Half Century

David Gugerli and Daniela Zetti

When we started working on computer history, the motivation for our research was the history of the second half of the 20th century.[1] We had questions about the change of organizations and companies, and certainly about the transformations of social interaction. Unsurprisingly, computers played an important role in this story. Sometimes, computers represented the longed-for hero, sometimes a murky Mephisto. In both cases, their role kept on changing over our 20 years of research. What remained stable was the emphasis of our interest in historical problems and solutions. When employees of banks, universities, the post office, or government bureaucracies had to reorganize their business because they could not cope with the impositions of growth, they moved their transactions into computers, invented new procedures, and restructured their communicative routines. To understand this historical shift, we read what contemporary engineers, administrators, politicians, and users had to say about points of sale, interrupts, databanks, addresses, and autonomy.[2] All this was tangible for us in archived project outlines, minutes of working groups, memoranda, and reports to clients, while books on computer history offered us a *grand récit* that could not clarify the specific problems we were interested in.[3] In fact, the big stories in computer history books rather obscured the sophistication of what was happening in those negotiating zones where the police or the check management (to name just two historical actors) allowed the linking of architectural issues of organizations and machines—and that is: of many machines and their reconfigured (im)material components.

When we reread a selection of books and papers for this chapter, we were above all interested in the shifts of historiographical problems in their connection to socio-technical change. We structure these problems by asking a handful of questions: Do developers of computing technology produce genuine historic narratives? How can the historicity of computing be organized and what role does it play in solving organizational difficulties? Have efforts to record and document computer history influenced its representational possibilities? Can the expansion of computer history into the past and present be interpreted as a shift of attention toward the digital age? How and where did historians begin to classify their own work, and what happened to this organizing effort when the great media hype of the 1990s virtually obliterated history?

Developers' History?

Unprecedented technologies are always difficult to justify. Early advocates of computers had to be willing to explain the sterling future of their machines to all who would listen as well as (and especially) to those who remained skeptical. With the past as the argumentative resource, very little could be achieved under these circumstances and history was of no use. However, estimates about the future were usually overtaken by the present more rapidly than expected. To remain credible, these estimates had to be continually projected into a new future, which is to say, revised. At such times, for all its optimism, computer technology was unexpectedly provided with a historical narrative. For instance, computer developers Murray L. Lesser and John W. Haanstra claimed in late 1956 that "originally" computers were created to process administrative tasks on a quasi-industrial scale. The two engineers did not specify the details of this origin. But their account was definitely situated in the past, could be addressed narratively, and was invoked precisely when a new future was about to be produced as was the case at IBM. Lesser and Haanstra maintained that future data processing would no longer be done in conveyor-belt fashion, with elaborately sorted, neatly stacked punched cards. The goal was emancipation from batch processing of uniform data and an attractive future for computers with random access to novel drum memories. This required creating a past that could be evoked by the word "originally." The development of computers gives rise to historical interpretations.[4]

Hence, the first histories of computers came with strategic acts of distinction of computer developers. Furthermore, it became the by-product of computer operations: in data centers, it was inevitable that old and new machines would come together. Printers and punching devices for punch cards, which had been connected to the Hollerith machines, were also used by data processors and shared space with fast drum memories, well-known magnetic tape stations, and brand-new hard disks. The old *functioned* alongside, in front of, and behind the new (or vice versa), leading to operational expense and annoyance. In addition, the data centers *documented* the history of their operations. Log files recorded the activities "in the computer," and eventually "in the network." They reminded "users" and "administrators" of which operations had been "completed successfully" and which had not. Computers did not forget their operational past and marked records with "time stamps," logged "interventions" by components and users, and recorded who had changed the valid version of an application program or the operating system. Whenever operations stalled or the system crashed, the chronicle kept in log files became the prerequisite for getting all interrupted processes running again.[5]

While developers needed an origin for their futuristic concepts, and data centers recorded their operational activities, the computer industry developed an entire range of protohistoric orientation aids. Several forms of self-historicization were discernible. The most popular and perhaps most obvious was genealogical numbering. Konrad Zuse's Z1, Z2, Z3, and in particular Z4 offer one such example. In naming this series of machines, which emerged out of the ruins of the Third Reich, Zuse wished to genealogically connect with the electronic and programmable computers of the postwar period.[6] The four Harvard computers known as Mark I–IV also relied on this family history technique. Machine numbering at IBM was somewhat more complex.[7] This was due on the one hand to the substantially larger product range, and on the other hand to the fact that some machines were assigned a number but were never built and remained paper tigers. Nonetheless, such machines not only brought an exuberant diversity to the devices but also dramatized the genealogical classification system. Thanks to the gaps and discontinuities in succession, the catalog reflected the promise of the future in the performance classes then available. The discontinuous numbering of the machine classes showed that IBM customers were on the right track and that development was also making remarkable leaps forward.[8]

Genealogies suggest entrepreneurial stability and technological variety, may identify performance classes, and document resilience of the industrial world. In the computer world, genealogies facilitated communication between those who manufactured, sold, programmed, and acquired computers, assuring these actors and the machines a place in the annals of both the computer industry and the data processing centers. This labeling practice was also popular with programming languages and was evident early in the case of FORTRAN. Preliminary versions of FORTRAN were already in existence by 1954, but it was not until 1957 that a compiler was declared sufficiently market-ready to be delivered together with IBM 704 machines. The fact that FORTRAN was followed in 1958 by a language called FORTRAN II meant two things: FORTRAN now had to be called FORTRAN I, and the expectation was that FORTRAN III lay within the realm of expectations. But every genealogical legitimation also entails risks. FORTRAN III never did see the light of day, whereas FORTRAN IV was published between 1962 and 1966, in different versions. It followed that the FORTRAN world was going to have to be reorganized to replace the obsolete genealogical system with a generalizable chronological system. As it happened, the further development of FORTRAN IV, labeled x3.9-1966, was named FORTRAN 66 and immediately ISO-certified. FORTRAN 77 did not appear until 1978. A successor planned for 1982, long vaguely referred to as FORTRAN 8x, was certified as FORTRAN 90 in 1991. All this did little to clear up matters but at least made chronological sense.[9]

Much more abstract than the genealogical and chronological classification systems was a third, quasi-scientific classification method known as Moore's Law. It completely brushed off software and certainly ignored the machine type, manufacturer, application type, and architecture of a computer. It was exclusively keen on the annual growth rate of processor performance. According to Gordon Moore, an electrical engineer employed by Fairchild Semiconductor, this rate was doubling approximately every 1.5 years. It enabled Moore to put the computer industry, which was particularly susceptible to disruptive twists and turns, on a continuous path of growth that sprang from the observable past, influenced the present, and helped steady the future. But there was one thing Moore's Law could not do: it could not allow any compartmentalization.[10] Each processor was equal to its predecessor and its successor. Processors could be included in his geometric series, and some may have become better known than others. But Moore's curve makes no allowance for milestones, discontinuities, or exceptional delays. Everything runs a little too smoothly.

Anyone looking to argumentatively bridge a longer period, and who required breaks and qualitative leaps to do so, and yet still wished to speak for the entire industry, was forced to resort to an archaic principle that classified computers according to their basic electronic components. One referred to electromechanical, electronic, and transistorized generations of computers. This enabled an argumentative flexibility that required neither flagging interruptions in homogeneous growth rates nor familiarity with the genealogical and chronological circumstances of individual companies or programming traditions.[11]

Organizing Historicity

The quest for historicity in the field of computer development is not just a professional hazard for historians. Historicity has arisen for many different reasons. No matter how interested *digital societies* (i.e., associations, user communities, or developer communities) are in developing a different future, they have always had a vital interest in historical orientation and rely heavily on monitoring, reporting, and documentation, and thus on archives.[12] Ever since these societies formed, they have created backups and printed collections of commands and instructions. They have produced and circulated evaluations in committees, collected and commented on conference papers, and subjected them to critical peer review.[13]

Digital societies began to reckon with their past early on and used it as an argumentative resource. But producing a future that was explicitly different from the past was also playing with fire, especially in computer-based public spheres. Anyone who could say how things had been in the past or "originally" soon came under suspicion of no longer wishing to shape the future. As a result, it took considerable time for some computer specialists to use their memories of original intentions not merely as a launch pad for future designs, but to classify the state of the art. In so doing, they adopted (mostly at the end of their career or when switching from industry to academia) a culture of argumentation relieved of the declamation that had served the computer industry as a source of legitimacy. The history that is the product of developer discourses is frequently a sign of the development of its representatives and communities.

The case of the British computer scientist Brian Randell provides an excellent illustration of such an "origin" in the history of the computer. Randell was among the first computer specialists to investigate the origins of digital computers. From 1957 to 1964, he was employed by the Atomic Power Division of the English Electric Company, where he developed compilers for ALGOL 60. He then worked at IBM's T. J. Watson Research Center in Yorktown Heights, New York, and later in San Jose, California, on computer architectures, operating systems, and system development methods. Randell was thus used to being at the technological forefront when, in 1969, he moved from the area of industrial development to academic teaching and research at Newcastle University. He may have found another library there, like the one he remembered from his studies at Imperial College. In any case, while preparing his inaugural lecture at Newcastle, he stumbled on some work by the Irish accountant Percy Ludgate. Written many decades earlier, the work dealt with issues relating to programming machines. The approach struck Randell as surprisingly familiar.[14]

Randell's interest in Ludgate made history a source of distinction for computer science as a nascent subject. He wished to ensure that the (prestigious) history of science, rather than the (mundane) history of technology would one day professionally deal with the matter. Randell may well have assumed that computer science would first have to be stabilized as an academic subject (so it could then be discovered by the history of science). Or, conversely, perhaps he assumed that the thoroughly contentious academic professionalization of his field could be accelerated if it had a decent past.

Randell's anthology of basic texts, published in 1973, obviously had classical claims. However, what was presented under the title *The Origins of Digital Computers* did not reckon, as Darwin did, with a development *principle* that produced different types of computers. Instead, Randell was searching for heterogeneous contexts of origin from which the present-day, "modern digital computer" emerged.[15] An audience interested in technical details about the emerging computer sciences encountered neither a natural history nor a theory for explaining contemporary diversity, but instead the documentation of a forgotten but diverse past intended to stabilize what had been achieved to date. To this end, Randell embarked on a quest for elements from which a concept of the modern computer could be assembled. His quest began with Charles Babbage's and Ada Lovelace's soon-to-be-famous "analytical engine," for which he found a description from 1837, and ended with another British machine, the Electronic Delay Storage Automatic Calculator (EDSAC), which went into operation in 1949. Randell regarded this year as the start of the present-day, following a more than century-old, highly complex, and somewhat curious development history, whose protagonists often had no idea how "it" could be done or how to solve their problems with elegance.[16]

The particularly striking thing about Randell's collection was neither the beginning nor the aim of the history. What is surprising is that Alan Turing was treated as a nonentity. Initially, as Randell later explained, he had not even planned to mention Turing. After all, Turing had never actually developed a machine. In 1936, he had written a very theoretical design for a general-purpose machine, but to the best of Randell's knowledge, nothing concrete or tangible had emerged from it. In the first edition of *The Origins*, at the insistence of his British colleagues,

Donald Michie wrote two cursory pages about Turing. Although at the time Randell himself was also and very cheerfully ensconced in theory research at Newcastle, he did not think that his work would not (as in Turing's case) lead to anything concrete. For sure, Turing was somehow worth mentioning in 1973. But due to his exclusively theoretical interest, he was not considered one of the forefathers of digital computers.[17]

Nor as such was he suitable material as a figurehead of computer science. Indeed, his place among the ranks was only assured after confirmation of the suspicion that Turing had met John von Neumann during the war, and that Max Newman and Tommy Flowers had used Turing's design to build an electronic computer at the British Post Office Research Station. Encrypted messages from the German Wehrmacht had been decoded in legendary Bletchley Park. This suspicion was confirmed, despite the British government's policy on secrecy, and was published as a sensational bit of news in 1976 by Randell at a conference on the history of computers at Los Alamos.[18]

Randell's new findings made it possible to say, contrary to expectations, that Turing's work had produced something useful. Moreover, perhaps the undisputed American computer supremacy was not the product purely of the nuclear physicists at Los Alamos and their atomic bomb. At any rate, the prevailing genealogy of machines and inventors began to look a little shaky. Turing and von Neumann could no longer be consulted. Konrad Zuse, who always considered himself as the first among the first, was terrified by Randell's revelations in Los Alamos, and remained silent. The next logical step then was to consult old documents, i.e., to engage in historical research.

The discovery of a British computer linked to Turing during the Second World War unsettled the family history of computers and their inventors. But the updating of Turing as a figure was useful, not only for British scientists.[19] In every category of apparatus—from mainframes, to midicomputers, to microprocessor-based machines—Turing appeared to have contributed principles of transnational importance. The *Turing machine* linked the computer with the fundamental cultural technologies of reading and writing, and the *Turing test* made it possible to evaluate the relationship between humans and machines. Moreover, the *historical*

Alan Turing,[20] who had managed to avoid even his mother's biographical grasp, had become accessible to computer history, all the while remaining surrounded by an aura of secrecy, enigma, and of outsiderness. He was therefore ideally suited for both the undocumented beginnings and the anthropological interpretation of the calculating human *vis-à-vis* his machine.[21]

The fact that the history of the computer was a zone of conflict had obvious consequences for its function as a source of orientation and argumentation. National claims to the creation myth of the computer would now have to be reassessed. But Randell's discovery was above all grist for the mill for academic computer scientists. As theoreticians, they both preceded and followed the machine. Turing's 1936 paper "On Computable Numbers" predated Colossus. Von Neumann's "First Draft of a Report on the EDVAC," published in 1945, came after the development work. Both a documentable machine history and an academically founded computer science were needed; both would come about.[22]

This evidence of theoretical relevance could easily have served to establish the ultimate supremacy of theory over practice. Yet, computer science as a field was wisely very tentative about moving in this direction. The scramble over Turing, von Neumann, Zuse, and "their" computers could much better serve the discipline if "the machine" in general was accorded proper respect—venerated as a museum piece and eventually put out to pasture. In short: the monumental computer (sometimes called Colossus) had to become both the aim *and* origin of development. This would grant the theory the desired relevance and lead to a technologically attractive union of all elements of the computer community.

Randell's collection was revised and reprinted in 1975 and again in 1982. *The Origins of Digital Computers*, the history of their early development, also changed. The book was simultaneously historically and professionally effective. In his 2012 essay "A Turing Enigma," Randell recounted the story of how he selected texts and authors and in so doing (with no small amount of mischief) reinvented the history. One year later, the BBC filmed Randell at the opening of an exhibition on the history of computers.[23] Behind him was Colossus, which had just been elaborately reconstructed, and in front of him was a neat laptop, which he was using to give a presentation. What might be seen as a physical arrangement between a colossal past, a successful rehabilitation of the theory, and the current projection culminated in a grand connection history between the secret and the published, the remembered and the existing, between software and hardware.[24] It was a history that was accessible to the computer world because it easily and elegantly combined academic mathematical theory with industrial practice; and yet, as was the case between Charles Babbage and Ada Lovelace, as well as between Alan Turing and Tommy Flowers, it reflected a successful collaboration involving shared tasks. It was a museum staging that repeatedly combined veneration of saints and the desire for enlightenment with entertainment.

Recording and Documenting

The discovery of Britain's Colossus computers in 1976 was the apex of the search for the historical beginnings of computer science. That it came to light only then was easily explained by the fact that some of the computer development had been kept secret during the war. The past intelligence work done during the Second World War haunted the increasingly professionalized computer science of modern times. It made the later discovery explosive, but it also provided a plausible explanation as to why entire computer complexes could vanish and be forgotten.

The existence of the Colossi called into question the inevitable loss of milestones in computer development. History worked differently than assumed in computer development and application and could, using the right methods, raise computer science to an entirely new level. The discovery that computers had been built in Britain and then dismantled was only possible owing to Randell's use of historical research methods.

Contemporary computers and computer science were not without precedent but possessed a narrative history that was the direct result of Randell's inquiries and research. The basis for his computer discovery was still isolated windfall, objects, and circumstantial evidence, as well as people who had developed computers. Randell increasingly disregarded the physical and abstract classification systems of computer development and abandoned the repositories carefully managed by information technology and transformed into digital spaces. The value of his research, indeed, the result, was the validation of configurations and relationships that were at odds with the familiar development stages and milestones of computer science. If the revelation of the once-vanished computers took the professional community and the interested public by surprise, it was at least as surprising that the future of the computer could be shaped within the work and argumentative space of the history of science, of all things.

The founding of a society to promote the history of computers was also broached in 1976 at the Los Alamos conference, right after the unveiling of the first new electronic computers. William Aspray, one of the first historians to study the history of computers, described the beginnings thirty years later. In Los Alamos, Arthur Norberg, a historian of science and technology from Berkeley, met printer manufacturer Erwin Tomash and his wife, Adele, at a "picnic away from the rest of the participants," where they talked about computer history for a long time. Norberg tried to persuade them to fund a museum for computer history.[25] He must have been thinking about the computer collections of the Smithsonian Institution in Washington, the Science Museum of London, and the Deutsches Museum in Munich. Digital Equipment Corporation (DEC), a Massachusetts-based computer manufacturer, also established a museum of computer history. The Tomashes rejected the idea of a museum because they thought the costs were unaffordable.[26]

However, in talking to "industry friends" over the next few months, they managed to fundraise a significant amount. An archives and research center was founded in 1977. In 1980, the Charles Babbage Institute (CBI), as it was called, was moved and affiliated with the University of Minnesota to bolster its academic orientation. The CBI provided bibliographies of computer history, created an archive, and established an oral history program under director Norberg.[27] A book series reprinted texts on computer development since Charles Babbage. In 1985, for example, the Moore School Lectures of 1946 were published as volume 9 of the series. In the first two decades, historians interviewed 330 contemporary witnesses of computer development and, together with them, revisited bygone times of societies in which computers played an increasingly important role.[28]

This brief review of the paths and detours taken by the CBI when it was founded in the late 1970s reveals why the historiography of the computer was so quickly established in the United States. With the financial support of computer manufacturers, the decision was made to link the CBI to university research structures and to use methods from the humanities that brought to light texts, memories, and narratives as raw material for later investigations. With a firm focus on research into the history of the computer, an astonishing amount of output on paper and tape was archived and produced. Of particular note is that digital documentation was not emphasized in corpora building and that one object of investigation—software development—was omitted. Moreover, CBI thus remained remarkably close in its research practice to the program that the digital community had pioneered in the 1960s and 1970s in its quest for the origins of its discipline. The computer was at the center of these efforts, even more so than in contemporary computer science. The focal point was not the heterogeneity of existing computer models, but a machine that was conceived of as universal. The British "inventor" Charles Babbage, the patron saint of the foundation, institute, and book series, stood for what, around 1980, was the hopeful marriage of the history of industrialization and mathematics under the common umbrella of a technological history of the computer.

AN UMBRELLA FOR THE DIGITAL AGE

The first historians to deal with the volume of interviews, old lectures, and papers, sources of mathematical history, electrical engineering, and what developers were thinking about the computer and its components came from the humanities. Paul E. Ceruzzi was an Americanist, and later a historian of technology. William Aspray was trained in mathematics and philosophy and history of science. Nancy Stern was a historian of technology. They edited the historical documents and narratives into prehistories of the computer—and of computing.[29]

In 1982, an article by Stern appeared in the *Journal of the Society for the History of Technology*, in which she "propose[d] to examine the entrepreneurial activities" of J. Presper Eckert Jr. and John William Mauchly. Between 1943 and 1951, Eckert and Mauchly had developed four computers: ENIAC, EDVAC, BINAC, and UNIVAC. After going through many boxes of archives and conducting several interviews, Stern concluded that judging success was not an easy thing to do. Although the early market launch of UNIVAC—a commercial digital computer—had put the two developers far ahead of their competitors, the real achievement was ushering in a new phase in computer history. Large companies had not been agile enough to commercialize digital computers, and the two engineers had willingly taken on the necessary risks. On the other hand, Eckert and Mauchly had been extremely naive and anything but successful as independent entrepreneurs. Despite Stern's focus on personalities and things, a "Colossus shock" was readily discernible.[30] Stern was in search of a novel and unique selling proposition for the early American digital computers, whose pride of place had been so thoroughly brushed aside by Randell. Moreover, by addressing the question of the innovative capacity of large companies, Stern took into consideration the contemporary early 1980s. By this account, UNIVAC was a forerunner of the personal computers that spread quickly around 1980.

Finally, in *Reckoners: The Prehistory of the Digital Computer, from Relays to the Stored Program Concept, 1935–1945*, Paul E. Ceruzzi neatly summed up the 1980s' interpretation of the prehistory of the computer. Before 1935 there had been machines that could calculate and process information, "but they were neither automatic nor general in capabilities. They were not computers." In the 1930s, when people used the word "computers," they were still referring to human beings. It was only after 1945 that the term "computer" definitively came to denote a machine that did calculations. From then on, computers developed and improved, and became cheaper and smaller.

In a monograph titled *John von Neumann and the Origins of Modern Computing*, published in 1990, William Aspray focused on a decade that he considered significant for computer history. He began with 1943, the year in which von Neumann had written a Princeton colleague that in England he had developed "an obscene interest in computation" and would be coming home changed, "a better and impurer man."[31] Up until his death in 1957, von Neumann was making lasting contributions to the design, application, theory, dissemination, and legitimation of computing.

Aspray based his investigation on, as he put it, the first careful evaluation of the rich archival legacy of von Neumann and his peers. He was thus confident that the book presented a comprehensive and complete rehabilitation of the many different contributions that von Neumann had made to the history of computers and that it attracted its readers' attention to a historically based understanding of the concept of computing.[32]

That same year, Aspray also published *Computing before Computers*, to which Ceruzzi contributed chapters on relay calculators and electronic calculators. In this case, the period of investigation was dramatically extended, and at the same time it was clear that, for Aspray, the term "computing" meant not only computer-based calculation, but calculation in general, which had always been technologically based. Martin Campbell-Kelly wrote a chapter on punched-card machinery. The area of computer science was represented by Allan Bromley and Michael Williams. Their chapters covered "early calculation"—computers in antiquity and in the Middle Ages, the abacus in Europe and "the Orient," and slide rules, as well as Charles Babbage's "difference engine" and his "analytical engine." Analog computing devices, such as planimeters and computers for air defense, rounded out the ancestral gallery. In his introduction, Aspray stated that the purpose of the book was to not forget a "rich history that extends back beyond 1945. Since antiquity societies have had a need to process information and make calculations, and they have met this need through technology." The prehistory of calculating machines and societies was so easily forgotten for two reasons: a ubiquitous discourse around the computer revolution and the information age, and public awareness and the rapid development of the computer since the Second World War. The computer had led to fundamental changes "in the way we conduct business, perform scientific research, and spend our leisure time."[33]

The efforts of the researchers affiliated with the CBI to promote professionalization were aimed at establishing a computer historiography that included industry, mathematics, and technology. At the same time, developers' discourses were refined, as in the 1986 book *IBM's Early Computers*. Historian of science I. Bernard Cohen contributed the foreword, in which he enthusiastically strove to persuade readers of the qualities of the book as a rare investigation of corporate history.[34] All too often, he wrote, histories of technology-based companies turned out merely to be company histories. Technological development was of peripheral interest. Now, all the authors of *IBM's Early Computers* were documenting for the record a long era of development activity in the computer industry. In other words, in the 1980s, the developer community continued to occupy itself with history.[35]

The period covered by *IBM's Early Computers* began after the Second World War and ended in the early 1960s. The authors—Charles J. Bashe, Lyle R. Johnson, John H. Palmer, and Emerson W. Pugh—examined management decisions at key (to them) moments to show how the computer had transformed IBM, the former manufacturer of punched-card accounting machines. They had scoured their own institution for earlier IBM-made computers and had made a discovery that to Cohen, a historian of science in the mid-1980s, seemed so distant and exotic as to warrant mention: the universal programmable computer of the 1960s and 1970s had had precursors.[36]

The precursors of the general-purpose computer were not distinguished by serial numbers but by their fields of application. In their search for predecessors of IBM's general-purpose computers, the authors encountered colleagues who had envisioned the future of their products exclusively either in science and engineering or in business, and who had increasingly forgotten this history. The final chapter of the 700-page book laid out the "architectural challenges" confronting IBM as an up-and-coming computer manufacturer from the 1940s to the 1950s. Here, "architecture" was understood to refer both to the organization of a company into departments and to the assembly of a computer from components. The double meaning of the architecture and the complex challenges it posed within the company were embodied by System/360, one of the "unified line" of products announced by IBM in 1964. According to Bashe et al., it was intended specifically to abolish the distinction between "computers for science and computers for accounting" within IBM. Bashe et al. had published the prehistory of the IBM system and the system at IBM.[37]

In Stern's book, as well as in those by Ceruzzi and Aspray, and particularly in *IBM's Early Computers* by Bashe et al., the widespread success of personal computers and small start-up companies prompted questions about the industry's power to innovate. The PC stimulated the search for criteria that would make it possible to distinguish the development stages of "the computer." Which brings us to another contemporary book that made a claim in 1984 that the authors considered sensational: even the PC had a history.

Paul Freiberger and Michael Swaine portrayed the development of the personal computer quite differently from the way Stern, Ceruzzi, Aspray, and Bashe et al. had tried to do, namely, as the story of a revolution carried out by a few and later more and more heroes. Charles Babbage served the two journalists as the conventional starting point of the narrative about the computer's unprecedented ascent. A few pages later other inventors followed: Hermann Hollerith, Alan Turing, and John von Neumann. Turing was accorded an appearance as the conceptual inventor of the universal computer, the bright mind who (theoretically) put his concept into a programmable electronic machine. For Turing "envisioned a machine designed for no other purpose than to read coded instructions describing a specific task and to follow the instructions to complete its own design."[38] It was, as Freiberger and Swaine claimed, John von Neumann who first turned the vision into reality. "The instructions became programs, and [Turing's] concept, in the hands of another mathematician, John von Neumann, became the general-purpose computer."[39]

Here, the prehistory of the computer, which had caused so many headaches for developers and historians, was formulated regardless of the state of research. But Freiberger and Swaine were not following the discussions of the computer historiography either. Their story began only—or rather, already—on page 17, which is where they first mention DEC. Here they let readers know that DEC would challenge giant IBM with smaller and cheaper computers. Accordingly, the first hero to make an appearance in the book was an innovative DEC employee named David Ahl, whose mission would prove epochal because, according to Freiberger and Swaine, he was the first to rebel against existing structures. He eventually left DEC, frustrated by what he considered to be the many wrong decisions made by management. To Freiberger and Swaine, he and many others had done exactly the right thing: "Had the personal computer revolution waited for the mainframe computer and minicomputer companies to act, it might still lie in the future."[40]

Titled *Fire in the Valley*, this compilation of Freiberger's and Swaine's Californian business acumen considered the IBM PC (launched in 1981) as a replica of earlier personal computers. IBM's success in the PC market was consequently not based on the company's own efforts at innovation. IBM's marketing savvy as explained in *Fire in the Valley* was the reason why, at the start of the 1980s, even the creative start-up companies themselves saw their sales market in office applications and showed that the marketing of history was the order of the day. Freiberger and Swaine's account of the mavericks conquering the world from Silicon Valley struck a chord that would be sounded repeatedly later. Despite all the claims of having invented or found a completely new computer history, *Fire in the Valley* is one of a piece with other hagiographic accounts of inventors.[41] It also supports the popular narrative of computer historiography. The book makes it plain that computer history had become a resource that could be used almost arbitrarily to serve the next revolution. In the mid-1980s, PC narratives had to suppress the fact that the PC industry, too, had a penchant for strict formatting and command sequences, fought over processes and protocols, and loved bureaucracy.[42] In other words, the history of entrepreneurial decision-making, which at the time was being painstakingly reconstructed in the case of IBM, was left out.

Historiographical Batch Processing

A few years later, in a 1988 essay titled "The History of Computing in the History of Technology," historian Michael S. Mahoney produced a constructive and critical review of the current historiographic program. From the very outset, he left no doubt that computer historiography itself was part of the development of information processing. "We live in an 'information society,' an 'age of information.' Indeed, we look to models of information processing to explain our own patterns of thought."[43] However, the history of computers did not come into the events and discussion of the history of science and technology. In contrast, wrote Mahoney, a look at the research literature on computer history showed that the people who wrote it were hardly addressing the questions that historians of technology were asking. He suggested integrating computer history into the history of technology from two directions: "The history of computing should use models from the history of technology at the same time that we use the history of computing to test those models."[44]

Mahoney summed up "computing's present history" as a form of batch processing.[45] He sorted all the literature he knew according to the way the authors dealt with history and computers. One pile consisted of books purporting to be historical research literature. The literature on computing consisted of investigations that focused, first, on hardware and, second, on the prehistory and early history of the computer. Contrary to the self-perception of its authors, however, the latter was only casually historical, at least insofar as it addressed the history of the 20th century or wished to take a broader perspective.[46]

A second, sizable pile consisted of accounts written by people who had once been involved in the developments they described. The format here varied considerably. Mahoney was familiar with a variety of publications ranging from routine reports, collections of groundbreaking essays and retrospectives, to biographies of people and machines, to corporate histories. Mahoney included Brian Randell's work in this pile as well as Bashe et al.'s *IBM's Early Computers*. These "insider" histories were full of expertise and "facts and firsts." Mahoney was critical of the authors accepting as technical givens what a more critical investigation might have concluded to be choice and decision making: "This literature represents for the most part 'insider' history. While it is first-hand and expert, it is also guided by the current state of knowledge and bound by the professional culture. That is, its authors take as givens (often technical givens) what a more critical, outside viewer might see as choices."[47] Reading these reports, one loses sight of alternatives, because the authors subordinated everything to knowledge that they had exclusively at the time they were writing. They lost touch with the time they were writing about.

A third pile comprised texts that Mahoney labeled as "written by journalists." Since the 1950s, journalists had been writing material that captured the essence of people and institutions and portrayed them so vividly that they found an audience. Journalists had "an eye for the telling anecdote." But they also tended to emphasize the out-of-the-ordinary, the sensational, and the epochal about people and lines of research, which boosted the popularity of microcomputers and artificial intelligence.[48]

Mahoney found these reports difficult to distinguish from a fourth pile of texts concerned with the effects of computers on society. This literature tended toward facile pronouncements about the present that were difficult to distinguish from utopian dreams. But what was most obvious here was the simplification of history. Computers were removed from past periods to instrumentalize history. History was thus not examined, but it served the purpose of contemporary social criticism. "Some of this literature rests on a frankly political agenda; whether its modes and models of analysis provide insight depends on whether one agrees with that agenda."[49]

It was fair to say that the history of computing had opened a discussion space where different voices could be heard and different perspectives on politics, society, technology, business, and culture were examined in a way the Society for the History of Technology had tried and tested for quite a while. Mahoney's sorting unearthed resources that computer historians could analyze profitably: anecdotes, facts, technical knowledge, and political strategies. From the limitations he described, it was also very easy to make out what Mahoney found regrettably lacking in some or all of the books, which he regarded as sources to be mined for computer historiography: information about choices and alternatives, insight into the past and a glimpse of the ordinary, and not least, authors with a commitment to history.

In retrospect, Mahoney could have come up with yet another pile. It consisted of a small number of remarkable studies that made use of a professional arsenal of methods from other disciplines. Among these books that attracted well-deserved respect was Donald MacKenzie's *Inventing Accuracy*. Its subject, nuclear missile guidance, referred to the computer-historical past of ballistics and the atomic bomb. The question was simple, the answer demanding: What must be computed for and by autonomous flying missiles—whether digitally in advance or during flight with electromechanical devices? However, due to the diverse body of knowledge at play in this technology, there are neither dominant theories nor outsized personalities, neither genealogical nor teleological inevitabilities.[50] Rather, inventing accuracy has always been a question of carefully selected and combined forms of knowledge and technical processes.

Inventing Accuracy transcends machine boundaries, disciplinary boundaries, attentional foci, and groups of actors to show how a black box was conceived, a technical revolution developed, a weapons system transformed, and technical facts reconstructed. With remarkable aplomb, MacKenzie has since written several essays and books in which society and computers are woven into a finely textured fabric. By refraining from dealing with only one particular form of computing, MacKenzie succeeds in presenting very differentiated sociotechnical problem areas—from "computer-related accidental deaths" to the autopoietic effect of mathematical descriptions on the dynamics of the financial markets.[51]

Paul Edwards' *The Closed World: Computers and the Politics of Discourse in Cold War America* appears to have been written much closer "to the computer." But the impression is wrong. What Edwards delivers is an "implicit critique of existing computer historiography." It is not about progress and revolution, he maintains, but rather about contingency and multiple determination. Technological change should be treated as a technological choice, "tying it to political choices and socially constituted values at every level." Thus, technology is depicted as a product consisting of complex interactions "among scientists and engineers, funding agencies, government policies, ideologies, and cultural frames."[52]

Both MacKenzie and Edwards approached "the computer" concerning its operational requirements and effects; these can be both distributed and centralized. However, they are never distributed or centralized for obvious, quasi-natural reasons, but rather as the result of a political, strategic, or cultural choice. The computer disperses, becomes ubiquitous, and forms a system that is highly integrated within the social context of action where computing, sorting, and interaction occur. The analysis thus also reflects a context that was shaped by the development dynamics of computer-aided networks and distributed computing.

"Taking God Out of Mathematics and Putting the Body Back In"[53] was Brian Rotman's goal in a 1993 essay on the ghost in Turing's machine. Rotman boldly pointed out the corporeality of abstractions and symbolic operations, even those of mathematics. He did so with explicit reference to the work of Alan Turing. For the history of mathematics, this was thrilling and changed a lot. In contrast, the history of computers has hardly changed the course of the history of technology in general. Even though Edwards and MacKenzie effectively demonstrated how closed disciplinary worlds could undergo productive change by importing new perspectives. The substantial interest of many representatives of the humanities in corporeality, materiality, and artifacts would also have helped pave the way for a new history of technology. The fact that so little happened was due both to the historical context and the hesitancy of the history of technology as a field.[54]

FUTURE WITHOUT PAST

In the mid-1990s, history changed far too quickly for historians. Together with the explosive growth of the World Wide Web, a vision emerged of "humans" always "surrounded" by "media and technology." Following Mahoney's motto "hype hides history," nearly everything that might have constituted a productive computer history in the promising MacKenzie/Edwards/Rotman style tumbled back into oblivion.[55] What was now reported about virtual space in media theory, philosophy, and literary studies was all geared toward establishing norms, an identity, an aesthetic sensibility, and a future orientation. If you believed what you heard, then everything was evaporating—history, society, theory, space, and of course the library. Cyborgs were crowding into cyberspace, this "specific elsewhere"; the piano was becoming a hyper tool. Even the Hollywood dream factory was vanishing into the Internet. Technology-oriented culture and media scientists focused fully and ahistorically on what had "always" been after the invention of movable type or what had been the case with information "ever since" Shannon.[56] This enabled mediation of the space between the latest newswire reports as plain text and Heideggerian aphorisms, between media mythologizing of the Big Bang and allusions to the current convergence of all media.

People were not astonished when, in 1993, Friedrich Kittler situated "the last historic act of writing" in 1978. He based his claim on his reading of a 1990 issue of *DOS* magazine. At the time, "a team of Intel engineers laid a few dozen square meters of drawing paper on empty garage floors in Santa Clara to plot out the hardware architecture of their first integrated microprocessor."[57]

Under these circumstances, there was nowhere for computer history to retreat to. Computers were too heterogeneous to serve as stable historical objects for embarking on yet another journey through the past. By now, the computer was somehow physically, conceptually, and procedurally everywhere simultaneously, a moving target that left many traces but whose replication was no longer palpable. Admitting that it had never been so would have undone the original narrative of the highly integrated universal machine.

In 1996, Martin Campbell-Kelly and William Aspray published their book *Computer* and put several obstacles in the way of an otherwise freewheeling history of computing. Subsequent editions met the string of new futures classifying computer history into different "computer ages" and continued undauntedly to string together further materials for computer historiography. It was required reading for anyone who needed historic references for things that in the 1990s were common knowledge in the history of computers, in the household, and at the office. At the same time, these were things that generally were considered completely foreign to the history of science or the history of technology. So, while readers might have been searching the book for clues to issues like "operating systems" or "batch processing," they also encountered anachronistic attributes—for example, where a "Mrs. John W. Mauchly" found her way into history as the wife of a professor, but without her own first name and evidence of her mathematical education. The book was thus yet another source to be mined. You had to plow through it and "process" it, so at least you knew right away what should be done now. For example, it was quite clear that no research had yet succeeded in prompting a rethink of the commonplaces about programmers. The book was reissued many times without corrections but simply extended to include the most recent period.[58]

The history of computers thus remained for more than two decades a largely irrelevant topic for the history of technology. What was obvious in the history of technology was of no interest whatsoever to the history of computers. When the history of technology began to innovate and to address artifacts and discourses, computer history tried to free itself almost in vain from machine genealogies and interviews with contemporary witnesses. However, neither the history of computers nor the history of technology could profit from other historical subdisciplines as a source of protection. In 2001, when the eminent corporate historian Alfred D. Chandler looked back on the just-concluded 20th century as an era of electronics, historians of technology should have taken note immediately. But the journal *Technology and Culture* waited a full five years before publishing a review. Obviously, computer historiography had failed to keep Chandler reliably informed about its research aims and findings.[59] Not that the aims were any secret: "Until recently, the history of computing in these fields has been written in terms of the machine and its impact (revolutionary, of course) on them. The emphasis has lain on what the computer could do rather than on how the computer was made to do it," stated Michael S. Mahoney on a somewhat sober note at the end of his career.[60]

According to Mahoney, the operational prerequisites and impact of historical computers could perhaps still be decentered by focusing on "communities, or bodies of shared disciplinary practices, who embraced the new device and helped to shape it by adapting it to their needs and aspirations" instead and then following the history of these groups. This acknowledges the role of the user, evokes the popular interdisciplinarity of science policy, and makes concessions to the science and technology studies community, which had been recounting such histories in its lectures for more than thirty years in the place of philosophy of science, which had long since been retired.[61] It was also another attempt to finally transcend the border between the analog and digital eras and thus to get closer to the original, but probably also present-day, universal computers. The focus on computers remained intact, even when people began to speak of communities and software, and no longer was it museum collections that had to be maintained, but websites.

Notes

1. For an earlier German version of this chapter see David Gugerli and Daniela Zetti, "Computergeschichte als Irritationsquelle," in *Provokationen der Technikgeschichte. Zum Reflexionszwang historischer Forschung,* ed. M. Heßler and H. Weber (Paderborn: Ferdinand Schöningh, 2019), 193–224. Giselle Weiss has provided us with a translation of this text; it served as a quarry for the following deliberations.

2. K. Girschik, "Machine-Readable Codes. The Swiss Retailer Migros and the Quest for Flow Velocity since the Mid-1960s," *Entreprises et histoire* 44 (2006): 55–65; Daniela Zetti, "Die Erschliessung der Rechenanlage. Computer im Postcheckdienst, 1964–1974," *traverse* 15, no. 3 (2009): 88–101; David Gugerli, "The World as Database. On the Relation of Software Development, Query Methods, and Interpretative Independence," *Information & Culture* 47 (2012): 288–311.

3. Paolo Bory and Daniela Zetti, *Digital Federalism. Information, Institutions, Infrastructures (1950–2000)* (Basel: Schwabe Verlag, 2022).

4. Gerhard L. Hollander, "Data Processing with a Quasi-Random-Access Memory," *Electrical Engineering* 74, no. June 1955 (1956); Murray L. Lesser and J. W. Haanstra, *The RAMAC Data-Processing Machine* (New York, 1957).

5. Geoffrey D. Austrian, *Herman Hollerith: Forgotten Giant of Information Processing* (New York: Columbia University Press, 1982); David Gugerli and Hannes Mangold, "Diskussionsforum: Betriebssysteme und Computerfahndung. Zur Genese einer digitalen Überwachungskultur," *Geschichte und Gesellschaft* 42, no. 1 (2016); Günther Sandner and Hans Spengler, *Die Entwicklung der Datenverarbeitung von Hollerith-Lochkartenmaschinen zu IBM Enterprise-Servern von 1887 bis 2000, in 10 Epochen* (Böblingen: Sandner, 2006).

6. Friedrich L. Bauer and Hans Wössner, The "Plankalkül" of Konrad Zuse: A Forerunner of Today's Programming Languages," *Communications of the ACM* 15, no. 7 (1972); Herbert Bruderer, *Konrad Zuse und die ETH Zürich. Zum 100. Geburtstag des Informatikpioniers Konrad Zuse (22. Juni 2010)*, Technischer Bericht / Departement Informatik. Professur für Informationstechnologie und Ausbildung (Zürich: ETH Zürich, 2010); Wilhelm Füssl, ed., *100 Jahre Konrad Zuse. Einblicke in den Nachlass* (München: Deutsches Museum, 2010); Fritz-Rudolf Güntsch, *Konrad Zuse und das Konzept des Universalrechners*, ed. Hans Dieter Hellige (Berlin: Springer, 2004).

7. Charles J. Bashe et al., *IBM's Early Computers*, 2nd ed. (Cambridge MA: MIT Press, 1986); I. Bernard Cohen and William Aspray, "Howard Aiken and the Dawn of the Computer Age," in *The First Computers: History and Architectures*, ed. Raul Rojas and Ulf Hashagen (Cambridge, MA: MIT Press, 2000); John A. Postley, "Mark IV: Evolution of the Software Product, a Memoir," *IEEE Annals of the History of Computing* 20, no. 1 (1998); Michael R. Williams, "Makin' Numbers: Howard Aiken and the Computer," *Technology and Culture* 41, no. 4 (Oct 2000).

8. Bashe et al., *IBM's Early Computers.*

9. John Backus, "The History of Fortran I, II, and III," *IEEE Annals of the History of Computing* 20, no. 4 (Oct–Dec 1998); Ian D. Chivers and Malcolm W. Clark, "History and Future of Fortran," *Data Processing* 27, no. 1 (1985); M. N. Greenfield, "History of Fortran Standardization," *AFIPS Conference Proceedings* 51 (1982); William P. Heising, "History and Summary of Fortran Standardization Development for the ASA,"

Communications of the ACM 7, no. 10 (1964); D. T. Muxworthy, "Standard Fortran. Short History," *Computer Bulletin* 16, no. 4 (1972). The vendor-independent programming language ALGOL 60, developed in a series of international meetings between 1957 and 1963, was designed from the outset to have an easy-to-follow chronological system. But the system was short-lived, and ALGOL 68 provoked a schism among programming language developers. In 1970, Niklaus Wirth (a member of the notorious ALGOL 68 opposition) published his programming language under the name "Pascal"—a move considered insubordination by the members of the programming guild who relied on international committee work. "ALGOL 68" did have a nice, scientific ring to it. But with "Pascal," and its powerful evocation of mathematics and history, Wirth had switched to a branding surely intended to be interpreted as the *invention of tradition* by a community reliant on the academic merits of its field. N. Wirth, *The Programming Language Pascal* (Zürich: ETH Zürich, 1970).

10. Paul E. Ceruzzi, "Moore's Law and Technological Determinism: Reflections on the History of Technology," *Technology and Culture* 46, no. 3 (2005).

11. Peter J. Denning, "Third Generation Computer Systems," *ACM Computing Surveys* 3, no. 4 (1971); John A. Postley and H. Jackabson, "The Third Generation Computer Language: Parameters Do the Programming Job," in *Data Processing*, ed. Data Processing Management Association (Los Angeles, 1966); Larry Press, Jeff Rothenberg, and Jim Carlstedt, "The Next Generation of Personal Computers: A Position Paper," *SIGPC Note* 1, no. 4 (1979); Michael R. Williams, "A Preview of Things to Come: Some Remarks on the First Generation of Computers," in *The First Computers: History and Architectures*, ed. Raul Rojas and Ulf Hashagen (Cambridge, MA: MIT Press, 2000); William Aspray, "Should the Term 'Fifth Generation Computers' Be Banned?" *Journal of Computers in Mathematics and Science Teaching* 3 (Spring 1985), 36–38. On generations in the history of computers see F. Hoof, "Das Problem der Generationen in der Computergeschichte," in *Zur Geschichte des Digitalen Zeitalters,* ed. R. Wichum and Daniela Zetti (Wiesbaden: Springer Fachmedien Wiesbaden GmbH, 2022), 33–52.

12. We use the term "digital society" as a form of self-description. It has been used by societies that delegate their functions and services to networks and computers. Over the last four decades "digital society" eventually replaced older forms of societal self-descriptions, such as the consumer, information, or knowledge society. David Gugerli and Daniela Zetti, *Digitale Gesellschaft (Rohfassung). Beitrag zum Historischen Lexikon der Schweiz. Preprints zur Kulturgeschichte der Technik 30* (Zurich: ETH Zurich, 2018).

13. David Gugerli, *How the World Got into the Computer. The Emergence of Digital Reality* (Zurich: Chronos, 2022).

14. Percy E. Ludgate, "On a Proposed Analytical Machine," *Scientific Proceedings, Royal Dublin Society* 12, no. 9 (1909); Brian Randell, "Ludgate's Analytical Machine of 1909," *Computer Journal* 14 (1971).

15. Brian Randell, "A Turing Enigma," in *CONCUR 2012 - Concurrency Theory*, ed. M. Koutny and I. Ulidowski (Heidelberg, Dordrecht, London, New York: Springer, 2012), 23.

16. Martin Campbell-Kelly et al., "Past into Present: The EDSAC Simulator," in *The First Computers: History and Architectures*, ed. Raul Rojas and Ulf Hashagen (Cambridge,

MA: MIT Press, 2000); David Nofre, "Review: It Began with Babbage. The Genesis of Computer Science by Subrata Dasgupta," *Technology and Culture* 56, no. 2 (Apr 2015); Randell, ed., *The Origins of Digital Computers: Selected Papers*, 2nd edition (Berlin, Heidelberg, New York: Springer, 1975 [1973]), viii–ix.

17. Randell, "A Turing Enigma," 24. It is noteworthy that colleagues always have a keen interest to strongly encourage others to quote authors they believe indispensable for the completeness of a text. Likewise, we have been asked to mention H. H. Goldstine, *The Computer from Pascal to von Neumann* (Princeton, NJ: Princeton University Press, 1972) and M. R. Williams, *A History of Computing Technology* (Englewood Cliffs, NJ: Prentice-Hall, 1985).

18. https://www.youtube.com/watch?v=WCeLTWPCEFI. Brian Randell, "Colossus: Godfather of the Computer," *New Scientist* 73, no. 1038 (1977); Randell, "A Turing Enigma"; Brian Randell, *Uncovering Colossus* (https://youtu.be/Yl6pK1Z7B5Q: BBC, 2013).

19. Randell, "A Turing Enigma," 32.

20. Sara Turing, *Alan M. Turing* (Cambridge: W. Heffer, 1959); Rolf Hochhuth, *Alan Turing. Erzählung* (Reinbek bei Hamburg: Rowohlt, 1987); Andrew Hodges, *Alan Turing: The Enigma* (London: Burnett Books, 1983).

21. Paul E. Ceruzzi, "Review: Turing's Man: Western Culture in the Computer Age by David Bolter," *Technology and Culture* 26, no. 2 (1985).

22. John von Neumann, "First Draft of a Report on the EDVAC," in *The Origins of Digital Computers: Selected Papers*, ed. Brian Randell (Berlin, Heidelberg, New York: Springer, 1945); Alan M. Turing, "On Computable Numbers, with an Application to the Entscheidungsproblem," *Proceedings of the London Mathematical Society* 42 (1937).

23. Randell, "A Turing Enigma."

24. Randell, "Colossus: Godfather of the Computer"; Randell, *Uncovering Colossus.*

25. William Aspray, "Leadership in Computing History: Arthur Norberg and the Charles Babbage Institute," *IEEE Annals of the History of Computing* 29, no. 4 (2007), 17.

26. Ibid, 7ff.

27. Computer historians produced research materials they did not find in archives by interviewing old practitioners of the computing field. A few years later Pierre Bourdieu criticized oral history as a *methode à la mode* by pointing to "the biographic illusion" and oral history's inclination to part from a continuum of subjectivity. See Bourdieu, Pierre, "The Biographical Illusion (1986)," in *Biography in Theory: Key Texts with Commentaries*, ed. Wilhelm Hemecker and Edward Saunders (Berlin, Boston: De Gruyter, 2017), 210–216; A. Petschow, "Zeitzeugeninterviews als Ego-Dokumente in der Stiftung Haus der Geschichte der Bundesrepublik Deutschland," in *Das dokumentierte Ich. Wissen in Verhandlung*, ed. C. Hämmerling and Daniela Zetti (Zürich: Chronos, 2018), 67–80.

28. Martin Campbell-Kelly and Michael R. Williams, eds., *The Moore School Lectures*, vol. 9, Reprint Series for the History of Computing (London, Los Angeles/San Fransisco, CA: MIT Press and Tomash Publishers, 1985); Arthur Norberg, "Use of CBI's Research-Grade Oral Histories from CBI," *Charles Babbage Institute Newsletter* 24, no. 3 (2002), http://www.cbi.umn.edu/about/nsl/v24n3text.pdf. The Smithsonian Online Virtual Archives provide evidence from an earlier oral history project. "The Computer Oral

History Collection was a cooperative project of the American Federation of Information Processing Societies (AFIPS) and the Smithsonian Institution. This project began in 1967 and was concluded in 1973." https://sova.si.edu/record/NMAH.AC.0196?s=0&n=10&t=C&q=henry+tropp&i=3

29. Aspray, "Leadership in Computing History: Arthur Norberg and the Charles Babbage Institute," 18.

30. Nancy Stern, "The Eckert-Mauchly Computers: Conceptual Triumphs, Commercial Tribulations," *Technology and Culture* 23, no. 4 (1982), 582.

31. William Aspray, *John von Neumann and the Origins of Modern Computing* (Cambridge, MA: MIT Press, 1990).

32. Ibid, xv.

33. Ibid, vii.

34. Bashe et al., *IBM's Early Computers.*

35. I. Bernard Cohen, "Foreword," in *IBM's Early Computers*, ed. Charles J. Bashe et al. (Cambridge, MA: MIT Press, 1986), xi.

36. Ibid, xv. Comment by William Aspray (email to the authors, June 6, 2023): "Cohen had become familiar with happenings at IBM because he was the paid historical consultant to IBM when IBM hired the famous designers Charles and Ray Eames to produce a traveling exhibit entitled The Computer Perspective, later appearing as a published catalog of the exhibit."

37. Bashe et al., *IBM's Early Computers*, 582.

38. Paul Freiberger and Michael Swaine, *Fire in the Valley: The Making of the Personal Computer* (Berkeley: McGraw-Hill, 1984), 6.

39. Ibid.

40. Ibid, 21.

41. Robert H. Reid, *Architects of the Web: 1,000 Days that Built the Future of Business* (New York: John Wiley & Sons, 1997); Daniel Ichbiah, *Die Microsoft Story: Bill Gates und das erfolgreichste Software-Unternehmen der Welt* (München: Heyne, 1993).

42. Freiberger and Swaine, *Fire in the Valley: The Making of the Personal Computer*, 297ff.

43. Michael S. Mahoney, "The History of Computing in the History of Technology," *Annals of the History of Computing* 10, no. 2 (1988), p. 1.

44. Ibid, 2.

45. Ibid.

46. Ibid.

47. Ibid.

48. Ibid, 3.

49. Ibid.

50. Donald A. MacKenzie, *Inventing Accuracy. An Historical Sociology of Nuclear Missile Guidance* (Cambridge MA: MIT Press, 1990).

51. Donald A. MacKenzie, *An Engine, not a Camera. How Financial Models Shape Markets* (Cambridge MA: MIT Press, 2006).

52. Paul N. Edwards, *The Closed World. Computers and the Politics of Discourse in Cold War America* (Cambridge MA, London: MIT Press, 1996), xiii.

53. Brian Rotman, *Ad infinitum. The Ghost in Turing's Machine* (Stanford CA: Stanford University Press, 1993).
54. H.U. Gumbrecht and K.L. Pfeiffer, "Materialität der Kommunikation" (Frankfurt a.M.: Suhrkamp, 1988).
55. Michael S. Mahoney, *Histories of Computing* (Cambridge MA: Harvard University Press, 2011), 120.
56. Sherry Turkle, *Life on the Screen: Identity in the Age of the Internet* (New York: Simon & Schuster, 1995); Jay David Bolter and Richard Arthur Grusin, *Remediation: Understanding New Media* (Cambridge, MA: MIT Press, 1998). Lev Manovich, *The Language of New Media* (Cambridge, MA: MIT Press, 2001).
57. F. A. Kittler, "Es gibt keine Software," in *Draculas Vermächtnis. Technische Schriften*, ed. F. A. Kittler (Leipzig: Reclam, 1993), 226. Here we ignore the role that computers have played in media studies, as well as the philosophical claim of media studies to history.
58. Martin Campbell-Kelly and William Aspray, *Computer: A History of the Information Machine* (New York: Basic Books, 1996).
59. Alfred D. Chandler Jr. *Inventing the Electronic Century: The Epic Story of the Consumer Electronics and Computer Industries* (Cambridge: Free Press, 2003); Nathan Ensmenger, "Review: Inventing the Electronic Century: The Epic Story of the Consumer Electronics and Computer Industries by Alfred D. Chandler," *Technology and Culture* 47, no. 3 (Jul 2006).
60. Michael S. Mahoney, "The Histories of Computing(s)," *Interdisciplinary Science Reviews* 30 (2005), 127.
61. Ibid.

Chapter 7

Visionaries and Transformative Product Launches in Computing and Information Business History

Shane Greenstein

Archetypes Associated with Transformative Product Launches

Transformative product launches are recurring and fundamental features of computing and information business history.[1] For example, in the middle of the 1960s, IBM obtained a dominant leadership position from the development, introduction, and support of System/360. In later eras, other organizations such as Apple, Google, Wikipedia, and Craigslist designed, introduced, and supported products that transformed their respective markets—with the launches of the iPhone, the search engine married to an ad auction, the crowdsourced online encyclopedia, and a platform for classified advertising. These transformations were not secret. News organizations covered the details and chronicled dramatic changes in the user experience and firm profitability. These new products became the benchmark against which subsequent new products were compared.

Many CEOs acquired extraordinarily high profiles from guiding and driving the launches if they did not possess them before. For example, the CEO of IBM during the launch of the System/360, Thomas Watson

Jr., achieved celebrity status, arguably outshining his father's.[2] In later eras, pictures of Larry Page and Sergey Brin graced the covers of magazines when their search engine became famous and propelled Google to become one of the most valuable companies in the history of Western capitalism. Before his untimely death, Steve Jobs garnered more than his share of celebrity attention by introducing the iPhone, gaining a status rarely achieved by any of the most famous entrepreneurs.[3] It is not an exaggeration to say these executives gained so much attention that their actions became the fodder for multiple books and cases on how to behave as a CEO.

A scholar new to computing and business history may wonder what to do with such a combination of unimpeachable transformative events and occasionally vulgar celebrity hyperbole. It's the type of puzzle technology historians have confronted many times and will recognize as the unavoidable challenge of "great mover history," a narrative mode that remains popular in trade books but falls in and out of fashion among professional historians. The benefits and challenges of focusing on an individual "mover and shaker" are well-known. Biographical details can complement the study of economic incentives for creating and selling new products. At its best, incorporating information about CEOs illuminates the origins behind the unique features of a strategy. It highlights when the processes and resources at the call of the CEO played an essential role and when the CEO's judgment made a difference.[4] Biographical details may shed light on salient events that models and statistics cannot capture, such as personality clashes between executives, idiosyncratic talents CEOs developed over years of experience, and flashes of unique insight.[5] Yet, so often, it is also poorly done. At their worst, CEO biographies can become hagiographic. Recent technological biographies have begun to frame accomplishments around a "superhero" trope as if the entrepreneur possesses extraordinary powers.[6]

This chapter proposes one approach to deriving insight despite these challenges. It draws on archetypes of visionaries. An archetype is a recurrent symbol or motif. An archetype can capture a type of person or team and their behavior and reappear as a story containing many of the same narrative elements in different contexts. We often recognize archetypes when we see them, but scholarship requires something more precise and generalizable than *ad hoc* recognition. For this chapter, a *visionary archetype* refers to an individual with two traits. First, he can perceive something about the market that others cannot and retains the courage of his convictions. Second, he possesses some executive authority to execute and implement his unique perception. This definition applies equally to all leaders and not exclusively to entrepreneurs or prominent executives at established firms.

This chapter uses this approach to compare transformational product launches in the history of computing and information services. Some of these launches have taken on an iconic status, including the launch of the IBM System/360, the IBM Personal Computer, the Netscape Web Browser, the Google Search Engine, Facebook's social media platform, and the iPhone's debut. Other prominent and transformational (albeit less iconic) product launches covered in his chapter are Encarta's CD-ROM encyclopedia launches, Craigslist's platform for classified advertising, Wikipedia's online encyclopedia, Nginx's web server for streaming, and the launch of Wi-Fi routers. These product launches represent a broad set of circumstances, users, and suppliers. Yet, all these product launches had visionaries driving their design, production, and deployment. The chapter shows how the archetypes help identify and catalog similarities and differences.

A taxonomy of four different archetypes of transformations emerges, illustrated in Table 7.1. **Visionary outsiders** comprise the first archetype. In this group, a firm led by an unknown CEO with a unique vision launches a product that leads to market transformation. Four examples illustrate it: Wikipedia, Encarta, Craigslist, and Nginx. The second group is a particular case of the first group. A **visionary academic**, previously unknown to established firms, leads a commercial firm transforming a market. Two examples illustrate it: Google and Facebook. Innovative **visionary insiders** comprise the third archetype. A leader of an established firm launches a product that creates so much new demand that it transforms the market and replaces what the established firm previously supported. The IBM 360 illustrates that archetype. The last group is a particular case of insiders, labeled **fast followers**. These visionary insiders react to pioneering visionary outsiders by launching a new product, and it transforms the market. The IBM personal computer illustrates that archetype. In each case, the narrative describes the vision behind the product launch and identifies the historical origins of the vision.

To close, the chapter analyzes three product launches in which combinations of visionary outsiders, academic outsiders, visionary insiders, and fast followers play crucial roles in transformations—the launches of the Netscape browser, the Wi-Fi router, and the iPhone. The analysis illustrates how to use the archetypes to cut through complications in each case. The archetypes help construct a succinct narrative for why specific details played crucial roles in outcomes.

Several conclusions emerge from this approach, as the chapter shows. First, visionary archetypes are not a standard feature of product development models, but comparing many examples suggests they could be. Related, understanding who has such knowledge explains why the assessment of opportunities and new product launches varies across market participants, with some participants predicting product launches with accuracy and others considering them unexpected. While an outsider visionary's actions may look random or unexplainable to the insiders, the activities of the outsider have a rationale for those who know the relevant details. In every case, this is crucial to explaining the subsequent transformation. Finally, these premises yield explanations that require extensive historical facts about the origins of visions. Those origins also have an important implication: they partially explain why visionary outsiders can differ in their intuition and judgment from visionary insiders, even though the insiders seem to have irreplaceable advantages from being established and calling on their organization's resources.

Conceptions of Outsiders and Insiders

The chapter builds a taxonomy of visionaries who act from a position as outsiders or insiders. That taxonomy draws on a core-periphery market leadership model, a staple of industrial organization. The model delivers a precise definition for insider and leader, outsider and non-leader. In these core-periphery models, market leadership is assigned by a simple metric, higher market share, which proxies for numerous factors

Table 7.1. A Simple Taxonomy of Transformational Product Launches

Status	Visionary founder(s) or Executive(s)	Important partners	Product launch	Market	Year of launch, quick success?	Yrs to develop?
Outsider	Jimmy Wales	Ward Cunningham	Wikipedia	Online Encyclopedias	2001, no	1–2
Outsider	Bill Gates and Craig Bartholomew	Funk & Wagnall	Microsoft Encarta	CD-ROM Encyclopedias	1993, yes	1–2
Outsider	Craig Newmark and Jim Buckmaster	None	Craigslist	Online ads for matching services	1999, yes	1–2
Outsider	Igor Sysoev	Open source community	Nginx	Streaming servers	2004, no	2–3
Academic Outsider	Sergey Brin and Larry Page	Andy Bechtolsheim, Eric Schmidt	Google search	Web search & targeted advertising	1998, no	3–4
Academic Outsider	Mark Zuckerberg	Sheryl Sandberg	Facebook	Social networking & targeted advertising	2004, yes	1–2
Insider	Thomas Watson Jr.	Thomas Watson Sr.	System 360	Compatible mainframes	1964, yes	3–4
Fast Follower	Frank Carey and Bill Lowe Jr.	Following the Apple II and other early PCs	IBM personal computer	Personal computers	1981, yes	1–2

(continued)

Table 7.1. *(continued)*

Status	Visionary founder(s) or Executive(s)	Important partners	Product launch	Market	Year of launch, quick success?	Yrs to develop?
A blend of academic outsider and insider	Mark Andreesen and Jim Clark	Eric Bina and Apache Webservers	Netscape	Browser	1994, yes	1–2
A blend of outsider & fast follower	Michael Dell followed Steve Jobs, and many prototypes	Lucent and IEEE 802.11 committee	Dell-branded Wi-Fi routers	Wi-Fi routers and wireless local area networks	2001, yes	1–2
A blend of insider and fast follower	Steve Jobs followed BlackBerry, Windows CE, and Symbian	AT&T Wireless	iPhone	Smartphones and iTunes	2007, yes	1–2

relevant to market leadership.[7] Most importantly, a high market share also demarcates outsider or insider status before a new product launch, with the insider possessing a high market share while outsiders have little or none. Equating insider status with leadership also implies how to measure persistence or change in leadership—i.e., market share is the primary benchmark.

Notice the difference compared to prior definitions. The concept of "outsiders" and "insiders" plays a vital role in the history of scientific ideas in which a paradigm determines the discussion of specialists. In this conception, associated with Thomas Kuhn, "insiders" construct the paradigm and act as gatekeepers of participation. It casts science as a social activity in which existing hierarchies resist change to established paradigms if it threatens established prestige and reward structures built around a specific scientific paradigm.[8] Outsiders come with scientific backgrounds and training that lead them to different perceptions and develop distinct fact bases to support new theories. This approach emphasizes how outsiders with new ideas gain wider acceptance and legitimization.[9] For example, this approach has been fruitful in explaining the rise of new ideas about biological mechanisms in life sciences and medicine that depart markedly from existing practice. It also explains why some (promising) ideas meet resistance (when outsiders advocate for them).[10]

The analysis of product launches borrows an essential insight from this prior scholarship, i.e., that different backgrounds and experiences can foster distinct perceptions about the potential for new products, and new modes of thinking make gaining acceptance among crucial participants challenging. But there is a critical difference. While acceptance or rejection of new ideas and science may be a part of realizing commercial goals, typically, it is subservient to the imperative to generate more revenue than expenses. Moreover, the key participants are not other rivalrous or allied scientists but, instead, are business partners or financiers who want to profit from the partnership. For product launches, the assumption of a core-periphery setting is essential. None of these product development decisions occur within a greenfield structure of participants.[11]

This study builds on classic literature focused on new product introductions,[12] particularly the analysis of novel product introductions that lead to "creative destruction" and "disruption."[13] These studies and analyses identify rationales for why insider and outsider visionaries take different approaches to exploring and assessing proposed product launches.[14] This line of research stresses economic incentives to experiment to gain new information and exploit it to profit from it. Some of those mechanisms will arise in some of the discussed episodes. However, much of this literature on "creative destruction" and "disruption" extends into areas far removed from computing and the internet. Exceptions are disk drives,[15] photolithographic equipment,[16] ethernet,[17] integrated circuits,[18] and Internet Service Providers.[19] In other words, while this chapter draws on several historical examples of transformative product launches, many examples herein are novel for the literature.

Visionary Outsiders

We start with outsiders because this definition is novel. Before launching a product, the outsider visionary claims to perceive something that others do not. That interacts with another critical feature, their status as an unknown player inside the market in which they aspire to launch a product. All four examples here are drawn from the history of transformational products that delivered information. Wikipedia and Encarta delivered encyclopedias in new formats, while Craigslist altered the modal delivery of classified advertising. The example, Nginx, is the web server that provides most streaming services, a form that has transformed video entertainment.

Online Encyclopedia. In the modern internet, the iconic outsider is Jimmy Wales, the founder of Wikipedia. Before this experience, he had dropped out of a PhD program in financial economics to found and operate several web-based ventures during the dot-com boom. He had no experience in the encyclopedia business nor with a not-for-profit organization. After gaining prominence, Wales pursued for-profit ventures in other organizations that deployed Wikis, such as Wikia.[20]

Wikipedia fits any potential definition of a transformational product launch. It is commonly held responsible for motivating Encyclopedia Britannica to cease the publication of books in 2012 and become an online and CD-ROM-based service exclusively.[21] In every middle- or high-income country without a censored internet today, Wikipedia is a top twenty site and has been for over a decade and a half.

Wikipedia's launch was circuitous. Wales and his employee, Larry Sanger, failed to develop an online encyclopedia by commissioning articles in 2001, an effort they called Nupedia. After taking this approach for a year and not getting far, they began to use a Wiki, and after that, they adopted the moniker Wikipedia. The Wiki was invented by Ward Cunningham in 1995 to support the development of documentation of open-source software. Sanger suggested it after observing its usage. One appeal of a Wiki was functional: it quickly enabled revision from contributors, just as it had enabled revision of software manuals from multiple contributors to open-source software. The second appeal was its open-source licensing, which cost nothing to use and allowed Wales and Sanger to adapt and modify the format of a Wiki for their purposes, namely, to compile encyclopedia articles.

Wales imbued this format with a mission. He asked contributors to make verifiable contributions and accompany them with external sources of information. While allowing contributors discretion, he asked them not to offer original opinions but, instead, to aspire to give a neutral point of view. Wales watched how these principles played out in practice, then tweaked them to make the crowdsourcing work effectively in the online encyclopedia. For example, if contributors desired to take the neutral point of view, then contributors often had to argue over how a view was characterized. However, rather than argue over which was right, contributors often had conversations over how to add another paragraph. That changed arguments into questions about portraying a controversial view in settings where information was unobvious and subjective.[22]

What did Wales learn? Oversimplifying for expositional convenience, there were two types of contributors. One was the regular and identifiable individuals who used their real names, and the other was the infrequent and anonymous volunteers. The frequent contributors often became devoted to high-level tasks such as editing and restructuring articles that interested them. Wales gave them additional editorial privileges and discretion to invite others. The irregular contributors came in various forms, but they typically corrected small details or other errors they spotted while reading an article. Articles would settle into forms where a few contributors felt ownership.[23] The combination resulted in articles that only some individuals could have composed. In the best case, the whole was something different and more impressive than the sum of its parts.

After only a year of growth, i.e., by 2003, it was clear to Wales that Wikipedia required formal protections, and he established the Wikimedia Foundation as its owner, protecting its not-for-profit status. Wales learned something that had escaped others: with a not-for-profit mission, many contributors gladly gave their time and insights to articles about science and history, something they would not have done if anyone profited from advertising. This differentiated Wikipedia from the plethora of question-answer sites that did not pursue the mission of an encyclopedia and sold advertising. Once it became more prominent, Wikipedia became a focal site, which became self-reinforcing.

Wikipedia was far from the only attempt at an online encyclopedia (Hill, 2013).[24] That comparison reinforces the uniqueness of Wales's vision for the online encyclopedia. Almost certainly, Wikipedia would not have succeeded with any other vision.[25] Wales soldiered on and maintained the site for several years with only volunteers. He could afford two regular employees, paid out of modest donations. By 2006, Wikipedia had become so prominent that it needed a more formal organizational structure. Wales helped raise money from foundations and establish the new organization, which was moved to San Francisco.

In sum, the best explanation for the successful product's prominent features lies in the founder's unique vision. Wales launched the project, guided it in the early years before all its features were established, and found a viable approach to scaling it. That effort transformed how the world thinks about an online encyclopedia.

CD-ROM-Based Encyclopedia

Before Wikipedia, there was Encarta, the product launched by Microsoft in 1993. This launch was also a watershed event. While several firms had earlier attempted to launch a CD-ROM-based encyclopedia, none had yet to make significant sales. In contrast, Encarta reached over a million sales units in 1995 and changed perceptions about the potential for PCs in the home.[26] It was held responsible for reducing US sales of the venerable Encyclopedia Britannica *in half in three years.* The decline shocked contemporaries and motivated frantic forecasts about the future deployment of digital technologies replacing established firms.[27]

The vision behind Encarta combined the perceptions of two people, Bill Gates Jr., the CEO and co-founder of Microsoft, and Craig Bartholomew, who would organize the launch of Encarta and headed the division until its closure. Although Bartholomew had briefly worked in publishing before joining Microsoft for another project, neither he nor Gates had worked at a firm producing encyclopedias. Moreover, in 1991, Microsoft was over a decade old and regarded as an outsider by other leading encyclopedia firms, such as Britannica, World Book, and Collier's.

Gates considered a CD-ROM-based encyclopedia a pet project to which he had repeatedly devoted time and energy despite other more lucrative opportunities for the company. Gates believed that the product would be valuable. He viewed a CD-ROM encyclopedia as a "pull product" for PC demand, and, as such, a successful launch could come back to help Microsoft's bottom line in the long run by growing the home market for IBM-compatible PCs. Like many others, his first view was to take an encyclopedia's content and put it on a CD-ROM.

The project began in 1987 but failed to generate successful partnerships with existing encyclopedia companies. By 1991, the board expressed skepticism and canceled funding for the project. All that remained of those efforts in 1991 was one licensing deal to use content from Funk & Wagnalls, a defunct children's encyclopedia maker.[28]

A comparative new employee at the time, Bartholomew wrote an internal memo to Gates after the cancellation was announced. After reading his message, Gates decided to revive the project, putting Bartholomew in charge, which the board approved. The CD-ROM encyclopedia would build on the content of Funk & Wagnalls but tailor the content to the unique strengths of the CD-ROM, such as its ability to transmit sound and video and include internal links between different entries.

Being a CEO's pet project did not insulate Encarta from a strict product-development budget. When it was revived, Encarta's funding was contingent on market response. Initially, only four and a half developers worked on the project, and personnel growth was constrained until its product launch. These resource constraints shaped the choices of the design, which was severely constrained by the limited memory of existing CD-ROMs.

It also shaped how Bartholomew structured the marketing, which leveraged existing relationships with third-party software retailers.[29] Bartholomew believed that the product had to sell without the aid of a person to explain the product to potential buyers. He instructed the designers to include startling videos, sounds, and pictures so these could be featured in product displays. These videos included the first landing on the moon and the meeting of Rabin and Arafat on the White House lawn. The former caught the attention of many parents who had lived through the event in 1969, and the latter caught the attention of many buyers who had witnessed the signing only a few months earlier on television. Moreover, it signaled that Encarta could display historical and current events with video, attributes not typically associated with a printed encyclopedia.[30]

As it happened, Encarta's first iteration failed fast. The team learned from their errors and fixed some features for its second release. The second release in the fall of 1993 was commercially successful, selling a hundred thousand copies in the 1993 holiday sales season. After the second launch, the high sales led Gates to authorize a budget to hire more editors, translators, and staff for broader distribution. Encarta grew from there, selling a million copies in 1995. This was far out of scale with book-based encyclopedias, where yearly encyclopedia sales reached no more than several hundreds of thousands.

Contrast the vision behind Encarta's design with what happened at Encyclopedia Britannica, which simultaneously pursued a CD-ROM-based encyclopedia for its content. Britannica regarded its existing text as the source of value for users. Rather than fit the limited space of one CD-ROM, they used multiple disks for the extensive text and pictures. Since it did not work on one disk, the designers did not use internal links between entries, as Encarta did. They included many images from their library but no videos or sounds. Once they had an offering, Britannica further encountered issues with distributing their product within their current sales force, who fretted over the impact of product cannibalization on their compensation. In summary, Britannica chose design features and distribution outlets that remained consistent with its existing business before any alternative demonstrated a different set of choices. Even after a competitor illustrated the appeal of an option, Britannica stayed with its intentions. As a result, Encarta handily won market share in head-to-head competition with Britannica's offering. Bleeding cash, Britannica was sold in 1996.

This example illustrates how the most succinct explanation for Encarta's success is the vision that shaped it. Gates and Bartholomew had a conception of the CD-ROM encyclopedia that appealed to users, carefully used the limited space on the medium, and adapted to the third-party distribution channel. This example also illustrates the ambiguity of the definitions of insider and outsider. Gates was deeply familiar with the personal computer industry, just as Bartholomew knew publishing. Both were outsiders to the encyclopedia market, but characterizing them as outsiders is too simple. We will return to this ambiguity again in other examples.

Online Classified Ads

Craigslist offered a new conception for organizing local classified advertising, and it took advantage of the growing World Wide Web. It departed from the typical design elements that had prevailed in local newspapers. Local newspapers attracted readers with content and sold ads. Craigslist dropped the content, the aspect to which every newspaper devoted most of its efforts. Craigslist eventually became a focal online site for local classified ads, displacing revenue at local newspapers, previously the focal place for many similar ads. One estimate found that Craigslist transferred $5 billion of annual revenue from newspapers to former ad buyers between 2000 and 2007.[31]

Founded by the programmer Craig Newmark in 1995, Craigslist took on its explosive growth after Newmark hired Jim Buckmaster in 1999, quickly elevating him to CEO and passing most decision-making responsibility to him. The eponymous name remained, but contemporary observers knew the vision combined input from both executives. Both had extensive programming experience in computing and the internet, but neither had worked in newspapers.

Together, they found a workable architecture for the organization and the website. The listings were organized around a simple departmental structure, with minimal text for local classified ads. There were generally no charges for listings other than in a few small categories like apartment rentals in Manhattan.[32] Craigslist contained no news, columns, or other content, nor did the organization spend funds marketing its site. The organization hired only programmers and grew to a national scale with approximately forty employees. In short, the vision for Craigslist was stark and simple: it costs little to operate and nothing to use much of the time, and it delivers a basic functional service to a broad set of users in all major cities.

The service changed little over the next two decades, a fact that could be interpreted in several ways.[33] In one interpretation, the founders implemented a foresighted vision to forestall any alternative competitor. In an alternative understanding, they had a rigid outlook and continued as long as revenues covered their costs. In either case, Newmark and Buckmaster generated more than enough revenue to meet their own goals for personal wealth. When questioned about their profitability, they boasted about giving away gains to users. Taken at face value, those interviews support the second interpretation.[34] It is also consistent with later behavior, in which the founders stopped searching for additional dollars of revenue after finding success.[35]

Streaming Web Server

Today, Nginx delivers a massive fraction of the traffic on the internet, much of it devoted to streaming. Streaming has restructured many services and supported the rise of several prominent transformational services, such as YouTube and Netflix. For our purposes, we will focus on the transformation of web servers. Nginx displaced Apache and IIS.[36]

The evolution of Nginx is, once again, a story of an outsider. Igor Sysoev resided in Russia at the outset of his efforts. Sysoev toiled alone on creating a high-volume web server that could handle volumes of requests far beyond the capacity of the two existing web servers, Apache and IIS. Among the key features Sysoev would invent, Nginx worked alongside Apache and IIS, the other two most popular web servers, rather than wholly replace them.

Sysoev experimented for several years after but generated little adoption. He started to use open-source approaches in 2004. That led to suggestions that changed the features of Nginx. The software improved tremendously from the recommendations of many other online participants. By 2007, Nginx had begun to grow into the primary server for the streaming economy. By 2011, it had become the dominant leading web server for streaming, rising to prominence with frontier services such as YouTube and Netflix.

After it started to gain adoption, Nginx faced resistance from the most prominent commercial internet users, which required enterprise-level services from a reliable firm. So, in 2011, Nginx Incorporated was established as a profit-oriented organization to offer services to business users needing enterprise-level reliability. This organization charged fees for its services, and Sysoev served as CTO. With this evolution, Nginx became a freemium service.

After F5 networks acquired Nginx in 2019, yielding a payout to Sysoev, this narrative gets as close as any story to a "rages-to-riches" Horatio Alger tale in modern times. However, Sysoev put a modern twist on the trope in this instance, going from an isolated Russian programmer to a leading technologist of frontier service.

Overall. The visionary archetype of an outsider provides a narrative for understanding transformation. For instance, Wikipedia designed a new approach to give discretion to many contributors while working within the current World Wide Web. Encarta changed the product to a CD-ROM but worked within the existing distribution system for third-party software. Craigslist did not reinvent the World Wide Web but revised a plan for classified ads as a matching service on the internet. Nginx redesigned the functions of a web server but not any other part of the internet and worked with those who wanted to place classifieds.

Another essential element is the time involved. Though contemporary news organizations portrayed some of these outsiders as overnight successes, that portrayal is misleading in one sense while accurate in another. All these products took many years to develop *before* launch, needing time to build capabilities and refine their product features to meet user needs. Yet, all these products continued to evolve *after* their launch. Some showed signs of transforming the markets quickly (e.g., Encarta, Craigslist), while others transformed the market only after many years (e.g., Wikipedia, Nginx).

Finally, notice how these narratives vary from the conventional history of an entrepreneur's journey.[37] For example, Wikipedia is a not-for-profit organization with a non-profit mission, aiming only to cover its costs, not to make a significant profit. Craigslist does not fit a conventional profit-maximizing model of an entrepreneurial organization. Nginx used crowdsourcing to grow, and only after many years did it adopt a for-profit structure.

Visionary Academic Outsiders

Academics have played essential roles in many transformations. In this archetype, academics become aware of the commercial possibilities for some invention or technology they studied and already knew deeply about. The academic is typically unknown to commercial firms in the market, and their perceptions about creating value depart from the views of industry insiders. In every case below, an insider will help the outsider overcome the challenges associated with outsider status.

The vision for an academic is linked to their origins. It may arise from the unique gifts of the individual attracted to academic life or the circumstances affiliated with their academic experience. The former explanation stresses that the probability of success rises after selecting a type of entrepreneur with scientific acumen. The latter description highlights that success increases after experience in an academic environment exposes (future) entrepreneurs to lessons not otherwise found in commercial markets or outside of academia.

The two examples in this section are among the most valuable and transformational product launches in the history of information services.

The first is Google, the search engine that altered user search on the World Wide Web and developed an incredibly profitable form of targeted advertising. The second is Facebook, the most widely used social media service globally. It also created a state of targeted advertising almost as valuable as Google's.

Ad-Supported Search Engine. The experience of Larry Page and Sergey Brin led to the Google search engine, the dominant online website. Their product launch indeed led to transformational change. It displaced all online portals and today ranks among the most profitable entrepreneurial firms in the history of Western capitalism.

As Stanford graduate students in computer science, Page and Brin had conducted a lab experiment with search algorithms, which they initially called Backrub. First invented in the summer and fall of 1995, the algorithm was seen as original by the duo's thesis advisors, who let them implement it on a server for users at the university. Over time, it gathered experience on the campus, and more usage helped Page and Brin refine the algorithm.[38] When both were but a couple of chapters short of their dissertation in 1998, the pair—with the help of the university—sought to license their algorithm to major portals at the time, including Yahoo! and Excite. They asked for a million dollars and did not get any buyers.[39]

The following events took numerous unexpected turns.[40] Soon after the failure to license, Brin and Page asked for a meeting with Andy Bechtolsheim, the retired CTO from Sun Microsystems, who had become an angel investor. One of their academic advisors arranged it. Bechtolsheim cofounded Sun while he was a PhD student. By this point, he had transitioned into a new role, advising many firms. Bechtolsheim became enthusiastic after one meeting and wrote a check immediately for one hundred thousand dollars, which catalyzed Google's founding soon after that.[41]

The technical basis for the algorithm was original and unlike any other at the time. It ranked web pages by their relevance to the keywords in the query. Accordingly, popular tales about Brin and Page, such as the best-selling book by Walter Isaacson,[42] associated Google's success with the technical proficiency of the algorithm. That characterization plays to the stereotype of the isolated geniuses who wake up from their academic bubble to find riches in commercial markets. That is, however, a misleading characterization of how Google became a lucrative commercial success.

The success took time to arise. First and foremost, the founders experienced some initial success licensing their software but quickly encountered difficulties managing the organization. By 2001, their venture capitalists insisted on bringing in professional management, hiring Eric Schmidt, a computer scientist with extensive industry experience at Novell. In addition, Schmidt brought others to build the organization.[43]

More subtly, the business was not merely licensing an ever-improving algorithm. From 1998 to approximately 2005, Google ran two related but different companies using the same search algorithm at its heart. The first was the primary source of revenue at the outset, and the other became the dominant source of income after 2005. The first business was not aimed at consumers, looked different from today's, and was the first to succeed. It supplied the algorithm to portals for miscellaneous searches. Though it was popular with these business customers, it did not have much to do with the firm's later high valuation. However, it was important for the opportunity it opened. It led Google to see a significant fraction of online search queries.

The second business was consumer-oriented, providing the algorithm to anybody who landed on the URL google.com. Initially, it showed some banner advertising, if it showed any at all. After 2001, it began experimenting with click-through ads next to organic searches, evolving the processes to select ads. Finally, after developing a scaled auction process, the services started to resemble today's search engine.

Even in 2001, the portals who licensed the algorithm did not view their licensing deals as helping a future rival. They had no reason to think so. The searches on Google were considered a peripheral activity, something users did as a last resort when a portal did not index the desired solution. Yahoo!, Excite, and many others did not view Google as a competitor and willingly made deals with Google to send searches their way in exchange for fees. In addition, in 2001, Google had yet to figure out how to generate ad revenue from a consumer service. This was the only firm that figured out how to generate significant ad revenue from a consumer service.

Brin and Page had a unique vision for advertising-supported search, treating the technical basis behind search as a separate and distinct activity from those associated with advertising. Google put a firewall between its search and revenue-generating advertising organizations. In other words, separating technical and commercial aspects was a distinct feature of the organization and the service experienced by users. Different search algorithms have experimented with allowing advertisers to bid for better placement in their organic listing in response to searches. Brin and Page had declared as graduate students many reasons why mixing the two would fail. They stuck to that vision for many years after founding their firm.[44]

The ad auction was invented by several Google employees, who studied other attempts to gain revenue from advertising and learned lessons from those failures. They learned how to invent and operate a quality-weighted position auction for ads that appeared in a fixed location, coordinated with the keywords that triggered the organic search results on google.com. These ads responded to the queries users put in the search bar, so they were designed to meet user needs. The designers worried users would not return to the search engine after viewing ads. Hence, the designers sought to make these ads respond to user needs. Users wanted to see them or, at a minimum, did not get annoyed after viewing them because they were relevant to their query.

Developing the auction required significant, costly effort and imaginative new designs. These efforts were underappreciated by all others in the industry when the prototypes initially emerged in 2002. It combined two traits—an automated method for generating revenue from many auctions and a display and weighting scheme that encouraged relevant ads to be placed higher. Industry confusion was understandable at the time. Even after the initial rollout of the prototypes, this auction did not generate much revenue, so its capacity for supporting a large scale of ads was not on display. Most online ad buyers, such as automobile companies, resisted using it. It took Google several years to develop a division to support ad buyers with tools, such as lists of common search terms, and generate a large and regular ad business. Though invented earlier, other portals and potential rivals were unaware of the breakthrough until 2005, when the scale of ads became visibly wider.

While this experience illustrates the academic outsider visionary, it also demonstrates a limitation of the archetype. A proper understanding of the unique features of the visionaries, in this instance, would need to encompass actions long after the initial product launch. The complete vision for the most lucrative parts of this business only emerged over time, years after its founding in 1998. Moreover, more than the technical breakthrough alone is needed to account for its transformational success.

Social Network

Another prominent example of an academic visionary was the entry of Facebook. Developed for his Harvard classmates, Mark Zuckerberg took control of Facebook in 2004, spreading it from one campus to another over the next two years, spreading it outside of campuses after that. In a few years, it displaced its direct predecessors, Friendster and MySpace, and its transformational impact went well beyond that. In many countries, it is the most popular internet destination. With the growth of its advertising, it also has become one of the most profitable firms in the history of Western capitalism.

As with Brin and Page, the academic setting provided opportunities and challenges. For Facebook, the academic environment served as the setting for the initial trial and scaling of the service. However, these origins did not guarantee success at a national scale or reduce the risks of finding ways to generate revenue with private buyers. The open question for Facebook in 2005 and 2006 was whether the network effect would spread outside campuses, depending on whether non-students would find it appealing. Again, learning was only possible by investing considerable resources to support the service outside a locale.[45]

The academic archetype provides a helpful way to characterize Zuckerberg's earliest actions, but it has limitations. First, it would take years before Facebook achieved the scale required to sell advertising. That came after years of experimenting with many applications, such as games and news feeds. Moreover, engineering a national service of this form was expensive and challenging. Sheryl Sandberg, who joined Facebook in 2008, receives credit for initiating and implementing those efforts.[46] With her prior experience at Google, she could be described as an insider when she joined. (She obtained some celebrity from this position and wrote a best-selling book about how women executives can succeed in business,[47] which took her to a new level of fame.) In addition, scaling Facebook required enormous investments in back-end computing to support a larger-scale site. As of this writing, Facebook is building its ninth data center in Prineville, Oregon. Finally, as later events demonstrated, it required seemingly simple but innovative changes in the interface, such as the invention of the thumbs-up.

Facebook also had to invent new forms for targeting advertising to its users. This involved creating an automated ad auction that resembled Google's scale and precision but differed in the information about the target. Facebook found a way to identify types of users based on their social activities and enabled advertisers to bid on those types. Once again, much of this came long after the product's initial launch and its transition from the academic setting into the commercial one.

In summary, the academic outsider archetype helps analyze the product launch into mass markets but provides less insight into the commercial evolution and potentially even less into the construction of the most lucrative parts of the business. Just as with the description of Google, the latter requires considerable detail about the individual's outlook and the assembled team within the organization, and that team contained many experienced executives who had status as insiders.

Overall. The central feature of the academic outsider visionary is not academic accomplishment as measured by the academy. Brin and Page never finished their PhDs, and Zuckerberg his undergraduate degree. Instead, for the academic visionary, academic life enables them to grapple with a technological frontier. It is too much to say the academy serves as an insulated cocoon. It provides a somewhat protected environment to experiment with and scale a product for a limited set of users. Still, the academic experience helps gestate new technology without competitive concerns.

In the two examples of academic visionaries, each required substantial work to transcend the academic experience and achieve transformational commercial success. Insiders helped immeasurably with the transition. In the above examples, insiders such as Eric Schmidt and Sheryl Sandberg played crucial roles in commercialization activities.

Visionary Insiders

Are there any examples from the history of computing where an insider visionary transformed markets after launching successful products? In the history of computing, Thomas Watson Jr. occupies the iconic position of insider visionary. The event leading to his fame, the launch of the System/360 in 1964, led to a re-conception of design for all major providers. In addition, it cemented IBM's market leadership for many years. As a result, it is one of the most chronicled events in computer history.[48] By comparison, only a summary is offered here, mainly to stress the continuity in the critical elements of the archetype.

Compatible Mainframe Computer

Watson Jr. inherited leadership at IBM from his domineering father, Thomas Watson Sr., after he died in 1956.[49] Watson Jr.'s independent outlook resulted in frequent clashes between them. Once solely in charge, Watson Jr. pursued and implemented a new vision for the company, building on the strong sales and distribution processes Sr. had established for numerous businesses. Watson Jr. became affiliated with selling mainframe computer systems and experienced large sales with the 1400 System. From this leading position, Watson Jr. chose to launch an ambitious new product, the System/360, the first mainframe with modular compatible components, betting the company on that launch. Compatibility, in this case, meant a user could buy a small system and develop many applications, and later upgrade to a large mainframe and still use the same software applications.

Watson Jr. shepherded the development and the rollout of System/360 throughout the organization. Among other advances, the system enabled users to upgrade their systems without losing many valuable investments in system-specific software. As a result, it became a seminal product introduction in business computing, generating widespread adoption and reinforcing IBM's leadership and dominance many years after the product launch. However, the System/360 went to the technical frontier of the time, particularly in its design of a series of compatible products, so it had to address new engineering problems, and success was not guaranteed.

For this chapter, it is essential to be precise about where Watson showed a unique vision compared to competitors and his firm's previous accomplishments. The product development process confronted many risky engineering challenges, and breakthrough engineering has received considerable attention in retrospect.[50] Addressing those challenges required extensive funding, and failure would have resulted in financial ruin for the company.

Insider status was vital because it allowed IBM to use its own employees to attempt to re-engineer the system rather than make a modular change in one component, preserving the marketing of a whole system. This was the norm for selling to business users, who regarded mainframes as risky and complex machines. In addition, insider status correlated with tangible resources that played essential roles in the product's launch, such as frontier knowledge about building equipment that stored large amounts of information in electronic format. IBM designers also had experience with designing and producing previous generations of equipment. They could undertake the development of such a complex redesign only because they had experience with the prior generation of computing memory, processing, compiling, software implementation, and hardware servicing.[51]

Insider status also contributed to the success of the distribution of this product. The innovations in engineering required complementary changes in the organization for servicing the equipment. The distribution of computing equipment and applications had already received considerable attention at IBM before the launch of System/360. An entire organization had been set up with compensation for sales and maintenance, which generated responsibility for the usage and produced a reputation of reliability for the previous generation of equipment. The sales force also had many relationships with businesses that used computers, specifically its 1400 series, a rollout with which Watson Jr. was also deeply familiar. It also had an established service organization to help users with issues. These marketing resources played essential roles in fostering the adoption of a new pioneering product.

In summary, the product launch of the System/360 illustrates a leading firm taking enormous financial and marketing risks on a frontier-stretching new design. Moreover, the launch built on existing R&D, production, and distributional assets, coordinating them towards an ambitious new product. As a result, it became the benchmark against which all future designs were compared.

This example illustrates two crucial differences between the outsider and insider archetypes. First, insider status enabled the incumbent firm to bring unique resources to bear on the product launch and related commercialization activities. In addition, the insiders brought an outlook learned from experience as the leading firm, and that helped implement a vision that had been challenging to reach. In either case, these were difficult for others to replicate. It also illustrates one other key metric for success. Not only did the new product lead to a loss of market share among existing players, but it generated a re-conception of design against which all future designs were compared.

Fast Followers

An outsider visionary may have a significant impact after introducing a new product, primarily if it generates a reaction from insiders, transforming the market, even though the insider is not the first to have the idea. The insider in this archetype is often called a "fast follower."[52] As the label implies, the archetype describes an established firm and insider that imitates the actions of another firm, typically an outsider who first pioneered a new product. The follower makes additional changes, such as adding some differentiated implementation or improvement to the product.

The fast-follower archetype can be portrayed as an economic morality tale, with the fast follower as either an opportunistic copycat of the pioneer's product or a wise innovator making additions to a flawed pioneer's product. The former narrative describes the pioneer as taking commercial risks but gaining little reward. In the latter narrative, the fast follower draws on their experience to improve on an early actor's missteps, which implies that the fast follower deserved his gains.

IBM's introduction of the personal computer is an excellent illustration of a fast follower. Once again, this product launch has been chronicled extensively, and it certainly was transformational. Its launch restructured the personal computer market and made IBM the leading firm supplying personal computers for the next half-decade. Its launch motivated many businesses to adopt a PC for the first time. It also motivated enormous numbers of firms to develop applications. If the PC division at IBM stood alone after three years (i.e., in 1984), it was larger than all but two other computer firms, DEC and the other parts of IBM. After 1985, clones of the designs began to erode IBM's market share and eventually came to dominate the market, using the label "IBM compatible." By the early 1990s, the leadership for IBM-compatible systems had visibly passed to Intel and Microsoft.

Personal Computer

What led IBM to be a fast follower? By the time of the PC launch in 1981, several other firms had begun selling pioneering systems of personal computers for years.[53] The largest selling of these pioneers was the Apple II, led by the entrepreneurial team of Steve Jobs and Steve Wozniak. Ironically, Jobs and Wozniak were also regarded as fast followers within their community. They imitated numerous pioneers in personal computing, such as Commodore, Radio Shack, and MITS. The Apple I and II stood out for their unique combination of extraordinary design and commercial focus.[54]

IBM's management knew that Apple's products had begun to appear in customers' offices to whom they sold mainframes. That was not concerning, but Apple's success motivated additional designs of software applications, including VisiCalc, the first spreadsheet. The appearance of office applications directly touched core services offered by IBM and represented a recognizable threat to any top-level executive. The strategic equivalent of the fast follower was explicitly discussed as a competitive response, and some executives inside IBM doubted IBM could develop a personal computer quickly enough to execute any fast-following strategy.[55]

Who had the vision for the IBM PC? That can be attributed to the CEO of IBM, Frank Carey, and Bill Lowe, the leader of the small-systems division. Carey arranged for Lowe to present a proposal to the board. Lowe's presentation was masterfully constructed to appeal to the executives at IBM. It included marketing projections for the number of PC units not to exceed the number of mainframe units. Such a low estimate did not match market realities, but Lowe did not want to offend the sensibilities in the room. Years of experience within IBM have made Lowe a savvy presenter.[56]

Lowe's plan involved developing a product within a year and doing so with an "open design." Not all components would be produced internally at IBM, a departure from an established practice, which Carey permitted to get the product to market quickly. The design came with expansion slots for additional hardware components and encouraged and anticipated software applications from non-IBM firms. At the time, Carey stated that he would field any phone calls from other parts of IBM to stop interference from a division that thought they should provide a component that Lowe refused to source internally. Carey did this to bring the product to market quickly.

IBM's personal computer was not dramatically more functional than those it imitated and was sufficient for many business customers. Although IBM had a large internal division for producing integrated circuits and a long history of using internal staff to design and build software, Lowe sourced the microprocessor from Intel and the operating system from Microsoft (decisions endlessly debated later by Monday morning quarterbacks). Lowe's group also deliberately used a modular design that could accommodate third-party software applications, which IBM invited because they anticipated IBM would be unable to provide adequate application software. Breaking with precedents, IBM published the specifications and requirements for compatible products.

The product launch was extraordinarily successful, partly because IBM brought all its marketing and distributional assets to bear on the launch, including running an extensive national advertising campaign and rewarding its sales force. In addition, several software firms and add-on cards were available. The explosive growth of sales to businesses motivated many software developers to build new software for the IBM PC, leading it to acquire enormous functionality from the cross-side network effects.

This example illustrates how a sense of morality shapes the analysis. Depending on the value judgments of the narrator, IBM had either gained profits that belonged to risk-takers or used its existing marketing strengths wisely in a new product that imitated those who lacked such assets. In either interpretation, it acquired status as the most dominant PC. Within a few years, most competing platforms had disappeared.[57]

Summary

The fast follower stresses the insider's ability to learn from the outsider, bring additional resources to bear on the setting, and direct new design elements toward customer needs. It explains how a pioneering entrant may catalyze change and restructuring but not survive or profit much from their pioneering actions. Instead, the insider learns from the outsider's actions and launches a product that transforms the market.

Drawing on Multiple Archetypes

No legal regulation or managerial business principle requires product launches to fit neatly into a stark taxonomy of visionaries. In practice, no visionaries care whether their actions illustrate any taxonomy. It is no surprise, therefore, that many product transformational launches blend the archetypes of insiders, outsiders, and fast followers. This discussion aims to illustrate how the archetypes manifest in such complex examples, typically involving participants with various statuses. For that purpose, this section discussed three iconic product launches from the history of computing: the launches of the Netscape browser, the wireless router, and the iPhone.

Commercial Browser

Mark Andreesen and Jim Clark, the cofounders of Netscape, the first firm to sell a commercial browser, illustrate one version of the academic archetype associated with transformational change. It also shows other aspects of the insider archetype with transformational change. Though unknown to young surfers today, this firm was undoubtedly transformational. Through 1995, 1996, and much of 1997, Netscape displaced Microsoft as the PC internet application software leader. Microsoft's response—which will not receive much attention in this summary—could be used to illustrate aspects of the fast follower. In any case, Netscape's product launch was catalytic, kicking off the dot-com boom of the latter part of the 1990s.[58]

The co-founders could not have been more different. Andreesen had previously toiled away at a University of Illinois research lab funded by the National Science Foundation and had developed the Mosaic browser for the young World Wide Web, which Berners-Lee had made available as shareware in 1991. Though several Unix-compatible browsers had been designed, Mosaic was the first to work on IBM PCs, the most popular potential client in use in 1993.[59] Andreesen was deeply familiar with the Web and, thus, was an insider concerning the growing commercial web and the newly privatized internet. At the same time, he was an outsider concerning any commercial software firm. As a result, when the firm Netscape first launched, attempts were made to market Andreesen as the next wunderkind of computing, as if he possessed talent comparable to Bill Gates.[60] Those statements became common when Netscape proliferated and less frequent as the browser wars raged.[61]

In contrast, Clark was a former Stanford faculty member and experienced entrepreneur known for leaving his faculty position to cofound Silicon Graphics. With that history, he already had a reputation for successfully being a commercial outsider at the frontier of computing who had transitioned into a commercial insider. He and Andreesen had conversed about browsers during the winter of 1994, just after Andreesen had graduated from college. After considering several alternatives, the two had become convinced that making and selling browsers could become a viable commercial venture. Once they settled on this option, they acted decisively, visiting the venture capitalists Clark knew well.

Could Andreesen have gotten any financial backing by himself as a visionary for a new product? The answer is no, at least at the level funded by Kleiner, Perkins, Caufield, and Byers, which enabled Netscape to deliver a beta version of their software in six months. This venture capital firm was willing to take on the risks because Andreesen had partnered with Clark, who had a positive track record for commercializing technical innovation. He also could articulate strategic visions in forms and language that venture capitalists appreciated.[62] By himself, Andreesen's vision encompassed only the technical aspects and design principles of making a browser, where he had extensive experience at the university with improving Mosaic. Andreesen had little knowledge that suggested he understood how to build an organization to support and commercialize software at a large scale.

The commercialization strategy for Netscape combined elements of an insider's and outsider's point of view. Netscape released its beta version in November 1994 and its first commercial release in February 1995. Netscape offered the browser for no price but charged businesses for licenses. That dovetailed with the firm's aspiration to displace Mosaic and attract users from among the number of non-academic Web users, which grew after the internet privatization in 1993. At its founding, there were approximately 25 million users, and Clark stated that he assumed usage would double by the time they shipped their product. Instead, their approach generated new adoption from both home users and academic users of browsers, and together, it led the product to be adopted rapidly and at an unprecedented scale. That started Netscape's life with a strong initial network effect, which supported gaining revenue from businesses.[63]

In this case, a visionary entrant used their academic experience to enter a commercial market with considerable expertise and insight into building and operating a product that non-academics want. While academic experience may provide little knowledge about distributing to commercial actors or managing and growing an organization, those issues become more salient after the first product launch.

Much has been written about the reaction Netscape induced in Microsoft. The fast-follower archetype illustrates some aspects of how Microsoft approached its browser development and commercialization efforts.[64] That said, no archetype alone could explain the complexity of the subsequent competitive events, and space constraints limit a complete treatment.[65]

Wireless Routers and Wi-Fi

It is hard to argue with the transformational impact of wireless routers, known by their branded name today as Wi-Fi. Wi-Fi router equipment started from almost nothing at the turn of the millennium and grew into nearly 12.5 billion dollars worldwide in 2020.[66] It displaced many wired local area networks and enabled a wide range of new uses for mobile devices.

The Wi-Fi router's development illustrates aspects of the fast-follower archetype and aspects of outsider and insider archetypes. That is because the events leading up to the development of Wi-Fi routers involve the vision of Michael Dell, the CEO of Dell Computer, an insider who served as a fast follower. He was following Steve Jobs, the newly returned CEO of Apple Computer, who, at that point, was both an insider in some markets and an outsider in others. In turn, Jobs reacted to events at the committee for standard IEEE 802.11, which issued a new protocol standard for "wireless local area networks" in 1997 and announced intentions to upgrade it in 1999. In turn, the committee had built its protocol after years of numerous prototypes and experiments with designs of wireless routers.

The initial launch of consumer-oriented routers began dramatically. In July 1999, Jobs took the stage at the MacWorld convention in New York. He introduced the audience to Apple Airport—the first mass-market Wi-Fi product. Famously, Jobs took out a hoop during the presentation to convince the audience that there were no wires between the wireless router and the computer. Viewed with hindsight, this event marked Apple's launch of the first Wi-Fi router using design 802.11b, the second design of the IEEE 802.11 committee, which improved the first (and flawed) design released in 1997. In addition, the Airport did something no prior wireless product had done: the laptop's expansion card was priced at $99 and came in a branded product from Apple. The base station was sold separately for several hundred dollars, making it affordable for a household. Finally, Apple distributed the entire system, backing it with its guarantees.

The Steve Jobs of 1999 had ambiguous status, but Apple's position was not ambiguous in a conventional core-periphery model. Apple had a low market share in 1999 and occupied a peripheral status in the computer industry. Many analysts thought the firm would fail soon. Jobs had returned to Apple only as part of a last-ditch attempt by the board to revive the firm after sales for its PCs had dropped to what many predicted would soon become unsustainable levels of sales. Due to two decades of history in the PC industry, the brand was well known, and so too was Jobs, particularly after his time at Pixar.[67] In other words, Apple looked like an outsider when assessed by market share, but by experience, Jobs was an insider.

The partnerships behind the Airport also designated the effort as coming from an outsider. Apple got the card and base station components from a division of Lucent in the Netherlands. Lucent was the equipment-making spinoff from AT&T, and though it was young, it had inherited insider status from its parent. Yet, the division of Lucent that became important in this episode had joined Lucent as part of the acquisition of NCR and did not occupy a central place in the organization's strategy for product development. It had no history of supplying into a supply chain for a consumer-oriented product. In that sense, they were outsiders to the consumer-oriented communications equipment market, and their vision carried little weight within their own company.

In another sense, Lucent's employees were insiders. They knew about the 802.11 standards because some of their employees had participated in the IEEE 802.11 committee. One of them had been the chair of the committee and had negotiated and maneuvered it to finally issue a standard. All of this had taken place on their initiative, not part of any strategy directed from headquarters. Moreover, they were keenly aware of alternative protocols that sought to address the same needs—such as HomeRF—and wanted their designs to prevail so they could supply equipment.

How did Michael Dell become involved? Lucent's cards for Apple laptops and the Apple Airport system worked well and sold well. Yet,

Lucent's product did not serve PCs from suppliers other than Apple, leaving a large part of the IBM-compatible PC market uncovered, and this was more than 90 percent of PCs in use at the time. They used a Windows operating system from Microsoft. This gap was not lost on Michael Dell, CEO of Dell Computer. Though he had started the firm in his dorm room in 1984, by 1999 his firm was the most prominent PC provider. Dell provided close to a quarter of all IBM-compatible systems. By any definition, he was an insider.

Michael Dell learned about the announcement of the Apple Airport shortly after the event in New York and contacted Lucent. According to the account from a Lucent employee, Dell was "furious" with Lucent because Dell was not the first to experiment with a wireless product release. Dell had to be gently reminded that, years earlier, he had explicitly declared to Lucent that he was giving up on their efforts to develop wireless modems.[68]

Lucent and Michael Dell subsequently signed a deal, and Dell arranged for Microsoft to modify Windows appropriately. Making Wi-Fi compatible with Windows XP was the main challenge for the team at Lucent. To do that, Lucent and Microsoft cooperated in changing the design of Windows XP, and a new version was released in 2001. It supported IEEE 802.11b in a Windows-based system. Like Apple, Lucent made a hardware card for an external slot in a PC, and Dell provided the entire system.

Dell's behavior fits the archetype of a fast follower. What did Dell, as a fast follower, bring to the market? Brand, distribution, design capabilities, and access to production facilities are all insider resources. After launching, Dell branded an entire system of wireless modems and laptops and became the world's leading provider of Wi-Fi products for a short time.[69] In light of the subsequent events, Apple's and Dell's actions were catalytic, jump-starting a process of introducing wireless modems into mainstream usage.[70]

This example also illustrates how morality plays of fast-follower archetypes shape analysis. Depending on the value judgments of the narrator, Apple and Lucent were risk takers, and Dell had gained profits that belonged to pioneers. Alternatively, Dell used its existing marketing strengths wisely to profit from a new opportunity and deserved the profits generated. In either interpretation, it acquired unquestionable status as a leading provider of routers.

Smartphone. There is no doubt that the launch of the iPhone was transformational. It was immediately popular.[71] It eventually led to the displacement of almost every cellular phone provider, leaving only two dominant smartphone providers. It led to a re-conception of design for phones among all providers in developed countries. Only one subsequent entrant, Android, achieved a similar scale of success.

The entry of the iPhone could illustrate entry from a visionary insider, or it could illustrate the actions of a fast follower. For the former interpretation, consider that the launch of the iPhone came directly from the vision of Steve Jobs, who returned to Apple in 1997. By 2007, he had revived both sales of its PCs and introduced the iPod and iTunes (a few years earlier). That looks like a visionary insider. But that is too simple. By 2007, numerous other smartphones had already been developed and sold in large numbers, suggesting Jobs had the benefit of learning from that previous experience. That makes him closer to a fast follower. For example, Blackberry had large sales for its unique design of a keyboard device with secure software operating inside a firm. Systems using the Symbian operating system had large sales, especially in Europe, and those supported a growing set of applications. Microsoft also worked on Windows CE for over half a dozen years, developing numerous partnerships to push out the operating system for applications in various phones.[72]

Jobs was aware of these other efforts, thought the iPhone was a crucial project for Apple, and was intimately involved in all the details. He wanted the iPhone to have backward compatibility with iTunes, Apple's music service, which the organization had developed to support its iPods, a music-playing device that sold in large numbers. At the launch, he described the iPhone as a device that could play music and make a phone call, highlighting the backward compatibility with the iPod and the phone's new features. The marketing strategy anticipated migrating users from their music devices to the iPhone, clearly using the insider advantages associated with an installed base of users of iTunes and iPods and branding as a trustworthy, innovative firm.

These events also involve the archetype of an outsider. Though more well-known in computing and fast achieving celebrity status in technology markets, Jobs and Apple had never sold a mass-market product that made a phone call except for one previous failed joint venture.[73] The design also marked a significant change from the norms at the time. While the iPhone was not the first smartphone by any stretch, the iPhone involved a large re-conception of crucial design elements, such as a touch screen, data intensity, and locus of computing on the handset.[74] In this sense, it resembles some of the fast-follower examples, except that Apple was not particularly fast by any definition. Others had been involved for years.

Being well-known in 2007 should be considered as something other than an unambiguous advantage. Apple needed partnerships with voice and data carriers to distribute the iPhone and make it operational.[75] However, by 2007, Jobs was intensely disliked by many other executives in the phone carrier business and, more importantly, mistrusted. Many business partners did not possess any appetite for sharing risks with him. When Jobs went to US carriers to negotiate deals for carrying the iPhone, he encountered unyielding refusals for partnerships from Verizon, Sprint, and T-Mobile, leaving only one primary US carrier, AT&T.

AT&T was willing to accept the terms that came close to those Jobs wanted only if Jobs gave them an exclusive deal for two years. AT&T aspired to move its market share with the exclusive agreement, selling the product alongside other smartphones and cellular phones that used AT&T's new 3G service, which could carry data at volumes not feasible with 2G. Jobs would have preferred wider distribution, which firms such as Motorola, Nokia, and Samsung enjoyed by taking the terms offered by carriers. As it turned out, both parties got what they wanted in the short run, as the adoption of the iPhone proliferated in its first year, and some customers switched to AT&T solely to gain access to the iPhone.[76]

The conception that informed the initial product launch changed after the enthusiastic reception from users. There were no applications in the first year, and initially, there were no plans for any developer support. Even though Apple had ample experience supporting app development for its personal computer and selling products over the Web on its iTunes platform, Jobs made no public reference to them in the initial product launch. The iPhone with applications came only after Apple's initial sales and after the other top executives at Apple persuaded Jobs that the installed base was big enough to support the cost of building tools and kits at Apple to help application development. Jobs had been reluctant to authorize such an expensive program. Nevertheless, the addition of applications turned out well for Apple. Many developers were willing to make applications after seeing the potential for many users.

As a result, the app store launched a year after the initial launch of the iPhone. The launch of applications played to the advantage of an insider, requiring significant resources, lots of experienced programmers, and an organization with regular employees supporting application development. The apps also had to be distributed somewhere, and iTunes had given Apple ample experience to understand how to design, build, and operate an online platform. Their online app store borrowed from that experience and built upon it. While this sequence of actions mitigated the risks, the second step is the key one. Apple restructured the market by attracting and selling applications.

The rest is well known. Google purchased Android in 2005, improved it, and released it within an open-source regime. The first Android phones came out in 2008. The iPhone and Android systems attracted app developers, creating platforms with functionality that appealed to users and exceeded that of the pioneering Blackberry, Windows CE, and Symbian systems. Users would settle on iPhone and Android, leading to a displacement of the pioneering platforms. Those two remained dominant for the next decade and continue to be, as of this writing.

In summary, and like many other examples, the visionary has a vital role in launching the product that brings about a transformation. In this case, the development of the smartphone with applications took place in multiple steps. As an insider, however, each step took advantage of and was hindered by Jobs's status—those advantages and hindrances made for marked differences from visionaries at entrepreneurial firms. That shaped aspects of the fast follower, too, as Jobs was learning from the many prior attempts at smartphones but tailoring Apple's designs to take advantage of Apple's installed base of users of iPods and Apple's strengths in online distribution.

Conclusion

This chapter provides an approach to analyzing the launch of transformative products and connects it to the influence of visionary CEOs. It argues that visionary archetypes of insiders and outsiders play crucial roles in bringing about transformational product launches. Archetypes of visionaries characterize the behavior of principal decision-makers. An archetype provides a systematic way of blending narratives with analysis and explaining why decision-makers behaved as they did before, during, and after a transformational product launch.

This analysis supports one broad conclusion. Transformations differ based on whether insiders or outsiders lead them. For example, visionary outsiders enter as entrepreneurs with nothing but their products and, unsurprisingly, put all their efforts into perfecting them to achieve their objectives. In contrast, insiders build on existing brands and distribution processes and choose competitive strategies in light of their strengths. Generally, insiders and outsiders use different approaches to product differentiation.[77]

Several additional implications deserve highlighting. For one, different origins for visionaries partially explain why the visionary differs in intuition and judgment and pursues opportunities with other resource and strategies. Second, these narratives of outsiders contain elements of surprise for insiders as the product launches, even when it has been gestating for a long time. That undermines models founded on complete information or common knowledge where all participants share the same information. Third, only a few of these visionaries acted alone, and none without organizations around them. Many have essential partners from outside their organization, contrary to myths of visionaries as lone geniuses.

Characterizing the competitive confrontation is the next step in the agenda. The chapter partially developed this point by identifying fast followers as an archetype but did not fully describe competitive actions. For example, since there is more than one archetype to describe "innovation from the edges," no single archetype of "innovation with incumbency" forestalls such entry. Further research could seek to identify the forms of these competitive interactions. In addition, nothing in the origins of an outsider precludes an insider from acquiring their firm for competitive purposes. What circumstances make such an outcome more or less likely?[78]

This chapter also raises questions about competition policy. The variety of narratives aligns with policy prescriptions to let a thousand flowers bloom to foster unusual product launches that catalyze transformative and innovative change. That folds back the open questions one layer. What market structures enable visionaries with various origins to experiment with commercial markets? Are some structures better able to support visionaries than others? This, too, is an open question for future work.

Finally, because the examples are selected for being transformational, it is unsurprising that the narrative arc usually ends with eventual triumph. For related reasons, the narratives share a similar redemptive tone. But that narrative arc comes with the selection bias inherent in studying winners with hindsight. Such a bias is difficult to correct without extensive recall of all the other visionaries who failed, and such details are oft interred with a visionary's bones. How would analysis change if all those details about the research community and prior precedents could be brought to bear on each example? That open question also emerges from this chapter.

NOTES

1. I am grateful to Bill Aspray, Tim Bresnahan, Innessa Colaiacovo, Jim Cortada, Roberto Fontana, Joshua Gans, Daniel Garcia-Swartz, Rebecca Henderson, Scott Stern, and Feng Zhu for many conversations. Nobody other than me is responsible for the errors herein.

2. Thomas Watson Jr., and Peter Petre, *Father, Son & Co.: My Life at IBM and Beyond* (New York: Bantam Books, 1990).

3. Walter Isaacson, *Steve Jobs* (New York: Simon and Schuster, 2011).

4. See, e.g., William N. Thorndike Jr., *The Outsiders: Eight Unconventional CEOs and Their Radically Rational Blueprint for Success* (Cambridge, MA: Harvard Business Publishing, 2012); and Walter Isaacson, *The Innovators: How a Group of Hackers, Geniuses and Geeks Created the Digital Revolution* (New York, Simon and Schuster, 2014).

5. See, e.g., Franco Malerba and Luigi Orsenigo, "The Dynamics and Evolution of Industries," *Industrial and Corporate Change*, 5, 1 (1996): 51–87; Franco Malerba, Richard Nelson, Luigi Orsenigo, and Sidney Winter, "'History-Friendly' Models of Industry Evolution: The Computer Industry," *Industrial and Corporate Change*, 8, 1 (1999): 3–40; and William J. Baumol, *The Micro-theory of Innovative Entrepreneurship* (Princeton, NJ: Princeton University Press, 2010) for recent treatment.

6. Thomas Haigh and Mark Priestly, "Innovators Assemble: Ada Lovelace, Walter Isaacson, and the Superheroes of Computing," *Communications of the ACM*, 58, 9 (2015): 20–27.

7. The phrase "core-periphery" is a potential source of confusion. This chapter's definition of a core-periphery model draws on conventional economic approaches to industrial organization, stressing that the firm with the highest market share is at the core. All others are at the periphery. It has nothing to do with standing in a social hierarchy, caste, or class system. Racial, ethnic, and educational backgrounds play no role in these models.

8. Thomas Kuhn, *The Structure of Scientific Revolutions* (Chicago: University of Chicago Press, 1962). Also, see the formulation in M. J. Mulkay, *The Social Process of Innovation: A Study in the Sociology of Science* (London: The Macmillan Press, 1972).

9. Scholars have identified several general paths (Gino Cattani, Simone Ferriani, and Andrea Lanza, "Deconstructing the Outsider Puzzle: The Legitimation Journey of Novelty," *Organization Science* 28, 6 (2017): 965–992). Efforts typically involve demonstrating a novel accomplishment and appealing to the shared sense of values supporting scientific experiments. The literature also stresses the importance of the presence of individuals who serve as "bridges" (between established social paradigms and outsiders' outlooks) by providing funding, recognition, and additional social connections (Gino Cattani, Simone Ferriani, and Paul D. Allison, "Insiders, Outsiders and the Struggle for Consecration in Cultural Fields: A Core-Periphery Perspective," *Amer. Sociol. Rev.* 78, 3 (2014): 417–447). Finally, scholarship identifies scientific and commercial actors' mechanisms to build institutions that facilitate acceptance (Huseyin Leblebici, Gerald R. Salancik, Anne Copay, and Tom King, "Institutional Change and the Transformation of Interorganizational Fields: An Organizational History of the US Radio Broadcasting Industry," *Administrative Science Quarterly*, 36, 3 (1991): 333–363). This framing has also been extended to knowledge communities outside of science (Hani Safadi, Steven L. Johnson, and Samer Faraj, "Who Contributes Knowledge? Core-Periphery Tension in Online Innovation Communities," *Organization Science* 32, 3 (2021): 752–775.

10. This type of scholarship focuses on the challenges and success of "Rebels, Mavericks, and Heretics," the title of one of the books on the subject by Harman and Dietrich. See Oren Harman and Michael R. Dietrich, *Rebels, Mavericks, and Heretics in Biology* (New Haven, CT: Yale University Press, 2008); Oren Harman and Michael R. Dietrich, *Outsider Scientists: Routes to Innovation in Biology* (Chicago: University of Chicago Press, 2013). For a contemporary example, consider the experience around mRNA vaccines, which only a few scientists investigated. This raises questions about why scientific researchers do not invest in risky research with the potential for a high payoff (both commercially and to society) and whether institutions to support research are biased against such research because it appears unusual or unconventional (Chiara Franzoni, Paula

Stephan, and Reinhilde Veugelers, "Funding Risky Research," in Josh Lerner and Scott Stern, eds., *Entrepreneurship and Innovation Policy and the Economy, Volume 1,* sponsored by the National Bureau of Economic Research (Chicago: University of Chicago Press, 2021); and Marco Iansiti, Karim Lakhani, Hannah Mayer, and Kerry Herman, "Moderna (A)," Harvard Business School Case 9-621–032, 2020.

11. In this sense, this essay is closer to the study of the early radio pioneers (Leblebici et al., 1991), where feasibility, social understanding among entrepreneurs with similar goals, and profitability played crucial roles in selecting surviving firms. A related theme arises in the study of Coco Chanel (Gino Cattani, Mariachiara Colucci, and Simone Ferriani, "From the Margins to the Core of Haute Couture: The Entrepreneurial Journey of Coco Chanel," *Enterprise & Society* (2022): 1–43, which studies how entering the commercial fashion and beauty market requires fostering the acceptance of new ideas while developing a successful business.

12. John Jewkes, David Sawer, and Richard Stillerman, *The Sources of Invention* (London: Macmillan and Co. Ltd., 1958).

13. Joshua Gans, *The Disruption Dilemma* (Cambridge, MA: MIT Press, 2016); Philippe Aghion, Celine Antonin, and Simon Bunel, *The Power of Creative Destruction: Economic Upheaval and the Wealth of Nations* (Cambridge, MA: Harvard University Press, 2021).

14. Joshua Gans, "A Theory of Visionary Disruption," Working paper, University of Toronto, 2022; Ajay K. Agrawal, Joshua S. Gans, and Scott Stern, "Enabling Entrepreneurial Choice," *Management Science* 67, 9 (2021): 5510–5524.

15. On disk drives, see Clay Christensen, *Innovators' Dilemma* (New York: Harper Collins, 1997). This research emphasizes the systematic differences in assessing new product launches with low capabilities at launch but high potential to gain capabilities over time.

16. On photolithographic equipment, see Rebecca Henderson, "Underinvestment and Incompetence as Responses to Radical Innovation: Evidence from the Photolithographic Alignment Equipment Industry," *Rand* 24 (1993): 248–270. This research emphasizes the systematic differences in incentives for product launches due to cannibalization and differentiated product strategies.

17. Urs von Burg, *The Triumph of Ethernet: Technological Communities and the Battle for the LAN Standard* (Stanford, CA: Stanford University Press, 2002). In this analysis, the views of the community of researchers and experimenting firms played a critical role in determining the consensus about the superior technical approach for resolving issues in local area networking.

18. Steven Klepper, "Disagreements, Spinoffs, and the Evolution of Detroit as the Capital of the U.S. Automobile Industry," *Management Science* 53, 4 (2007): 616–631; Steven Klepper, *Experimental Capitalism: The Nanoeconomics of American High-Tech Industries* (Princeton, NJ: Princeton University Press, 2016); Steven Klepper and Peter Thompson, "Disagreements and Intra-industry Spinoffs," *International Journal of Industrial Organization* 28, 5 (September 2010): 526–538. In the latter approach, the formation of new firms originates from disagreements about the direction of engineering experiments to support product development. This chapter characterizes the disagreements of the

descendants from Shockley Instruments as a series of "confrontations between insiders with distinct visions."

19. Shane Greenstein, *How the Internet Became Commercial: Innovation, Privatization, and the Birth of a New Network* (Princeton, NJ: Princeton University Press, 2015). The analysis stresses the interplay between government regulation of telephone companies, which protected bulletin board providers, and the earliest commercial Internet Service Providers, who gain from the boost in demand from the privatization of the Internet and the development of the World Wide Web.

20. Shane Greenstein, Rebecca Frazzano, and Evan Meagher, "The Triumph of the Commons: Wikia and the Commercialization of Open Source Communities in 2009," Harvard Business School Case KEL 464-PDF-ENG, 2009.

21. Shane Greenstein, "Reference Wars: Encyclopedia Britannica's Decline and Encarta Emergence," *Strategic Management Journal* 38, 5 (2016): 995–1017.

22. Shane Greenstein, Yuan Gu, and Feng Zhu, "Ideology and Composition Among Online Collaborators: Evidence From Wikipedians," *Management Science* 67, 5 (2021): 3067–3086.

23. Sam Ransbotham and Gerald Kane, "Membership Turnover and Collaboration Success in Online Communities: Explaining Rises and Falls from Grace in Wikipedia," *MIS Quarterly* 9, 1 (2011): 613–627.

24. Benjamin Mako Hill, "Almost Wikipedia: What Eight Early Online Collaborative Encyclopedia Projects Reveal about the Mechanisms of Collective Action," in *Essays on Volunteer Mobilization in Peer Production*, PhD dissertation, Cambridge, MA: Massachusetts Institute of Technology, 2013, https://mako.cc/academic/hill-almost_wikipedia-DRAFT.pdf

25. That conclusion is further reinforced by the experience of Larry Sanger, who had been the person who initially suggested using a Wiki. He grew frustrated at Wales's approach to decentralized control over the content, arguing that it led to too much low-quality content. Sanger left the project early and became one of its most vocal critics, harping on the lack of mechanisms to ensure high quality.

26. Between 1993 and 1996, Encyclopedia Britannica lost half its sales. See Greenstein (2016).

27. Randall Stross, *The Microsoft Way* (Reading, MA: Addison-Wesley Publishing Company, 1996); Greenstein (2016).

28. Even after the success of Encarta, others at Microsoft continued to question why it belonged in the firm and was within scope. See Greenstein (2016) for a description.

29. Greenstein (2016).

30. Bartholomew stated that the video of Arafat and Rabin shaking hands on the White House lawn was regarded as "revolutionary to get it in there."

31. Robert Seamans and Feng Zhu, "Responses to Entry in Multi-Sided Markets: The Impact of Craigslist on Local Newspapers," *Management Science* 60, 2 (2014): 476–493.

32. See https://www.startups.com/library/expert-advice/interview-with-craig-newmark-founder-of-craigslist- org, accessed August 2022.

33. Many services attempted to take share from Craigslist, and a variety of them did take minor amounts with high-margin service and additional features. These

services—such as dog-walking, home repair services, and vacation rentals—grew the market size for matching services, relegating Craigslist's role to listing essential services. However, until the emergence of Facebook Marketplace, none had taken substantial market share from Craigslist. That is an astonishing achievement.

34. https://www.startups.com/library/expert-advice/interview-with-craig-newmark-founder-of-craigslist-org, accessed August 2022.

35. As Craig Newmark recounted numerous times, his organization was contacted frequently by others with ideas and suggestions for opportunities to make changes that would generate additional revenue. However, such opportunities typically departed from the policy of charging nothing, so the organization refused them. In strategic terms, they were interested in something other than pursuing a high-cost, high-quality service.

36. For a longer description, see Raviv Murciano-Goroff, Ran Zhuo, and Shane Greenstein, "Unsung Software and Veiled Value Creation: Illustrations from Server Software," *Research Policy* 50 (2021): 1–31.

37. In the conventional narrative, the founders do not start their innovation journey by taking on the market's core. Instead, they target emerging segments of limited interest to the incumbent. Second, they enter with a value proposition (i.e., a business model) that differs significantly from the mainstream one. Third, as these entrants experiment with their offerings, they learn and grow resources that enhance their viability.

38. Greenstein (2015), chapter 13.

39. Isaacson (2014).

40. With the benefit of hindsight, many Monday-morning quarterbacks have wondered if the university needed to articulate why Stanford asked for so much money or if the firms had not appreciated what they were offered. In either case, it would be an understatement to notice that Page and Brin were lucky to experience such a disappointing outcome.

41. Bechtolsheim eventually invested millions of dollars in Google and became one of its largest non-employee stockholders.

42. Isaacson (2014).

43. Erik Schmidt and Jonathan Rosenberg, with Alan Eagle, *How Google* Works (New York: Grand Central Publishing, 2014).

44. Greenstein (2015), chapter 13.

45. Steven Levy, *Facebook: The Inside* Story (New York: Penguin Random House, 2020). In retrospect, it is impossible to say whether Zuckerberg spread his product from campus to campus because he had no alternative path available or because he knew how to do one thing well and did it until he ran out of campuses. As it turned out, it worked well for a young firm with limited resources that anticipated competing with an established incumbent.

46. Levy (2020).

47. Sheryl Sandberg, *Lean In: Women, Work, and the Will to Lead* (New York: Knopf, 2013).

48. Frank Fisher, John McGowan, and Joen Greenwood, *Folded, Spindled, and Mutilated: Economic Analysis and U.S. v. IBM* (Cambridge, MA: MIT Press, 1983).

49. Watson and Petre (1990).

50. Emerson Pugh, *Memories that Shaped an Industry: Decisions Leading to the IBM System 360* (Cambridge, MA: MIT Press, 1984).

51. Pugh (1984).

52. See also Greenstein (2020) for an elaboration of the competitive model that supports this archetype.

53. See Neil Gandal, Shane Greenstein, and David Salant, "Adoptions and Orphans in the Early Microcomputer Market: Chickens and Eggs, Hardware and Software," *Journal of Industrial Economics* 47, 1 (March 1999), 87–105.

54. James Cortada, *IBM: The Rise and Fall and Reinvention of a Global Icon* (Cambridge, MA: MIT Press, 2019); and Timothy Bresnahan, Shane Greenstein, and Rebecca Henderson, "Schumpeterian Economies and Diseconomies of Scope: Illustrations from the Histories of IBM and Microsoft," pp. 203–276 in Josh Lerner and Scott Stern, eds., *The Rate and Direction of Technical Change, 50th Anniversary* (Chicago: University of Chicago Press, 2012).

55. Bresnahan et al. (2012).

56. Cortada (2019).

57. Gandal et al. (1999).

58. Michael Cusumano and David Yoffie, *Competing on Internet Time: Lessons from Netscape and Its Battle with Microsoft* (New York: Free Press, 1998); Greenstein (2015), chapter 6.

59. In contrast, his research partner at the University of Illinois, Eric Bina, developed the web server to work with it. Apache, previously mentioned, eventually descended from Bina's efforts.

60. The VC who funded Netscape, John Doerr, was fond of describing how he expected a child to walk into his office with a revolutionary piece of software. Likewise, he was fond of describing his first meeting with Andreesen in such tones.

61. Chapman, drawing on various archetypes from sporting rivalry, blames Andreesen for using too much locker-room confrontational language in his public remarks and undermining his firm's competitive position by motivating Microsoft's employees to deliver a comeuppance (Merrill Chapman, *In Search of Stupidity: Over Twenty Years of High Tech Marketing Disasters* [Berkeley, CA: Apress, 2004]).

62. That reputation certainly mattered. Clark's business plan was remarkably minimal because of the esteem in which his funders held him. Immediately after its founding, the new venture hired every University of Illinois programmer involved in the Mosaic effort that aided the development speed and prevented any other competitor from doing the same: Greenstein (2015), chapter 4.

63. Greenstein (2015), chapter 6.

64. For example, Cusumano and Yoffie (1997).

65. Greenstein (2015), chapter 11.

66. "Wireless Router Market Revenue Worldwide," Statista, accessed June 2023).

67. These events also fit Jobs's self-conception after his experience at Pixar as an executive who stood between different markets and could act as an intermediary between them (Isaacson [2011]).

68. Wolter Lemstra, Vic Hayes, and John Groenewegen, *The Innovation Journey of Wi-Fi: The Road to Global Success* (Cambridge: Cambridge University Press, 2011), chapter 4, page 131. Links says Michael Dell "was furious that he had been beaten by Apple." Lucent executives had to remind Michael Dell that he had an opportunity to participate in discussions as early as 1992. However, Dell decided in 1993 to stop the talks because he concluded (incorrectly, as it turned out) that there was no market for the technology.

69. The delays did not hurt Dell much because the Apple and IBM-compatible PCs were heavily differentiated, and Apple's market size was small then.

70. The key events included the development of a certification process, a rebranding (i.e., from 802.11 to Wi-Fi), and the entry of many additional providers of Wi-Fi. The supply chain for Wi-Fi also underwent a restructuring a couple of years later when Intel designed a branded motherboard and included the 802.11 standard, calling it "Centrino." For analyses, see Roberto Fontana and Shane Greenstein, "Platform Leadership and Supply Chains: Intel, Centrino, and the Restructuring of Wi-Fi Supply," *Journal of Economic Management Strategy.* 30, 2 (2021), pp. 259–286.

71. Apple shipped almost six million iPhones before opening the App Store in mid-2008. Shipments skyrocketed in the third quarter of 2008 when Apple shipped nearly five million iPhones worldwide. See Daniel Garcia-Swartz and Martin Campbell-Kelly, *Cellular: An Economic and Business History of the International Mobile-Phone Industry* (Cambridge: MIT Press, 2022).

72. Garcia-Swartz and Campbell-Kelly, 2022.

73. The one exception was a joint venture with Motorola in 2005. They produced a "music phone," the ROKR, but the product failed to gain many customers. See Garcia-Swartz and Campbell-Kelly (2022), pp. 273–275.

74. Contrast the Blackberry and the Apple iPhone in a variety of features: from keyboard to touch keying; from small screen to prominent color; from minimizing bandwidth to allowing high bandwidth applications; from minimal web presence to high web support; from data-center to cloud-based support; from deep-carrier cooperation to tension with carriers; from protected e-mail to data plan for surfing. Those are significant differences. See Timothy Bresnahan and Shane Greenstein, "Mobile Computing: The Next Platform Rivalry," *American Economic Review, Papers, and Proceedings*, 104, 5 (May 2014): 475–480.

75. In other aspects of its business Apple had fruitful partnerships. For example, it had plenty of such partners in the inexpensive assembly of devices in Asia, notably a partnership with Foxconn.

76. The iPhone was introduced in June 2007 and sold 270,000 units in two days. It is generally believed that it sold a million units by September.

77. Product differentiation is a term from industrial organization. Businesses invest in processes and designs to distinguish their products from anticipated rivals, aiming to gain a competitive advantage and define unique value propositions to appeal to buyers.

78. Shane Greenstein, "An Archetype for Outsiders in Technology Commercialization," pp. 137–172 in William Aspray, ed., *Historical Studies in Computing, Information, and Society: Insights from the Flatiron Lectures* (Cham, Switzerland: Springer, 2020).

CHAPTER 8

Writing the Economic History of Computing

Daniel D. Garcia-Swartz and Martin Campbell-Kelly

THE ECONOMIC HISTORY OF COMPUTING

WHAT DOES IT MEAN TO APPROACH THE HISTORY OF COMPUTING—AND of high-technology industries more generally—from an economic-history perspective?[1] In this chapter, we argue that it does not involve merely accumulating business and economic facts from the past. Rather, economic history works with economic "models," which are somewhat simplified frameworks that provide an interpretation for what we observe in the real world. Although no model can capture all the details of the real world, a good model will capture the essential features of the phenomenon we are trying to explain. We share the Box-Draper view of models: all models are wrong, but some are useful.[2]

Second, the economic history of computing attempts to measure and quantify. But the quantification that matters is not disconnected from economic models—it is rather inspired by them. Doing economic history is not an attempt at measuring everything—when everything is relevant, nothing is relevant. Economic history quantifies the patterns that are important from the perspective of the models that inspire our research.[3]

Third, the economic history of computing explores the role played by institutions, organizations, and policies. Doing economic history is more than writing a collection of equations, finding the "equilibrium" of the model, and checking whether its predictions are borne out by the facts, although that may play an important role in certain contexts and for certain audiences. Writing the economic history of computing involves examining how different institutional and policy regimes have shaped outcomes in the real world—how such regimes can help explain, for example, why a domestic computer industry developed in certain countries but not in others.

Finally, the economic history of computing compares outcomes across industries, regions, and countries to identify causes and effects. This approach entails comparing, say, countries that were similar in several meaningful ways but nonetheless generated different outcomes; and then trying to identify the one factor—or set of factors—that varied across countries and that accounts for the differential outcomes.[4]

The chapter explores several economic themes including increasing returns to scale and transportation costs, the role of policies and institutions, knowledge spillovers, and the differences between closed and open standards. The historical applications include computer timesharing, the international computer industry in its early years, the genesis of information-technology geographic clusters, and several open and proprietary standards in the mobile-phone industry.

Several of the issues discussed in the chapter highlight the overlap between the history of information and the history of high-technology industries. For example, many economists have argued that informal exchanges of information facilitated by economic proximity—what social scientists label "knowledge spillovers"—explain the origins of high-technology geographic clusters. Further, the history of computing and the history of mobile phones have witnessed multiple battles between open and closed standards—and what distinguishes open from closed standards is precisely the amount of intellectual property (or information) that is made publicly available free of charge or at a reasonable price.

Economies of Scale in the Production of Computer Performance: The Computer Timesharing Industry

In economics, the concept of "returns to scale" refers to what happens to output when all inputs are increased proportionately. For example, if all inputs double, will output also double? If output doubles when all inputs are doubled, then there are constant returns to scale. If output less than doubles, there are decreasing returns to scale. If output more than doubles, there are increasing returns to scale.[5] When there are increasing returns to scale, average production costs fall as output expands, and the production process is said to benefit from economies of scale.

Several factors explain why a firm's average costs may decline as output rises. For one, there may be costs that do not vary with the level of output. These are usually referred to as fixed setup costs and are incurred by firms that must make initial investments to be able to produce anything at all. Further, division of labor and specialization become easier as output expands, bringing about potential gains in productive efficiency.[6]

In the history of the computer industry, computer timesharing provides a nice example of the role of economies of scale in the production of computer performance.[7] In the United States, computer timesharing started in the early to mid-1960s as a collection of academic experiments that allowed many users to interact simultaneously with a powerful mainframe, creating the illusion that each user was working on his or her own machine. Soon, private corporations perceived that there was money to be made in timesharing, and a "timesharing industry" started to develop around 1965. Table 8.1 provides an overview of the growth of the computer industry in the United States between 1956 and 1978. The table captures the impressive growth of the timesharing industry (labelled "interactive" in the table) between 1965 and 1978.

Table 8.1 reveals two important facts. First, in the first three decades of the computer industry, revenues from computer services and software grew much faster than revenues from computer systems. If, in 1956, software and services accounted for about 11 percent of industry revenue, by 1978 they accounted for almost 43 percent. Second, and more importantly, timesharing grew at a faster pace than the overall software and services sector. Timesharing, which barely existed in 1965, accounted for

Table 8.1. Total revenues from computer systems, services, and software in the United States, 1956–1978, in USD billion.

	SYSTEMS	SOFTWARE AND SERVICES						TOTAL
Year	System shipments	Traditional batch	Remote batch	Interactive	Software	Other	Total software and services	Total industry
1956	0.155	0.020					0.020	0.175
1957	0.245	0.025					0.025	0.270
1958	0.395	0.040					0.040	0.435
1959	0.495	0.090					0.090	0.585
1960	0.590	0.125					0.125	0.715
1961	0.880	0.180					0.180	1.060
1962	1.090	0.220					0.220	1.310
1963	1.300	0.260		0.005	0.005		0.270	1.570
1964	1.670	0.285		0.010	0.020	0.002	0.317	1.987
1965	2.060	0.340		0.015	0.050	0.005	0.410	2.470
1966	3.330	0.410		0.020	0.100	0.010	0.540	3.870
1967	4.030	0.480		0.050	0.175	0.030	0.735	4.765
1968	4.835	0.600	0.010	0.110	0.270	0.050	1.040	5.875
1969	4.919	0.740	0.050	0.160	0.340	0.080	1.370	6.289
1970	4.355	0.945	0.087	0.243	0.375	0.110	1.760	6.115
1971	4.275	1.075	0.115	0.335	0.415	0.145	2.085	6.360
1972	5.620	1.235	0.145	0.440	0.470	0.190	2.480	8.100
1973	5.945	1.405	0.205	0.630	0.605	0.255	3.100	9.045
1974	7.030	1.580	0.280	0.825	0.830	0.335	3.850	10.880
1975	6.663	1.740	0.350	1.025	1.090	0.415	4.620	11.283
1976	7.079	1.860	0.565	1.160	1.375	0.495	5.455	12.534
1977	8.913	1.935	0.840	1.410	1.730	0.575	6.490	15.403
1978	10.313	2.100	1.077	1.738	2.110	0.660	7.685	17.998

almost 10 percent of total computer-industry revenues by 1978. This was a remarkable achievement for an industry that is barely mentioned in the books on the history of computing.

Economists would argue that, if timesharing grew as it did, it must have made economic sense—and, in fact, it did. Timesharing grew as it did because of Grosch's Law.[8] In the early years of the computer industry, the pundit Herbert Grosch asserted that the average cost of computer equipment—or more precisely, computer performance—decreased as the computer's size and performance increased. Although Grosch's "law" was never published, it was usually expressed in the following form:

$$C = K P^{0.5} \qquad (1)$$

Here, C stands for a computer's monthly rental, P for the system's performance, and K is a constant. Alternatively, the "law" was expressed as:

$$P = A C^{2} \qquad (2)$$

Here, A is $\frac{1}{K^2}$. To put it simply, if we have two computer systems, S_1 and S_2, and S_2 rents for twice as much as S_1, then S_2 should deliver performance proportional to four times the performance of S_1. If S_2 rents for four times as much as S_1, then S_2 should deliver performance proportional to sixteen times the performance of S_1.

Grosch's Law says that there are economies of scale (or increasing returns to scale) in the production of computer performance. Note that equation (1) can also be expressed in terms of the average cost of computer performance:

$$\frac{C}{P} = K \frac{1}{P^{0.5}} \qquad (3)$$

Equation (3) says that, if Grosch's Law holds, the average cost of generating computer performance declines as the size (and performance) of computer systems increase. In a 1966 article, economist Martin Solomon showed that the "law" held reasonably well in the real world.[9] Solomon collected data on computer performance and computer monthly rentals for several IBM/360 models. He focused on the IBM family of systems introduced in the mid-1960s so that he could explore the relationship between performance and cost *for a given level of technology*. He estimated several regression models and interpreted his results as confirming the validity of the law.

Figure 8.1 shows the cost per million executions of a matrix-multiplication program for various IBM/360 systems, from the smallest (360/30) to the largest (360/75).

Figure 8.1 shows the essence of Grosch's Law: the average cost of performance declined as the size (and performance) of the system grew. Grosch's Law was important for timesharing because, by relying on large systems, timesharing companies could deliver levels of performance to individual users at a cost that was just a fraction of what users would have paid for the same performance had they rented their own (smaller) system. Table 8.2 provides an example.

For the five System/360 models studied by Solomon, the table shows the monthly rental and the performance (in millions of operations per second). We progressively raise the number of users starting at one for the smallest system (360/30) to keep the performance delivered to each

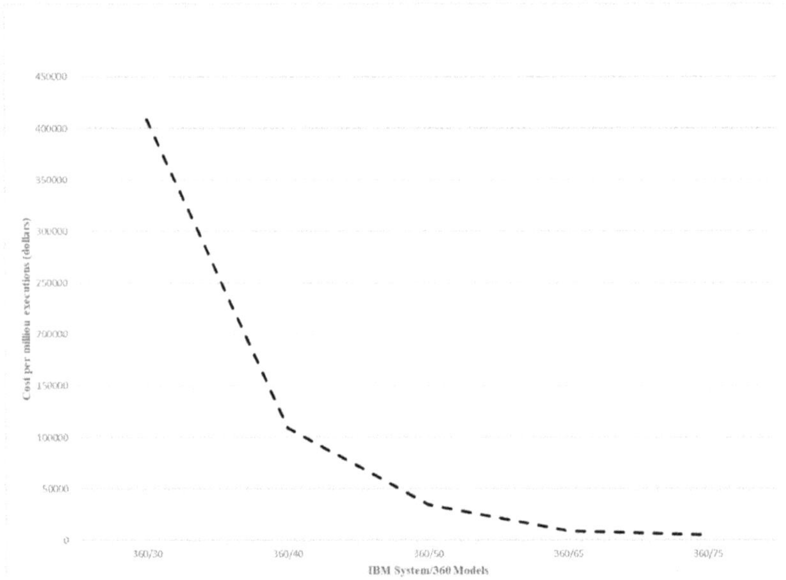

Figure 8.1. Cost per million executions of matrix-multiplication program for various IBM 360 systems (constructed by authors from data in M. Solomon, "Economies of Scale and the IBM System/360," Communications for the ACM 9:6 (June 1966), 435–440)

Table 8.2. Cost and performance of various IBM/360 models: monthly rental and million operations per second (matrix multiplication)

System	Monthly rental ($000)	Number of users	Performance (million ops per second)	Performance per user	Cost per user
IBM 360/30	8.00	1.00	2.45	2.45	8.00
IBM 360/40	17.00	3.74	9.14	2.45	4.55
IBM 360/50	32.00	11.91	29.13	2.45	2.69
IBM 360/65	50.00	47.66	116.59	2.45	1.05
IBM 360/75	80.00	85.96	210.26	2.45	0.93

user constant. If there were constant returns to scale in the production of computer performance, the cost per user would remain constant as we move up to larger systems. But it does not—the cost per user declines almost by a factor of ten from the smallest (360/30) to the largest (360/75) system. Timesharing made a lot of sense—and it continued making sense until the early 1980s, when the personal computer transformed the economics of computing.

Economies of Scale and Transportation Costs: Why a Domestic Computer Industry Developed Only in a Few Countries in the 1950s and 1960s

Economies of scale and transportation costs may explain why the early computer industry developed precisely where it did. Economic theory suggests that, when a firm is deciding how much output to produce, only marginal considerations matter: a firm produces a level of output such that, at that level, the revenue the firm obtains from an additional unit (marginal revenue) equals the cost of producing that unit (marginal cost). Fixed costs should play no role in this decision.

This is true, however, only after a firm has decided that it will produce any output at all. Fixed costs do matter for the decision whether to produce anything at all or not. If a firm has to make substantial investments in, say, research and development (R&D) before it can produce even a single unit of output, then the firm's expected profits over variable costs have to be large enough to cover the costs of the investments in the long run. If they are not, no investment will be made.[10]

Here is where the size of the market for a product plays an important role. In an industry in which fixed costs—and, equivalently, economies of scale—are substantial, a firm will only set up production plants in regions or countries that offer large markets for the industry's output. And if transportation costs matter enough, the size of the market for the firm's product may be determined mostly by the size of the domestic market.

The International Location of the Early Computer Industry

Studies of the American computer industry by Gerald Brock and of the British computer industry by Paul Stoneman suggest that fixed costs—and economies of scale—in the production of computer systems were sizable in the early decades of the industry. Although there were modest economies of scale in the manufacturing process itself, economies of scale in the early computer industry arose mainly from the large investments in R&D required to produce large computer mainframes. As the years went by, large investments in software creation for computer mainframes also generated scale economies.[11] There were economies of scale in the sense that the average cost of producing computer systems (of a given performance) declined as the number of systems produced increased. (As explained earlier, there were also economies of scale in producing computer performance, but here we interpret output as the number of computer systems of a given performance level, to keep things simple.)

Figure 8.2 shows the largest markets in the world in 1950 ranked by gross domestic product (GDP) in international dollars. Note that the size of the domestic market is determined by both population and GDP per capita, and GDP is a combination of both.

An economic model that emphasizes economies of scale and transportation costs appears to explain the facts quite well: during the 1950s, a domestic computer industry developed only in the United States, Britain, Germany, France, and Japan; and these were among the countries with the largest domestic markets in the world at the time. (There were some computer developments in other European countries, but they were more limited.)

Further, Figure 8.2 shows that, in 1950, the United States had, by far, the largest domestic market in the world. This helps explain why the domestic computer industry that developed in the United States had no equal. American computer firms were able to penetrate foreign markets and capture large shares of those markets, but foreign firms were unsuccessful in the American market.

Another piece of evidence confirms the importance of large domestic markets for the development of the early computer industry: when

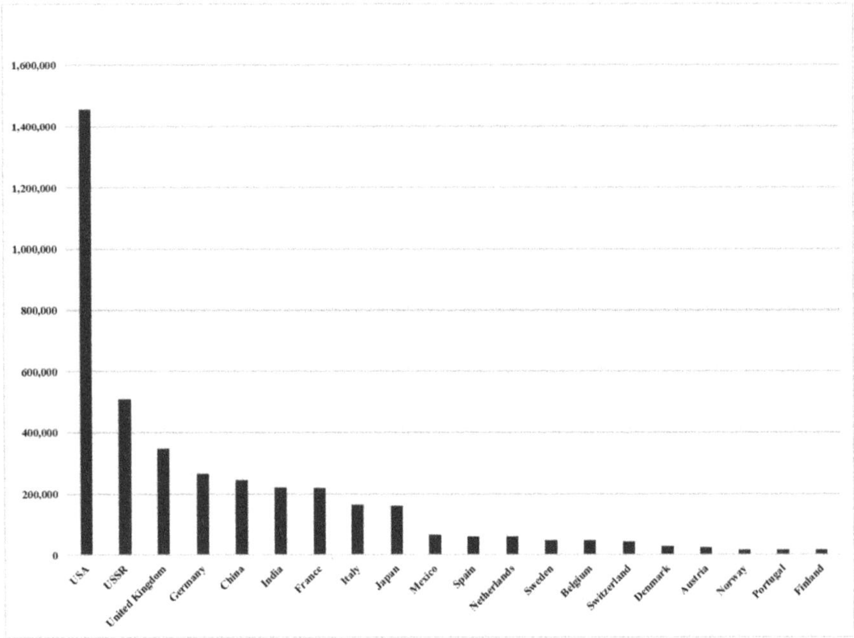

Figure 8.2. GDP in 1950, million 1990 international $ (Source: A. Maddison, Contours of the World Economy, 1-2030 AD (Oxford: Oxford University Press, 2007), p. 379)

companies such as IBM set up production plants abroad, they set them up only in the countries that had the largest domestic markets. IBM, for example, had sales offices all over the world but had production plants for computer mainframes only in Britain, France, Germany, Italy, and Japan.[12]

Potential Objections

A theory that explains the international localization of the early computer industry as driven by the size of the domestic market raises some potential objections.

The Pre-computer Business-Machinery Industry

The first is that there was a business-machinery industry in the countries where a computer industry developed, and that the business-machines industry had been around for decades before the advent of computers.

Would the presence of companies such as IBM, Burroughs, and NCR in the pre-computer American market—and of similar companies in Britain and France—not be a better explanation for the rise of a computer industry in the 1950s? Our response is that this is not really an "explanation" for the rise of a computer industry in certain countries. This objection simply raises the question of why a pre-computer business-machinery industry developed *precisely in those countries*, and here again the answer likely has something to do with the size of the domestic market for business machines.

The Rise of the Asian Tigers in the 1980s

A second objection is that, in later decades, the computer industry became globalized and small countries such as Taiwan and Hong Kong became important exporters of computer products. This objection simply confirms our theory. Countries such as Taiwan developed their own computer industry not in the 1950s, when mainframes were dominant, but in the 1980s, once personal computers had come to the forefront.[13]

The Role of Transportation Costs

A third, and stronger, objection is that transportation costs did not really matter in the early computer industry. In a comprehensive study of the (interregional and international) location of the early computer industry published in 1979, Servet Mutlu argued that the mainframe computer industry was an industry with a high value-to-weight ratio and that, consequently, transportation costs did not matter for location purposes, at least when computer vendors chose where to locate their plants within the United States. The idea is that, because a computer mainframe was a product that had a high dollar value per pound of weight, transportation costs added relatively little to the price of the product delivered at its final destination. Thus, the cost of transporting the mainframe from the manufacturing plant to the user did not influence the choice of location for the plant. Other studies have indeed confirmed that computers were, as of 1980, among the industries with high value-to-weight ratios (and low ad valorem air freight charges).[14]

That transportation costs may have played at most a small role in the regional localization of early mainframe manufacturers in the United States, however, does not imply that they played no role at all in the international development of the industry. Mutlu himself argues that, when IBM started expanding abroad in the pre-computer era, it did so to overcome tariff barriers and to save on transportation costs. Further, he points out that the size of foreign markets determined the location of mainframe manufacturing plants abroad. Regarding the expansion of IBM abroad, Mutlu writes: "Starting with the biggest markets, all the important markets came to have computer manufacturing facilities. . . . The size of the market was the most important consideration behind the location of new manufacturing facilities." The size of the domestic market mattered, he adds, because it "permitted production at or near optimum levels of plant, for most of the products."[15]

Just as the size of each domestic market mattered when American mainframe manufacturers located their plants abroad, it also mattered for the development of a domestic mainframe computer industry in a select group of countries—precisely those with the largest domestic markets. The argument that the size of the domestic market mattered suggests, first, that economies of scale mattered and, second, that transportation costs were not completely inconsequential.[16]

Large Markets without a Domestic Computer Industry

The fourth, and final, objection is the most interesting: although the USSR, China, and India were very large markets in the 1950s, no domestic computer industry developed in these countries. How are we to explain this fact? We address this question in the next section of this chapter.

Institutional Regimes and Economic Policies: Why a Domestic Computer Industry Failed to Develop in the USSR, China, and India in the 1950s and 1960s

The fact that, in the 1950s, no domestic computer industry developed in large countries such as the Soviet Union, China, and India suggests that a large domestic market was not enough to "generate" a domestic computer industry. The economic model "fails," but it is at least as useful when it fails as when it succeeds. When it fails, it suggests that factors other than the size of the domestic market mattered as well. Economist Douglass North was awarded the Nobel Prize in Economics for his study of the role of institutions in economic history.[17] In line with North's ideas, one could argue that, during the 1950s and 1960s, a native computer industry did not grow in all countries that had very large domestic markets but only in those that *also* had the "right" institutional regimes and economic policies.

But before focusing on institutions, we need to address two simple questions: Did countries such as the USSR, China, and India have the computer know-how in the 1950s and 1960s that could have laid the foundation for the growth of a domestic computer industry? In addition, were those countries wealthy enough (in per-capita income terms) to fuel demand for computers and foster the growth of a domestic computer industry?

The Issue of Know-how

Lack of computer know-how does not explain the absence of a domestic computer industry in the Soviet Union, China, and India during the 1950s and 1960s. There were several centers of computer research in the USSR in the early 1950s. In one of them, Sergei Lebedev and his team developed the BESM-1 computer, which became operational in 1953. BESM-1 was better than UNIVAC 1, according to certain performance measures, although not as good as Whirlwind I. Scholars of Soviet computers suggest that, in 1953, the USSR was barely two years behind the United States in computer technology.[18]

In the second half of the 1950s, the Soviets transferred computer know-how to the Chinese. With support from Soviet experts, the Chinese built two computers: Model 103, operational in 1958, and Model 104, operational in 1959. Model 104 was based on the Soviet BESM-2 and was described as the fastest computer in Asia at the time.[19]

Table 8.3. Advanced technologies in the United States, the USSR, Britain, France, Japan, and China

	United States	USSR	Britain	France	Japan	China
First nuclear reactor	1942	1946	1947	1948		1956
First atomic bomb	1945	1949	1952	1960		1964
First hydrogen bomb	1952	1953	1957	1968		1967
First jet plane	1942	1945	1941	1946		1958
First satellite	1958	1957		1965	1970	1970
First prototype computer	1946	1953	1949		1957	1958
First commercial computer	1951	1958	1952		1959	1966

During the 1950s, computer research in India was concentrated in two institutes, one headed by the statistician Prasanta Mahalanobis and the other by the nuclear physicist Homi Bhabha. Based on technical documentation for the ORDVAC and ILLIAC computers developed at the University of Illinois, the center headed by Bhabha built the first Indian computer, the TIFRAC, operational in 1959.[20]

More generally, Table 8.3 below shows that countries such as the USSR and China managed to develop a panoply of advanced technologies under Socialism, albeit with a lag *vis-à-vis* the West.

In short, the USSR was barely behind the United States in computer technology at the starting point. Further, although both China and India started with a lag of several years *vis-à-vis* American computer technology, the gap did not appear insurmountable. The absence of a domestic computer industry in the USSR, China, and India was not due to lack of computer know-how (or to lack of scientific and technical knowledge more generally).

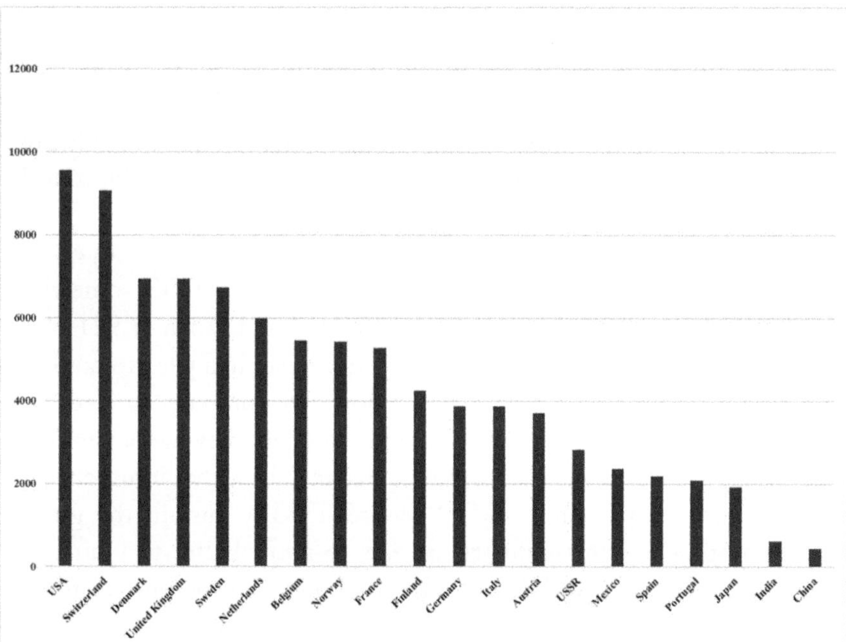

Figure 8.3. Per capita GDP in 1950 international $ (Source: A. Maddison, Contours of the World Economy, 1-2030 AD (Oxford: Oxford University Press, 2007), p. 381)

The Issue of Per-capita Income

Another potential objection to an institutional explanation is that the USSR, China, and India did not have per-capita incomes high enough to fuel the demand for computers and foster the development of a domestic computer industry. Figure 8.3 shows per capita GDP in international dollars in 1950.

The figure shows that four of the five countries where a domestic computer industry developed—the United States, Britain, France, and Germany—ranked very high not only in total GDP but also in per capita GDP. By contrast, two of the three countries we are focusing on here—China and India—ranked at the very bottom in per capita GDP. Critics could argue that, although the USSR, China, and India had large populations (and thus large domestic markets) in 1950, their economies were too poor and immature to sustain the growth of a domestic computer industry.

The idea that per capita GDP in the early 1950s was too low to sustain a domestic computer industry does not apply to the USSR. The Bolsheviks won the civil war that followed the revolution of 1917. In 1928, the First Five-Year Plan set in motion a process of "big push" industrialization. Economic historian Robert Allen has shown that, if the decade of the 1940s is excluded, per capita GDP in the USSR grew by 4.5 percent per year between 1928 and 1970, a remarkable rate by historical standards.[21]

Angus Maddison estimated that in 1950 the USSR had a per capita GDP of 2,841 international dollars. This put the USSR somewhere between Germany (3,881 international dollars) and Japan (1,921 international dollars) in 1950. Both Germany and Japan had per capita GDPs high enough to sustain the growth of a domestic computer industry.

China and India are different because, in 1950, they had a per capita GDP that was barely a fourth that of Japan. Thus, the argument that a certain level of per capita GDP was required to sustain the growth of a domestic computer industry appears to apply better to China and India than to the USSR. What we know about the economic history of China is that the level of economic development (and per capita GDP) in 1950 varied considerably by region: while many inland regions were mainly agricultural areas where people lived barely above subsistence levels, some areas on the eastern coast were reasonably mature economies with a substantial level of manufacturing development.[22] Substantial differences across regions existed in India as well.[23]

Although there is no room in a short chapter such as this one to develop these ideas any further, we should at least entertain the possibility that regions of China and India were economically mature in the early 1950s and might have supported the growth of a domestic computer industry.[24] This takes us back to the hypothesis that a domestic computer industry failed to develop in the USSR, China, and India due to institutions, organizations, and policies.

Institutional Regimes: Computers in the USSR

Between the late 1920s and the late 1970s, the USSR underwent a process of rapid industrialization, to the point that, by the 1980s, it produced about 15 percent of the world's manufacturing output.[25] However, although there were important computer developments in the Soviet Union during the 1950s and 1960s, no domestic computer industry grew in the country. Several factors explain this fact.[26] For one, in countries such as the United States, private corporations entered the computer industry in the 1950s; in the Soviet Union, by contrast, computers were designed and manufactured by scientists and engineers working for government ministries.

Further, although computers were originally conceived—both in the West and in the Soviet Union—as tools for performing high-speed numerical calculations (mainly for military purposes), in the West corporate data-processing needs soon became a key source of computer demand. In the Soviet Union, by contrast, commercial data-processing considerations did not fuel computer demand in the early years. In addition, because Soviet computers were developed by government agencies exclusively for military purposes during the 1950s, the designers and manufacturers of computers in the Soviet Union did not benefit from the kind of customer innovation, especially in software, that was a key driver of computer developments in the West.

Moreover, while innovation in computers in the West was driven also by intense competition among private corporations that constantly attempted to beat their competitors in price-performance, such competition was non-existent in the Soviet Union. In the West, private corporations made substantial efforts to acquire and maintain government and corporate customers, whereas in the Soviet Union, by contrast, lists of suppliers and customers were provided to computer manufacturers by central planners—and manufacturers were mostly focused on meeting CPU production quotas. Finally, computers in the West developed as "systems"—as collections of hardware, software, and services that interacted well with one another. In the Soviet Union, by contrast, the focus was almost exclusively on hardware to the detriment of other system components.

Institutional Regimes: Computers in China

In China's centrally planned economy, the organization of R&D activities followed the Soviet model.[27] Unlike what happened in the West, where private industrial corporations carried out their own R&D, in China there was a hierarchy of R&D organizations that was segregated from industrial enterprises. At the top of the hierarchy was the Chinese Academy of Sciences, which oversaw dozens of research institutes and universities. The second tier of the hierarchy was made up by a panoply of military-industrial research labs under the authority of several industrial ministries. The third tier comprised research institutes supervised by the industrial bureaus of regional governments.

From the beginning of the planning system, the Chinese government identified computers as a priority. But rather than allowing, or encouraging, private corporations to experiment with computer technology, the government set up the Institute of Computing Technology within the Chinese Academy of Sciences. This institute built the first two Chinese computers, Model 103, operational in 1958, and Model 104, operational in 1959. These vacuum-tube machines were copies of Soviet computers and were built with blueprints provided by the Soviets. After the breakdown of Sino-Soviet relationships in the early 1960s, the institute developed several generations of computers and became China's top research institute in computer technologies. The first transistorized computers were introduced in 1964–1965, with a lag of six or seven years *vis-à-vis* the United States. The first integrated-circuit computers were introduced in 1973, again with a substantial lag relative to the United States.[28] During the 1950s and 1960s, computers were viewed exclusively as tools for military and scientific applications. It was only in 1973 that large-scale production of computers for industrial and commercial uses was even considered.

More radical change happened only in the 1980s, after a new institutional framework was created: the so-called Science and Technology Enterprises. Like the high-tech ventures that drove technological change in Western economies for decades, these entities focused on turning the scientific and technological knowledge accumulated in the state R&D system into commercial products. The four entities that would become the leading organizations in the Chinese information-technology scene were all Science and Technology Enterprises: Stone, Legend, Founder, and China Great Wall Computer. Except for Great Wall, they were all non-governmental enterprises.

Institutional Regimes: Computers in India

India achieved independence from Great Britain in 1947. After independence, India did not adopt the Soviet model of full state ownership of enterprise but rather a mixed-economy model in which the state reserved certain economic activities for itself and allowed private firms to undertake others. Institutional and policy factors on both the demand and the supply sides prevented the development of a domestic computer industry in India before the 1980s.[29]

Most of the computer installations in India in the 1960s and 1970s were set up in academic institutions and research labs. Of the fourteen large computer systems installed in India in 1964, twelve were in R&D organizations.[30] (In 1964, the United States had almost 20,000 computer installations.[31]) Whereas in the United States, corporate data-processing needs fueled the demand for computers starting in the late 1950s, corporate demand did not take off in India, to a good extent because the first government of Indira Gandhi (1966–1977) imposed heavy regulations and restrictions on customers wishing to acquire computers.

Domestic private corporations were not allowed to enter computer manufacturing until the 1970s. The only domestic entity that manufactured computers in India was the Electronics Corporation of India Limited (ECIL), a government corporation established in 1967. By 1974, there were 224 computer installations in India. IBM accounted for 141 of them, the Indian subsidiary of the British ICL for 25, and ECIL for only 35.[32]

In short, government policies and regulations prevented private corporations from playing the kind of role they played in the West—as customers of the computer industry on the demand side and as vendors of computers on the supply side. By 1977, there were only about a thousand computer installations in India. It was only in the 1980s, during the second tenure of Indira Gandhi as prime minister (1980–1984), that the government changed its policies to encourage the development of a native computer industry. By that time, it was too late for the country to catch up with the pioneers in the manufacturing of computer hardware (although India did become a major exporter of computer services).

Taking Stock: The Role of Policies and Institutions

The literature thus suggests that the absence of private corporations on both the supply and the demand side of the computer market helps explain why a computer industry failed to develop in the Soviet Union, China, and India during the 1950s, 1960s, and 1970s. What did these countries miss because private corporations were absent from the supply side? First, they lacked the economic experimentation that is essential for success in high-technology industries.[33] Second, they lacked the incentives for innovation that come with intense competition among private firms. Third, they lacked the improved quality and falling prices that come about when private corporations compete to win customers.[34]

What did these countries miss because private corporations were absent from the demand side? First, they missed the customer innovation that drove progress in computer technology in countries such as the United States.[35] Second, they missed the feedback from customers that helped computer manufacturers improve existing products and invent new ones.[36] Finally, and perhaps most importantly, they missed the expansion of the market that happened in the West when computers started being used not just for scientific calculations but also for commercial data processing.

This explanation does not attempt to minimize the important role that government entities played in the United States (and elsewhere) as a source of demand for computers and semiconductors in the early years. The point, however, is that, by the late 1950s and early 1960s, civilian (corporate) demand had become the key driver of overall demand for high-technology products in the United States. In the USSR, China, and India, by contrast, the segregation between computer-related R&D and industrial needs continued for decades.

Note how economic models help us understand the origins of the international computer industry. First, the combination of economies of scale and transportation costs led us to focus on the size of domestic markets—and the model seemed to explain well the rise of a domestic industry in the United States, Germany, France, Britain, and Japan. And then, the failure of the model in countries such as the Soviet Union, China, and India led us to focus on the role of institutions and policies.

Economies of Scale, Transportation Costs, Knowledge Spillovers, and Historical Accidents: The Origins of Information-Technology Clusters

In an excellent literature survey, economists Steff Proost and Jacques-François Thisse argue that the essence of spatial economics is the trade-off between various sorts of increasing returns to scale, on the one hand, and various forms of transport costs, on the other.[37] In fact, we have just argued that the combination of scale economies and transportation costs offers a reasonable, albeit incomplete and controversial, explanation for the differential localization of the early computer industry across countries. In this section we focus on the geographic concentration of computer production not *across* countries but *within* them.

Figure 8.4. Computer-hardware establishments by county in 1974.

Figure 8.5. Computer-software establishments by county in 1974.

Geographic Clustering of Computer Hardware and Software

In the early years of the computer industry, the production of computer hardware and software agglomerated not only in specific *countries* but also in specific *regions* within each country.

Figures 8.4 and 8.5 rely on County Business Patterns data to show the concentration of computer hardware and software firms by county in the United States in 1974.

The figures reveal a few fundamental facts about the location of computer hardware and software firms in the early 1970s. First, most counties in the United States had neither computer-hardware nor computer-software establishments. Second, hardware and software establishments were concentrated in a few counties on the East Coast, a few on the West Coast, and a few in several other states, including Illinois, Michigan, Minnesota, and Texas. Third, software establishments were slightly more dispersed than hardware plants, and tended to locate in, or close to, large urban agglomerations.

It could just simply be the case that there were more hardware and software establishments in some counties because those counties were more heavily populated (and they had more establishments for all kinds of industries). To account for this, geographers and economists use measures such as location quotients. The location quotient for an industry *j* in a region *r* is defined as:

$$LQ_j^r = \frac{E_j^r / E^r}{E_j / E} \qquad (4)$$

Here, E_j^r is total employment for industry *j* in region r, E^r is total employment in region *r* for all industries, E_j is total employment for industry *j* in the country as a whole, and E is total employment for all industries in the country as a whole. The location quotient tells us how concentrated industry *j* is in region *r* relative to its concentration in the country as a whole. A location quotient greater than one for an industry in a county suggests that the industry is more heavily concentrated in that county than in the country as a whole. There are more sophisticated measures of an industry's geographic concentration but this one will suffice here.[38] Table 8.4 presents the location quotients for computer hardware in the top 20 counties in the United States between 1970 and 1986. Table 8.5 does the same for computer software between 1974 and 1986.[39]

Table 8.4. Location quotients for computer-hardware establishments, 1970–1986, top 20 counties based on 1970 ranking (counties with at least 3 establishments in 1970). Source: Computations by authors from County Business Patterns data files.

Fipstate	Fipscty	County Name	1970	1978	1986	Establishments in 1970
36	43	Herkimer County, NY	25.35	5.21	7.83	3
8	13	Boulder County, CO	19.35	4.76	7.96	5
36	7	Broome County, NY	18.34	2.59	1.93	7
25	17	Middlesex County, MA	17.38	7.19	7.33	42
6	85	Santa Clara County, CA	16.88	16.87	17.03	31
27	123	Ramsey County, MN	13.98	7.89	3.32	12
24	31	Montgomery County, MD	11.52	5.13	0.94	9
26	161	Washtenaw County, MI	10.99	11.60	6.65	4
12	9	Brevard County, FL	10.77	6.42	2.44	4
6	59	Orange County, CA	10.69	7.21	5.97	24
27	53	Hennepin County, MN	9.91	4.43	3.99	20
51	59	Fairfax County, VA	8.33	3.09	3.03	3
34	7	Camden County, NJ	8.31	2.79	1.25	7
4	13	Maricopa County, AZ	6.40	2.67	1.78	12
42	91	Montgomery County, PA	5.93	2.69	1.42	8
25	21	Norfolk County, MA	4.70	3.66	2.57	5
9	1	Fairfield County, CT	4.59	2.37	1.90	8
36	55	Monroe County, NY	4.22	1.71	1.50	5
6	73	San Diego County, CA	3.57	1.79	3.12	8
6	37	Los Angeles County, CA	3.42	2.62	1.98	51

Table 8.4 shows that computer hardware production was heavily clustered in certain counties in California, New York, and Massachusetts, among other states. California and Massachusetts were the states where the new computer firms were locating in the 1970s, and New York State reflected the persistence of the old stalwarts of the industry, such as IBM. Table 8.5 shows that computer software production was concentrated in

Table 8.5. Location quotients for computer-software establishments, 1974–1986, top 20 counties by 1974 ranking (counties with at least 3 establishments in 1974). Source: Computations by authors from County Business Patterns data files.

Fipstate	Fipscty	County Name	1974	1978	1986	Establishments in 1974
51	13	Arlington County, VA	19.47	14.29	7.47	20
1	89	Madison County, AL	15.84	7.42	4.69	17
51	650	Hampton City, VA	11.34	8.28	2.81	7
51	59	Fairfax County, VA	11.32	9.47	5.60	25
24	31	Montgomery County, MD	9.68	6.96	3.66	35
26	161	Washtenaw County, MI	7.32	4.45	3.64	10
25	17	Middlesex County, MA	7.08	4.23	3.93	64
17	43	Du Page County, IL	5.92	3.63	3.02	21
6	83	Santa Barbara County, CA	5.01	2.92	1.64	11
6	85	Santa Clara County, CA	4.91	3.32	4.47	37
6	113	Yolo County, CA	4.91	0.00	0.37	3
34	7	Camden County, NJ	4.63	2.14	1.38	14
12	9	Brevard County, FL	4.61	1.37	2.18	7
24	33	Prince George's County, MD	4.43	3.94	2.42	13
42	91	Montgomery County, PA	4.02	2.67	1.76	22
6	41	Marin County, CA	3.61	2.25	2.90	6
12	1	Alachua County, FL	3.54	1.18	0.49	3
34	27	Morris County, NJ	3.43	2.49	2.34	9
34	21	Middlesex County, NJ	3.38	2.35	2.49	7
8	41	El Paso County, CO	3.31	2.42	2.19	6

a few counties in Virginia and Maryland, close to Washington, DC, as well as in California and a few other states.

Why did hardware and software firms tend to locate in just a few counties? Why did they locate precisely where they did? Several economic explanations have been offered for this phenomenon.

Internal Economies of Scale and Transportation Costs

The first explanation is one we have already explored. When the production of a good—such as computer hardware or computer software—involves substantial fixed costs, firms will tend to concentrate production in very large plants.

But where will those large plants be located? If transportation costs matter, producers will tend to set up their plants close to large markets for their products. As explained earlier, Mutlu argued in 1979 that transportation costs were irrelevant for computer mainframes because mainframes were high value-to-weight products. But transportation costs likely mattered for computer software, not because "software products" were expensive to move but rather because it was costly to move people and information: in the early years of the industry, selling computer software required heavy interactions between vendors and customers, and thus proximity to customers was highly valued.[40] This would explain why software firms tended to locate close to Washington, DC, where many of their government customers were located. It would also explain why they located in, or close to, large urban agglomerations, where many of their corporate customers were based.

Knowledge Spillovers: Economies of Localization

An economic theory that combines economies of scale that are internal to firms and transportation costs, however, is not the only possible explanation for the origins of industrial clusters—and especially, information-technology clusters. An alternative theory emphasizes economies of scale that are external to firms but internal to the regions where firms tend to agglomerate. Since Alfred Marshall, economists have argued that scale economies that are external to firms—such as knowledge spillovers—are a crucial mechanism favoring the rise of industrial concentrations.

Marshall-style knowledge spillovers are the external benefits that accrue to people from living and working near academic and research centers, to highly skilled workers more generally, and especially to other people working in the same industry. Research and innovation are among the most concentrated activities in the world, and some of the most famous information-technology clusters developed near universities such as Stanford and MIT in the United States and Cambridge in England. In his book on the evolution of cities, economist Edward Glaeser has noted that, even in the era of the Internet, "ideas cross corridors and streets more easily than continents and seas."[41] Marshall-style externalities are frequently labelled economies of localization.

Knowledge Spillovers: Economies of Urbanization

The knowledge spillovers emphasized by Marshall benefit workers belonging to the same industry. The urbanist Jane Jacobs has argued that knowledge spillovers happen not only within the same industry but also, and fundamentally, across industries—and especially across industries that are somehow related to one another.[42] This explains the very existence of cities, which are environments full of diversity.

According to Jacobs, it is precisely such diversity that encourages people in cities to be creative and come up with new solutions to old problems. Whereas within-industry knowledge spillovers are frequently labelled economies of localization, cross-industry spillovers happening in urban environments are frequently labelled "Jacobs-style externalities" or economies of urbanization.[43] Jacobs-style knowledge spillovers would explain why software firms tended to cluster in large urban agglomerations, where diversity increased creativity.

The Knowledge Spillover Theory of Entrepreneurship

Economists Zoltan Acs and David Audretsch have developed a slightly different perspective on knowledge spillovers and their role in the genesis of industrial clusters. They have highlighted the role of knowledge spillovers as the driving force of entrepreneurship and have identified entrepreneurs as the medium through which such spillovers happen.[44] They have argued that, in geographic areas where knowledge is produced intensely, part of that knowledge cannot be fully exploited by the firm or organization that creates it. Knowledge thus spills over, benefiting other firms in the same geographic area.

In this view, entrepreneurs leave large and established companies to found start-ups that will exploit such knowledge spillovers. Thus, in regions where knowledge is heavily produced, we should see plenty of entrepreneurial activity. In the Acs-Audretsch model, firms in the same industry locate close to one another because knowledge-based established firms generate plenty of start-ups, and the start-ups tend to locate close to the parent firms that gave them birth. This applies well, for example, to the early semiconductor industry in Silicon Valley, where the origins of many companies can be traced back to William Shockley and the engineers that left Shockley Semiconductor to set up Fairchild Semiconductor.[45]

The essence of the "knowledge spillover theory of entrepreneurship," as the Acs-Audretsch model is usually called, is frequently conveyed through "genealogical trees," which track the connections between the parent organizations—including private corporations, government research labs, and universities—and the start-ups they gave birth to. Such trees have been constructed for several high-tech clusters in America (Silicon Valley, New England, Austin–San Antonio, San Diego, and Boulder), England (Cambridge), Canada (Calgary), and Denmark (North Jutland).[46]

The knowledge spillover theory of entrepreneurship helps explain not only the development of the semiconductor industry in Silicon Valley but also the early growth of the computer industry in Southern California. Historian Richard Sprague has noted that Northrop Aviation, a large computer customer, spawned as many as fourteen computer start-ups, all of which tended to locate close to the corporate entity that gave them birth.[47]

Historical Accidents

In studies published in the 1980s, economist W. Brian Arthur noted that, in processes such as geographic clustering, where increasing returns to scale and feedback effects play an important role, historical accidents matter—especially in determining why an industry locates in one of many potentially favorable places.[48] Two important such "accidents" likely had long-lasting implications for the geography of information-technology clusters in the United States.

The first accident involved William Shockley, who co-invented the transistor at Bell Labs with John Bardeen and Walter Brattain. Shockley moved from the East Coast to Palo Alto because Palo Alto was his hometown and he wanted to live close to his mother. Shockley then recruited a group of bright engineers from East Coast labs, and they became the foundation of what became the West Coast semiconductor industry.[49]

The second accident involved Fred Terman, who eventually became vice-president of Stanford University. He attended Palo Alto High School and later Stanford University, earning a degree in electrical engineering in 1920. He then moved to the East Coast to do a PhD at MIT under Vannevar Bush. After completing his doctoral program, Terman was offered a faculty position at MIT. Before the beginning of the MIT academic term in 1924, however, he travelled to Palo Alto to visit his family and there he contracted tuberculosis. After spending a year in Palo Alto recovering from his illness, he decided to stay on the West Coast because of the California weather and became an electrical engineering professor at Stanford. Terman later established the Stanford Research Park, where Hewlett-Packard and many other high-technology firms located. It has been argued that, without Terman and the Stanford Research Park, Silicon Valley may have never developed as it did (or, more precisely, where it did).[50]

Taking Stock: The Geographic Concentration of the Information-Technology Industry

A model based on (internal) economies of scale and transportation costs sheds light on why the production of information technology tended to be geographically concentrated not only *across* but also *within* countries. However, such a model is not the only explanation for information-technology clusters. Other economic models focus on external economies of scale, including knowledge spillovers, and historical accidents. A particularly promising approach is the knowledge-spillover theory of entrepreneurship, which views entrepreneurs as the conduits through which knowledge spills over from parent companies to start-ups in high-technology industries. The question of why the information-technology industry has tended to concentrate geographically (both across and within countries) is a nice example of how economists apply models to understand historical patterns.

Standards and Information Sharing: Open and Closed Standards in Computers and Mobile Phones

Economists view computers and mobile phones as system-type industries: the basic product or service supplied by the industry has different components that must interact smoothly with one another. The crucial components in the computer industry are hardware and software. In the cellular industry, they are handsets and base stations.[51]

Because computers and cellular phones are system-type industries, they are also industries with standards, which are the rules of interaction between a system's components. Economists and other social scientists usually distinguish between "closed" and "open" standards. The core difference between open and closed standards reveals another overlap between the history of information and the history of computing: whether standards are considered open or closed hinges on how much information is made available to implementers of the standard, either free of charge or at a reasonably low price.

Historically, closed standards were developed by a single company that retained ownership and control over the intellectual property at the core of the standard. Often, the standard creator was the sole supplier of technology for the standard, at least for a period of time. In the computer industry, IBM's System/360 is a good example. In the cellular industry, the first-generation standards in Germany, France, Italy, and Japan are good examples.[52]

The intellectual property at the core of open standards, by contrast, tended to be available to all comers, either free of charge or at reasonable rates. Sometimes, open standards were developed by a single company—the first-generation AMPS cellular standard created by AT&T is an example. The Unix standard, also developed at AT&T, is a good example for the computer industry. In the cellular industry, open standards were often developed by several entities that participated in committees in the context of standard-setting organizations—GSM, the second-generation European cellular standard, is a good example.

Economists have pointed out that there is a key trade-off at the core of the decision to make a standard open or closed.[53] Open standards usually foster adoption. They do so for two reasons. First, they encourage competition among technology providers within each standard, which lowers prices. Second, network effects (or demand-side economies of scale) kick in. Once customers start adopting a standard, more customers want to adopt it. The problem with open standards is that the creators of the standard relinquish control over the evolution of the standard, which often leads to fragmentation. In addition, it is frequently difficult for the creators or sponsors of an open standard to appropriate the benefits derived from the standard.

Closed standards, by contrast, are adopted much more slowly. Since closed standards tend to prevent competition within the standard, prices of technology for closed standards tend to remain relatively high. But, on the other hand, the creator or sponsor of a closed standard retains control over the evolution of the standard and usually appropriates more easily the benefits from the standard.

Here we focus on several interesting examples of the trade-off between closed and open standards—in the first generation of cellular phones, in the early smartphones (Symbian and BlackBerry), and in mature smartphones (Android and iOS).

The First Generation of Cellular Phones

During the 1980s, in the first generation of cellular phones, the contrast between closed and open cellular standards was stark. Closed standards were developed in Germany, Italy, France, and Japan. In Germany, Italy, and France, the electronics national champion in each country was in charge of creating the standard. In Japan, NTT developed the standard and worked with several Japanese conglomerates to develop products for the standard. Open standards, by contrast, were developed in the United States (AMPS), Scandinavia (NMT), and Britain (TACS).[54]

Open standards facilitated faster and wider adoption of cellular phones. First, cellular phones were adopted at a faster pace in the countries with open cellular standards. Figure 8.6 shows that, in the first group of standard creators (Japan, the United States, and Scandinavia), cellular density grew faster in Scandinavia and America—the regions with open standards—than in Japan.

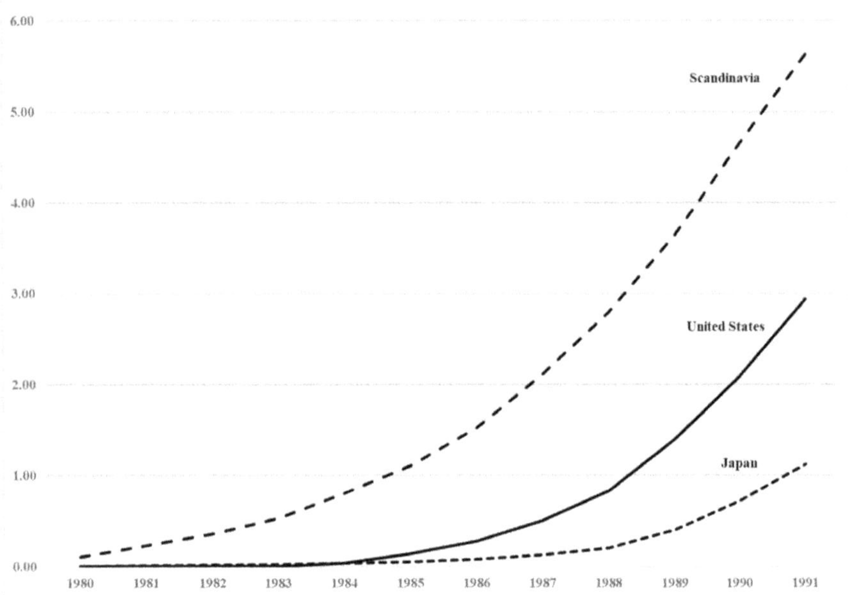

Figure 8.6. Cellular density in Scandinavia, the United States, and Japan, 1980–1991. Source: ITU database.

Figure 8.7 shows that, in the second wave of standard creators (Britain, Germany, France, and Italy), cellular density grew faster in Britain—the only country with an open standard—than anywhere else.

In addition, open standards were heavily adopted outside the regions where they were created. Figure 8.8 shows worldwide adoption of closed and open standards during the 1980s: the dominance of open standards in number of subscribers is clear.

The Early Smartphones: Symbian versus BlackBerry

The early smartphones were built on the technology of Personal Digital Assistants (PDAs). All the main smartphone operating systems in the pre-2007 period—Palm OS, Windows CE (later called Windows Mobile), Symbian, and BlackBerry OS—came from PDAs.[55]

In the early years of the smartphone market, vendors faced the choice between a closed system and an open system. Palm Computing and Research in Motion (BlackBerry) chose the "closed" route. In this

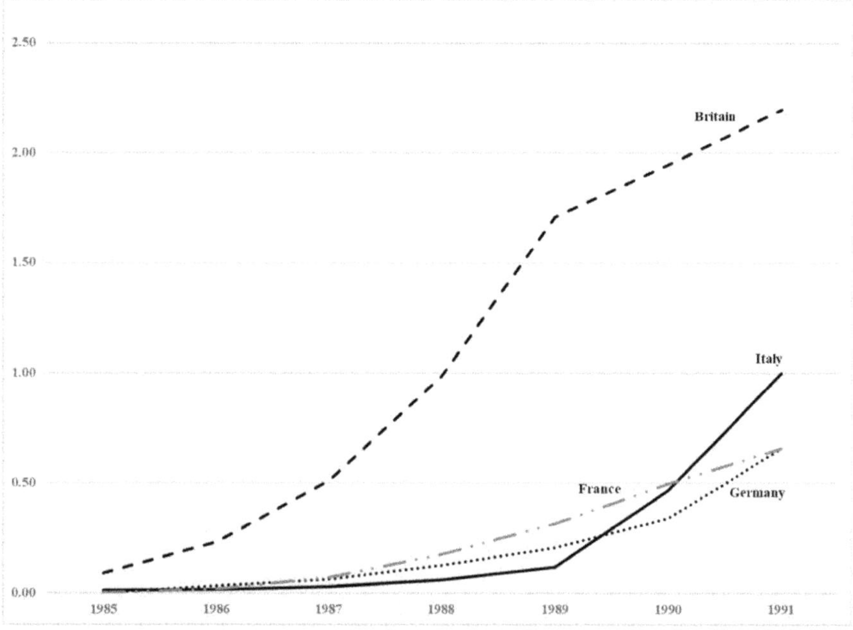

Figure 8.7. Cellular density in Britain, Germany, Italy, and France, 1985–1991. Source: ITU database.

model, the mobile operating system was closed-source, and its owner was a corporation that also functioned as the handset-maker. Put differently, the entity that owned the proprietary operating system sold an integrated bundle of hardware and software.

Nokia, Ericsson, and Motorola—the three key vendors of cellular handsets and infrastructure—partnered with Psion, a British computer company, to develop the Symbian operating system. The Symbian model was reasonably "open": although the mobile operating system was closed-source, the Symbian consortium licensed it for a price. Although Nokia, Ericsson, and Motorola sold mobile handsets that ran on Symbian, the mobile operating system was licensed to many handset-makers that were neither its "owners" nor its "sponsors." Further, different interfaces were developed so that handsets running on Symbian achieved a certain degree of product differentiation.

Although BlackBerry achieved success during the 2000s, especially in the United States, Symbian became, by far, the dominant smartphone

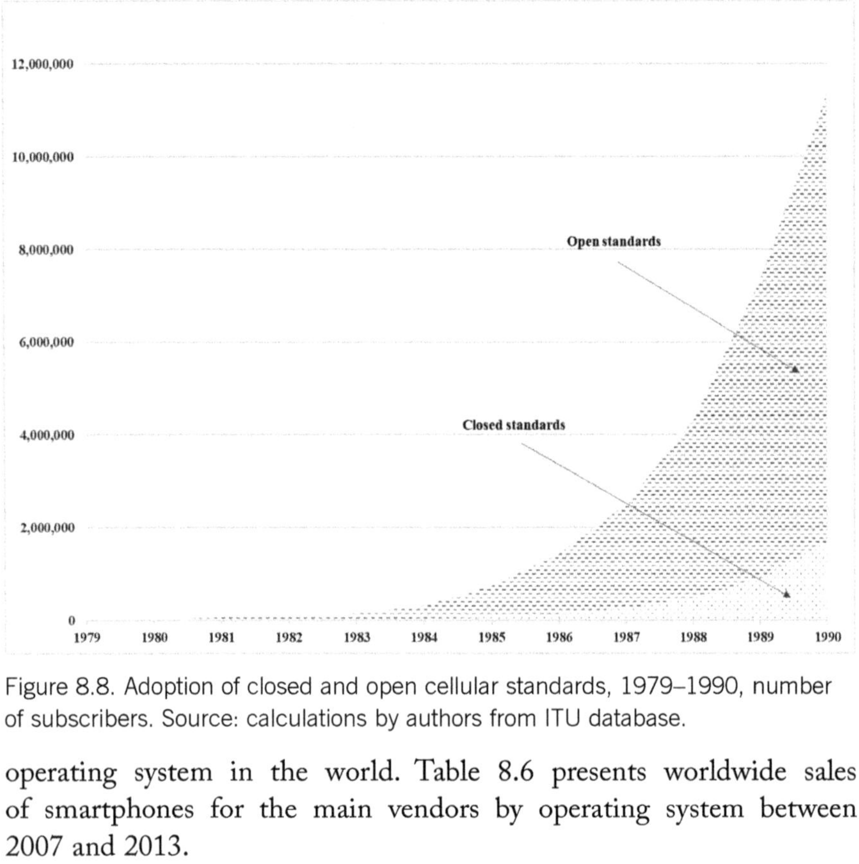

Figure 8.8. Adoption of closed and open cellular standards, 1979–1990, number of subscribers. Source: calculations by authors from ITU database.

operating system in the world. Table 8.6 presents worldwide sales of smartphones for the main vendors by operating system between 2007 and 2013.

In 2007, the year when the iPhone was introduced, more than 77 million Symbian smartphones were shipped. Neither BlackBerry nor Palm OS were even close: about 11 million BlackBerry smartphones were shipped that year. The early years of the smartphone market witnessed the resounding victory of the open system (Symbian) over the closed system (BlackBerry), to the point that, by 2007, Symbian was outselling Blackberry by a factor of about six.

Mature Smartphones: Android versus iOS

The year 2007, however, marked the beginning of the end for Symbian, BlackBerry, and all the other early smartphone vendors. In 2007, Apple introduced its first iPhone. In 2008, T-Mobile released the first

Table 8.6. Smartphone sales to end users by operating system, thousands of units, 2007–2013.

Year	Symbian	Other	RIM	Microsoft Windows Mobile	iOS	Android	Total
2007	77,684.0	14,863.4	11,767.7	14,698.0	3,302.6		122,315.7
2008	72,933.5	15,289.8	23,149.0	16,498.0	11,417.5		139,287.8
2009	80,878.6	10,432.1	34,346.6	24,889.8	15,027.6	6,789.4	172,364.1
2010	111,576.7	11,417.4	47,451.6	12,378.2	46,598.3	67,224.5	296,646.7
2012	47,203.0		34,210.3	16,940.7	130,133.2	451,621.0	680,108.2
2013	8,821.2		18,605.9	30,842.9	150,785.9	758,719.9	967,775.8

smartphone running on Android, the operating system supplied by Google.[56]

In 2007, owners and sponsors of the new mobile operating systems faced the choice between a closed and an open system. Consistent with its history, Apple chose the "closed" route, the same that BlackBerry and Palm Computing had chosen years earlier: Apple marketed the iPhone as a proprietary bundle of hardware and software. Google, by contrast, went the "open" route: Android was marketed as an open-source mobile operating system, developed by Google and its partners, and licensed to a wide swath of handset manufacturers.

Table 8.6 above shows that, once again, the open system won the diffusion battle: by 2013, about 758 million Android smartphones were shipped all over the world versus 150 million iPhones. As predicted by theory, however, Android's diffusion victory has hidden Android's problems: manufacturers that adopted the Android operating system have tended to tweak it for particular devices, which has made life more difficult for users and developers.[57] Further, it has been easier for Apple to appropriate the returns from its closed system than it has been for the Android handset manufacturers to appropriate the returns from the Android open system. It has recently been reported that, in the first quarter of 2023, Apple and Samsung accounted for 96 percent of operating profits in the smartphone market, with Apple alone accounting for 80 percent of total profits.[58]

Conclusions

What can an economic-history approach contribute to the overall history of information technology?

First, in the introduction to this chapter we argued that economics contributes a collection of models that help us interpret patterns in the history of high-technology industries, and these models generate hypotheses about the kinds of patterns we should expect to encounter in the real world. For example, one such hypothesis is that, when economies of scale and transportation costs matter, industries will tend to locate and develop in countries that have large domestic markets. The theory helps us both when it predicts well and when it fails. When it predicts poorly, it leads to additional questions, such as why a domestic computer industry failed to develop in the USSR, China, and India.

Second, we argued that the economic-history approach emphasizes measurement and quantification. But it is not measurement for the sake of measurement that matters. The models provided by economic theory give us clues as to which of the many numbers and figures we encounter in the real world are worth measuring. For example, in investigating the success of the commercial timesharing industry, a model that emphasizes economies of scale in the production of computer performance suggests that we should look carefully at the relationship between rental costs and performance measures for computers for a given level of technology.

Third, we argued that the economic-history approach is comparative history. Comparisons across countries and regions are useful because they allow us to better identify the factors that explain an industry's growth. For example, among the cellular pioneers, why did cellular density grow faster in the Scandinavian countries and the United States than in Japan? Among the second wave of standard creators, why did cellular density grow faster in Britain? One possible answer, which we have suggested here, is that open standards in Scandinavia, the United States, and Britain led to lower technology prices and faster adoption. Comparing how the industry evolved across countries raises many interesting questions, and a theory of open versus closed standards suggests potential answers to those questions.

Finally, the economic history of the computer industry suggests that institutional regimes and economic policies do make a difference. We have argued, for example, that a domestic computer industry failed to develop in the USSR, China, and India because these countries did not have the "right" institutional framework. Further, it should be noted that open standards in the computer industry—such as the Unix operating system—tended to splinter, whereas many open standards in the cellular industry did not. One possible explanation goes back, once again, to the importance of institutions: in the cellular industry, open standards such as NMT and GSM were defined by committees with the participation of multiple stakeholders. Such committees carefully specified the relevant interfaces, which helped the standards survive for a long time without fragmenting.

Notes

1. We acknowledge the superb research assistance of Nicolas Garcia-Vicente and helpful comments from Bill Aspray, Jim Cortada, and Shane Greenstein. All errors are our responsibility.

2. See G. Box and N. Draper, *Empirical Model Building and Response Surfaces* (New York: John Wiley & Sons, 1987), p. 424.

3. On the connections between quantification and theory, see especially R. Fogel, "The New Economic History: Its Findings and Methods," *The Economic History Review* 19:3 (1966), pp. 642–656.

4. For examples of comparative history and comparative economic history, see J. Diamond and J. Robinson, *Natural Experiments of History* (Cambridge: The Belknap Press of Harvard University Press, 2010); and T. Hutton, K. O'Rourke, and A. Taylor, *The New Comparative Economic History: Essays in Honor of Jeffrey G. Williamson* (Cambridge: MIT Press, 2018).

5. See, for example, W. Nicholson, *Microeconomic Theory*, 4th edition (Chicago: The Dryden Press, 1982), pp. 282–285.

6. See D. Carlton and J. Perloff, *Modern Industrial Organization*, 3rd edition (Reading: Addison-Wesley, 2000), pp. 35–40.

7. See M. Campbell-Kelly and D. Garcia-Swartz, "Economic Perspectives on the History of the Computer Timesharing Industry, 1965–1985," *IEEE Annals of the History of Computing* 30:1 (January–March 2008), pp. 16–36.

8. For a discussion of Grosch's Law, see M. Solomon, "Economies of Scale and the IBM System/360," *Communications of the ACM* 9:6 (June 1966), pp. 435–440; and W. Sharpe, *The Economics of Computers* (New York and London: Columbia University Press, 1969), pp. 314–322.

9. Solomon, "Economies of Scale and the IBM System/360."

10. For an intuitive discussion, see D. McCloskey, *The Applied Theory of Price* (New York: Macmillan Publishing Company, 1985), pp. 265–275.

11. G. Brock, *The U.S. Computer Industry: A Study of Market Power* (Cambridge: Ballinger, 1975), especially pp. 27–41; and P. Stoneman, *Technological Diffusion and the Computer Revolution: The UK Experience* (Cambridge: Cambridge University Press, 1976), especially pp. 45–65. For a detailed discussion of economies of scale in the early mainframe industry, see also N. Dorfman, *Innovation and Market Structure: Lessons from the Computer and Semiconductor Industries* (Cambridge: Ballinger Publishing Company, 1987), especially chapters 3–5.

12. To our knowledge, the most comprehensive study of the (regional and international) location of the early computer industry is the PhD dissertation of Servet Mutlu. See S. Mutlu, "Interregional and International Mobility of Industrial Capital: The Case of the American Automobile and Electronics Companies," University of California at Berkeley, 1979. On the foreign expansion of American computer companies, see

especially chapter XII. See also J. Cortada, *IBM: The Rise, Fall, and Reinvention of a Global Icon* (Cambridge: MIT Press, 2019), especially p. 111.

13. See M. Campbell-Kelly and D. Garcia-Swartz, *From Mainframes to Smartphones* (Cambridge: Harvard University Press, 2015), chapter 9.

14. See, for example, J. Dalton, "A Theory of Just-in-Time and the Growth in Manufacturing Trade," Wake Forest University, January 2013. For an excellent discussion of how transportation costs evolved in the second half of the 20th century, see D. Hummels, "Transportation costs and international trade in the second era of globalization," *Journal of Economic Perspectives* 21:3 (Summer 2007), pp. 131–154. Hummels presents a careful discussion of how changes in the weight-to-value (or value-to-weight) ratio affect transportation costs.

15. S. Mutlu, "Interregional and International Mobility," pp. 655–656.

16. For a theoretical discussion of the role of economies of scale and transportation costs in economic geography, see especially P. Krugman, "Increasing Returns and Economic Geography," *The Journal of Political Economy* 99:3 (1991), pp. 483–499; and P. Krugman, *Geography and Trade* (Cambridge: The MIT Press, 1992).

17. See the classic book by D. North and R. Thomas, *The Rise of the Western World: A New Economic History* (Cambridge: Cambridge University Press, 1973).

18. See, among others, R. Judy, "The Case of Computer Technology," in: S. Wasowski, editor, *East-West Trade and the Technology Gap* (New York: Praeger Publishers, 1970), chapter 5; and M. Cave, "Computer Technology," in: R. Amann et al., editors, *The Technological Level of Soviet Industry* (New Haven and London: Yale University Press, 1977), chapter 8.

19. See, among others, D. Audette, "Computer Technology in Communist China, 1956–1965," *Communications of the ACM* 9:9 (September 1966); and especially Z. Jiuchun and Z. Baichun, "Founding of the Chinese Academy of Sciences' Institute of Computing Technology," *IEEE Annals of the History of Computing* (January–March 2007).

20. See D. Sharma, *The Outsourcer: The Story of India's IT Revolution* (Cambridge: MIT Press, 2015), especially chapters 1–3.

21. R. Allen, "The Spread of Manufacturing," in: L. Neal and J. Williamson, editors, *The Cambridge History of Capitalism, Volume II* (Cambridge: Cambridge University Press, 2013), pp. 22–46.

22. On the economic history of China see, among others, B. Naughton, *The Chinese Economy: Transitions and Growth* (Cambridge: MIT Press, 2007); and D. Ma, "Economic Changes in China: The Role of Institutions and Ideology," in: S. Broadberry and K. Fukao, editors, *The Cambridge Economic History of the Modern World, Volume II: 1870 to the Present* (Cambridge: Cambridge University Press, 2021), chapter 5.

23. On the economic history of India see, among others, B. Gupta, "From Free Trade to Regulation: The Political Economy of India's Development," in S. Broadberry and K. Fukao, editors, *The Cambridge Economic History of the Modern World, Volume II: 1870 to the Present*, chapter 6; B. Tomlinson, *The Economy of Modern India: From 1860 to the Twenty-First Century* (Cambridge: Cambridge University Press, 2013); and T. Roy, *The Economic History of India, 1857 to 2010* (Oxford: Oxford University Press, 2020).

24. Although the years immediately prior to 1949 were a period of economic decline in China, it appears that, by 1952, many industries had reached levels of production that surpassed the pre-1949 peak. See especially C. Riskin, *China's Political Economy: The Quest for Development since 1949* (Oxford: Oxford University Press, 1988), especially Table 4.1, p. 54. For a historical comparison of GDP per capita in the coastal cities versus the inland regions of China, see especially F. Lemoine, G. Mayo, S. Poncet, and D. Ünal, "The Geographic Pattern of China's Growth and Convergence within Industry," Working Paper CEPII No. 2014-04, February 2014.

25. Allen, "The Spread of Manufacturing," especially pp. 22–24.

26. On computer developments in the Soviet Union, see especially G. Trogemann, A. Nitussov, and W. Ernst, editors, *Computing in Russia* (Braunschweig and Wiesbaden: F. Vieweg & Sohn Verlagsgesellschaft, 2001). See also S. Goodman, "Soviet Computing and Technology Transfer: An Overview," *World Politics* 31:4 (1979); R. Stapleton and S. Goodman, *The Soviet Union and the Personal Computer 'Revolution,'* Report to the National Council for Soviet and East European Research, 1988; S. Goodman and W. McHenry, "The Soviet Computer Industry: A Tale of Two Sectors," *Communications of the ACM* 34:6 (June 1991); and S. Gerovitch, *From Newspeak to Cyberspeak: A History of Soviet Cybernetics* (Cambridge: The MIT Press, 2004). A detailed discussion of the early development of information technology in the USSR is in J. Cortada, *The Digital Flood* (Oxford: Oxford University Press, 2012), pp. 238–276.

27. We draw from D. Simon and D. Rehn, *Technological Innovation in China: The Case of the Shanghai Semiconductor Industry* (Ballinger, 1988); and Q. Lu, *China's Leap into the Information Age: Innovation and Organization in the Computer Industry* (Oxford: Oxford University Press, 2000), especially chapter 1.

28. Details on the introduction of computers in China can be found in P. Guo and Q. Sun, "General Review of Computer Technology and Application in China," *Computers in Industry* 8 (1987), pp. 113–117. See also Cortada, *Digital Flood*, pp. 443–490.

29. Here we draw mainly from D. Sharma, *The Outsourcer: The Story of India's IT Revolution* (Cambridge: MIT Press, 2015). See also Cortada, *Digital Flood*, pp. 491–545.

30. Sharma, *Outsourcer*, p. 49.

31. See M. Campbell-Kelly and D. Garcia-Swartz, "Persistence and Change in the Information-Technology Industry," in: M. Kipping et al., editors, *Oxford Handbook of Industry Dynamics* (Oxford: Oxford University Press, 2022).

32. Sharma, *Outsourcer*, p. 49.

33. On the value of economic experimentation, see N. Rosenberg, "Economic Experiments," in: N. Rosenberg, *Exploring the black box: Technology, economics, and history* (Cambridge: Cambridge University Press, 1994), chapter 5.

34. On improvements in quality-adjusted prices for computers in the United States, see R. Gordon, *The Measurement of Durable Goods Prices* (Chicago and London: University of Chicago Press, 1990). On gains in customer welfare from technological change in computers, see E. Brynjolfsson, "The contribution of information technology to consumer welfare," *Information Systems Research* 7 (September 1996); and S. Greenstein, "From Superminis to Supercomputers: Estimating Surplus in the Computing Market,"

in: T. Bresnahan and R. Gordon, editors, *The Economics of New Goods* (Chicago: University of Chicago Press, 1997).

35. On customer innovation, see E. von Hippel, *Democratizing Innovation* (Cambridge: MIT Press, 2006).

36. On learning by using and customer feedback, see N. Rosenberg, "Learning by Using," in N. Rosenberg, *Inside the Black Box: Technology and Economics* (Cambridge: Cambridge University Press, 1982), chapter 6.

37. See S. Proost and J. Thisse, "What Can Be Learned from Spatial Economics?," *Journal of Economic Literature* 57:3 (2019), pp. 575–643.

38. For a more sophisticated measure that has been used often in the economic-geography literature, see G. Ellison and E. Glaeser, "Geographic Concentration in U.S. Manufacturing Industries: A Dartboard Approach," *Journal of Political Economy* 105:5 (October 1997), pp. 889–927.

39. In the tables, FIPS are the Federal Information Processing Standards codes, which the Census Bureau assigned to geographic units for over 30 years.

40. Economists Edward Glaeser and Janet Kohlhase have shown that the cost of moving goods declined substantially in the 20th century, but the cost of moving people did not. In their view, this explains why services have tended to locate in high-density areas but manufacturing industries have not. See E. Glaeser and J. Kohlhase, "Cities, Regions, and the Decline of Transportation Costs," NBER Working Paper 9886, July 2003.

41. E. Glaeser, *Triumph of the City* (New York: Penguin Books, 2011), p. 36.

42. J. Jacobs, *The Death and Life of Great American Cities* (New York: Modern Library, 2011), especially pp. 187–312.

43. Historians of the urban movement have argued that exchanges of information are at the core of the origins of cities around the world. See, for example, P. Hohenberg and L. Lees, *The Making of Urban Europe 1000–1994* (Cambridge: Harvard University Press, 1995), p. 47; and P. Bairoch, *De Jericho á México: Villes et économie dans l'histoire* (Paris: Gallimard, 1992), p. 133.

44. Z. Acs and D. Audretsch, *Innovation and Small Firms* (Cambridge: MIT Press, 1990) and D. Audretsch, *Innovation and Industry Evolution* (Cambridge: MIT Press, 1995).

45. E. Rogers and J. Larsen, *Silicon Valley Fever* (New York: Basic Books, 1984), p. 45.

46. See C. Mason, "Entrepreneurial dynamics and the origin and growth of high-tech clusters," in: C. Karlsson, editor, *Handbook of Research on Innovation and Clusters* (Cheltenham: Edward Elgar, 2008), chapter 3, especially pp. 33–36.

47. R. Sprague, "A Western View of Computer History," *Communications of the ACM* 15:7 (July 1972), pp. 686–692.

48. For a collection of Arthur's early papers, see W. B. Arthur, *Increasing Returns and Path Dependence in the Economy* (Ann Arbor: University of Michigan Press, 1994).

49. Rogers and Larsen, *Silicon Valley Fever*, p. 37.

50. Rogers and Larsen, *Silicon Valley Fever*, pp. 25–39.

51. See, among others, C. Shapiro and H. Varian, *Information Rules: A Strategic Guide to the Network Economy* (Cambridge: Harvard Business Review Press, 1998).

52. For a review of the early economic literature on standards, see P. David and S. Greenstein, "The Economics of Compatibility Standards: A Review of the Recent Literature," *Economics of Innovation and New Technology* 1 (1990), pp. 3–41. On the history of the standard-setting movement since the late 19th century, see J. Yates and C. Murphy, *Engineering Rules: Global Standards since 1880* (Baltimore: Johns Hopkins University Press, 2019).

53. See, for example, D. Garcia-Swartz and M. Campbell-Kelly, "Openness as A Business Strategy: Historical Perspectives on Openness in Computing and Mobile Phones," *Information Economics and Policy* 48 (September 2019), pp. 1–14.

54. D. Garcia-Swartz and M. Campbell-Kelly, *Cellular: An Economic and Business History of the International Mobile-Phone Industry* (Cambridge: MIT Press, 2022), especially chapters 1–2.

55. See Garcia-Swartz and Campbell-Kelly, *Cellular*, chapter 7.

56. See Garcia-Swartz and Campbell-Kelly, *Cellular*, chapter 10.

57. See, for example, S. Bose, "How to deal with Android fragmentation," *BrowserStack*, March 23, 2021, available at https://www.browserstack.com/guide/what-is-android-fragmentation.

58. See H. Walia, "Global Smartphone Market Declines 14% YoY in Q1 2023; Apple Records Highest-Ever Q1 Share," Counterpoint, May 5, 2023, available at https://www.counterpointresearch.com/global-smartphone-market-declines-14-yoy-q1-2023-apple-records-highest-ever-q1-share/. Google has obtained returns not from the mobile operating system *per se* but from ads, apps, and services supplied on the operating system.

Chapter 9

Writing Law and Public Policy into Computing and Information History

Philip Doty and William Aspray

This chapter discusses the importance of law and public policy in the writing of the story of the creation, use, maintenance, and spread of information and computing since 1930. This concern is important because public policy has often been no more than a background for many historians who have told this story.[1] Moreover, some people believe that government regulation outpaces technology and is therefore largely irrelevant in telling this history—a point of view that we strongly disagree with. While the focus here is on the United States, there is a similar dynamic at play between public policy and computing/information in other parts of the world.

Much of the writing on the history of computing focuses on either technological or business matters. We argue that public policy and law shape both information and information technologies, as well as the businesses associated with them. Even by adding law and public policy to the scholarly historical mix, we do not capture the complete story. One might want also to consider labor, race, gender, social institutions and practices, or culture to get a more complete or nuanced picture of computing and information history. Our focus in this chapter, however, is particularly on the interaction of law and public policy with computing and information. Although we provide some examples where law and policy have an effect in this realm, our goal is not to enumerate the most important examples as a shopping list or prescriptive guide for future historical scholarship.[2] Indeed, law and public policy interact with computing and information in many places, and our goal instead is to heighten awareness of this fact and suggest ways to bring legal and policy considerations into the telling of many of these stories.

How Computer and Information Historians Have Treated Legal and Policy Matters

To give a rough indication of how computer historians and information historians have treated legal and public policy concerns in the United States since 1930, we conducted a quick-and-dirty content analysis of two leading and recent "master narratives" of computer and information history: Thomas Haigh and Paul Ceruzzi's *A New History of Modern Computing* (2021) and James Cortada's *All the Facts: A History of Information in the United States since 1870* (2016).[3] Haigh and Ceruzzi explicitly discuss public policy topics 19 times in their book, Cortada 24 times. The policy coverage of these two books overlaps on only two topics: the government-funded computing system SAGE (1955) and the creation by the Defense Department of the ARPANET (1969).

Most of the references to public policy concerns in Haigh and Ceruzzi address either government-funded support for the creation of advanced hardware and software technologies[4] or government acquisition of such technology.[5] The only other three public policy matters they discuss are the Mansfield Amendment to the Military Procurement Authorization Act (1970), which limited the Defense Department to supporting only computing technologies that had military purposes; *U.S. v. Microsoft* (1998–2000), regarding antitrust enforcement; and *A&M Records v. Napster* (2000–2001), which involved copyright regulation of peer-to-peer file sharing.

A New History is good about identifying the role of government funding in computing in the early years but offers less coverage of this topic for later years. The book includes numerous stories about individuals' technological contributions to cybersecurity, RISC architectures, and graphical computing—all of which have government connections—but the authors do not discuss these connections. The book contains little about the role of government regulation in areas such as copyright infringement and antitrust, although, as noted above, it briefly mentions the antitrust case against Microsoft and the copyright infringement case against Napster. In all, their coverage is better in some areas, such as defense and antitrust (but not even very complete in those areas), and less so in certain other policy areas such as privacy and employment policy.

Like Haigh and Ceruzzi, Cortada sometimes discusses government acquisition of computing technologies such as tabulating equipment (1941–1943), or government support for research and education.[6] Many of the public policy topics covered by Cortada concern data collection, analysis, distribution, and occasionally regulation—especially from Franklin Roosevelt's New Deal but occasionally from later administrations.[7] Many of Cortada's policy topics, however, are about neither data collection nor funding, but instead about regulation: wartime censorship (early 1940s); import and export of technological goods, e.g., General Agreement on Tariffs and Trade (GATT, 1947); many instances of privacy legislation from the 1970s through the 1990s;[8] and several pieces of legislation and court decisions regarding copyright.[9] Cortada is strong on government use of information in carrying out its business and on privacy, less so on government support for technological research and development.

It is not our purpose to criticize either Haigh and Ceruzzi or Cortada—these are both excellent, path-breaking books—but instead use them to illustrate how computer and information historians more generally cover public policy matters. These two master narratives appear to be somewhat representative of the larger literatures of computer and information history.[10] Computer historians have been best considering political action when telling the story of national defense or of computers as tools of government bureaucracy. Often, other matters of regulation and law serve only as background impediments to technological advance, e.g., in the antitrust actions against IBM in the 1950s or against Microsoft in the 1990s; in the labor disputes over H-1B visas in the early 2000s; and in the trademark and copyright conflicts that occurred over domain names and file sharing in the 2010s. Cortada is unusual in the history of computing literature because of his extensive coverage of privacy regulation. Other important public policy questions related to computers and information, such as environmental impact or accessibility for the disabled, are only now beginning to receive attention from computer historians.

Public Policy Is Powerful in Shaping the Development and Use of Information Technologies

One of the reasons that historical accounts have given less than complete coverage to legal and public policy concerns may be that policymakers are not viewed as strong, rapidly responsive actors capable of participating in the fast-moving development of computing technology. This perspective comes partly from many technically oriented scholars' traditional view of the tools of policymaking. They equate policy action solely with the passage of laws by Congress signed by the president, regulations created and enforced by federal executive agencies, and case law arising from court decisions. All of these are, of course, important tools in the toolbox of governance, but these kinds of actions can be slow-moving—involving government hearings, report writing, multiple court cases, and more. What is much less well covered by these historical treatments is that government has a much wider range of tools, many of which are used frequently and effectively in the shaping and control of information technologies. Table 9.1, redacted from another of the authors' papers,[11] presents various ways in which the U.S. government might regulate technology, the ways in which the government can offer incentives and disincentives to shape the course of technological innovation, and methods the government employs that enables it to be more responsive to the pace of technological innovation.

Indeed, many technically oriented people believe that information technology fundamentally outpaces public policy and hence is less likely to be a major factor in telling the story of computer history. The authors of this chapter and others believe the situation is much more complex. The tools listed in Table 9.1 can enable the government to be involved in the control and regulation of fast-moving innovation in the computing

Table 9.1. Policy Means for Controlling Technology in the United States.

Ways in which the U.S. government regulates technology:

- bills passed by Congress
- regulations created by federal agencies
- case law produced by federal courts
- its bully pulpit (see below)
- government participation in technological development
- funding of technological research
- procurement policies for the federal government and government contractors

Government incentives and disincentives that shape the course of technological innovation:

- low-interest loans
- fees and vouchers
- subsidies
- contracts and grants
- tax breaks
- economic regulations
- liability and tort law
- information campaigns

Government approaches to enhance the current legal system as it tries to regulate technology:

- streamlining administrative rulemaking processes to reach final rulemaking quickly
- cooperative regulation between federal agencies and industry organizations
- specialized courts specifically created to address the arcane conflicts that arise when regulating technology
- sunset clauses in existing regulations so that they do not remain in effect after a technology has changed significantly
- delegating regulatory authority to independent organizations, e.g., the Internet Corporation for Assigned Names and Numbers (ICANN)

field, and Table 9.2 identifies four important ways in which policymaking can anticipate or outpace computing technology.

This section of the chapter has argued that law and public policy are indeed actors in the ever-evolving story of computing and information. It has also identified many of the wide array of tactics that are at the policymaker's disposal to shape these histories. The next sections of the

Table 9.2. Means by Which Policy Can Outpace Technology

Means	Examples
Governmental entities establish explicit technical standards (either anticipatory or *ex post facto*), usually in concert with actors in technology enterprises and others, for consumer products or products designed for internal governmental use	TCP/IP, ISDN, and V. 32, or Electronic and Information Technology Accessibility Standards under section 508 of the Rehabilitation Act of 1973
Government IT and IT procurement can lead technological development and deployment	Major IT projects in the U.S. Department of Defense or National Archives and Records Administration requiring particular types of computer and networking security and other system characteristics. They specify operating systems and additional functionalities, and demand characteristics such as interoperability, backwards compatibility, and ability to serve governmental and other users.
Government creates and aggregates markets for information services and products (Say's Law)	Packet switching, computer networking protocols (e.g., TCP/IP), email, remote login, e-commerce, supercomputing, and digital libraries
Government reshuffles policy domains to take a more comprehensive and coordinated approach to communication, commerce, and related matters	Formerly separated areas of technological activity and governmental regulation are forced to relate and integrate in novel ways, e.g., telephony, computer networking, TV and radio broadcasting, copyright, privacy regulation, and First Amendment law including matters such as free expression, film making, newspaper and other publication, tort liability, slander and libel, safe harbors for publishers of all types, and education

chapter identify some ways of writing computer and information history using specific examples of public policy and legal analysis.

Enriching Information and Computer History with Policy and Legal Research

In the first section of the chapter, we analyzed how leading computer and information historians have addressed some public policy concerns—identifying both strengths and gaps in their coverage. In the second section, we discussed the commonly held belief that technology outpaces policy (and the corollary conclusion that policy might be regarded as unimportant to computer and information history) and argued instead for policy's importance. In this section, we discuss how policy research often involves historical examination, the study of which can enrich and inform the analyses of computer and information historians.

Understanding the Study of Public Policy

Over the years, we have heard from our computer science and information science students, and sometimes from others, beliefs about the political process that are either incorrect or simplistic. They argue that *public policy is about expressing one's personal or one's organization's beliefs*; that it is the process of saying "here's what I think, or what my party or my organization thinks, and here's why everyone who disagrees with us is wrong." They believe that *public policy is generally about choosing one of two binary choices*, for example, those offered by right/left, conservative/liberal, Republican/Democrat, Trump/Republican in Name Only (RINO), and so on, when in fact there are often a multitude of reasonable positions regarding a particular policy conflict. Many believe that *public policy is about debate*; whereas the rhetoric of "debate" is often reductive and misleading, belying the complexity of a situation, implying that one "side" and only one side can be correct, and suggesting that there can be and will be one and only one "winner." These beliefs ignore the fact that some public policy issues are intractable, that both the issues and their resolutions are often emergent, and that compromise is oftentimes a mark of success rather than of failure or surrender. Finally, many believe that *"political" is a pejorative term*; it is not. Doing ICT politics is an area of human endeavor in which there needs to be intensive study to understand technologies, the multiple ways in which they are used, and the underlying values and reasoning of the many different interest groups that have a vested interest in a particular technology policy issue.

We believe that, for the computer and information historian, it is important to note that public policy is an intensely scholarly practice marked by workshops, hearings, conferences, national study groups, and careful book-length studies. These are carried out by government organizations (e.g., the [now-defunct] Congressional Office of Technology Assessment, the Government Accountability Office, or the Congressional Research Service), professional societies (e.g., ACM or Computing Research Association), trade associations (e.g., Information Technology Association of America), and other non-government organizations (e.g., the National Research Council, Brookings Institution, Center for Democracy and Technology, Cato Institute, and Electronic Privacy Information Center). Some historians have already drawn upon these rich resources, but there is still much more to be mined here.

In fact, it is worth noting that research involving technology and public policy goes on at two different levels. There are the policymakers who are attempting to create new policy, and various interest groups who are trying to influence policy being made, especially the kinds of organizations listed in the previous paragraph. They have strong interests in the particular technology policy instruments (laws, regulations, and case law) created. But there is also a group of academics, such as political scientists, legal scholars, public administration scholars, and economists, who are studying a particular policymaking episode not because they are trying to influence the outcome but instead because they are trying to understand the various factors that went into the decision-making and understand the impacts of the decision.

Developing a scholarly point of view does not mean asking researchers to extinguish their own political commitments and values. Instead, this scholarly practice aims to help policy researchers maintain, enrich, and express their own informed opinions. Computer historians may have had personal views about topics from the way in which computer history was usually practiced. They may have thought, for example, that

IBM was acting unfairly when it used its market position to stifle its mainframe competitors, or that the U.S. government was wrong when it prevented IBM from taking actions that restricted the development and spread of new technologies. But, unless one worked for IBM or for one of these competitors directly influenced by IBM's monopolistic actions, one was perhaps unlikely to get too incensed by these policy decisions. As computer history has increasingly turned to the study of the interaction of technology and individuals, e.g., algorithms that discriminate against people who are poor, female, or members of minority groups, however, such conflicts seem to become more personal—to resonate more strongly with scholars' and others' personal values.

The best practice of public policy as an academic pursuit, by political scientists and legal scholars, draws broadly on various academic disciplines, including philosophy, the social sciences, and history.[12] There are numerous examples of good political science and legal studies scholarship that rely explicitly on historical methods to explore and deepen our understanding of information policy. Many different policy areas arise in connection with information and information technologies: free expression, information and cyber security, privacy, surveillance and security, monetization of information, copyright, government information classification, and freedom of information, particularly since 9/11/2001. Here are three good examples of policy study selected from the many hundreds that are deeply informed by historical methods:

- Jessica Litman's 2017 monograph on *Digital Copyright* with its incisive and inquiring use of Supreme Court archives, official and unofficial records of the justices' deliberations, justices' memoirs, and other sources.[13]
- Simone Browne's 2015 monograph *Dark Matters: On the Surveillance of Blackness*, and its deep investigations of post-colonial studies of race and slavery; army and commercial archives in the early United States, England, and elsewhere; links between terrorist studies and the "othering" in historical and contemporary surveillance; and more.[14]

- Julie Cohen's 2022 book *Between Truth and Power: The Legal Construction of Informational Capitalism*, relying on close examination of legal institutions and the traditions of the English common law in the U.S. and beyond, based on a sophisticated historical investigation of how the law and networked digital infrastructures and technologies have co-evolved and helped define each other, most particularly in building our contemporary system of information capitalism.[15]

What Law and Public Policy Contribute to Computer and Information History

There are at least three ways in which computer and information historians can benefit from policy studies. First and foremost is that, because policy issues are understood in the serious policy literature as important policy conflicts of public interest, policy studies underscore the importance of understanding their historical bases. Second, given the understanding of policy issues as conflicts of important public interest, policy and legal studies identify and make explicit important sources of consensus and dissensus about these policy issues, and an ability to identify key policy stakeholders and actors related to these issues. Third is a willingness to recognize that many actors in the policy space have reason. That is, we recognize that reasonable people will disagree about what can and should be done about important policy issues. Further, the policy researcher recognizes that the policy system is meant, in part, to reveal and adjudicate among these conflicting perspectives and value judgments.

This last point requires seeing politics as a primary method for how a community gives witness to its values, and negotiates conflict over values, since reasonable people may disagree about values and about how to act upon them. At the same time, good policy and researchers develop and exhibit a grounded understanding of the *power asymmetries* that characterize modern society and an ability to act to help achieve a more just society, however imagined. Policy and legal research, especially its historical emphasis and embrace of multiple narratives, helps identify and understand voices, perspectives, and values often silenced, in the minority, and overlooked.

Finally, political science and legal studies are more mature academic disciplines than computer and information history, and can teach computer and information historians about the methods, concepts, and frameworks they use in their intellectual activities, including:

- Modeling openness and recognition of many modes of reasoning and values.
- Insisting upon fair and accurate representation of various legal and policy positions.
- Breadth of reading, including methodologically.
- Intimate knowledge of literatures related to public policy and the law, especially beyond technicist, "phases" models to include narrative, gender-sensitive, critical, and empirical social science methods.
- Use of primary as well as secondary literatures, including policy instruments such as the various kinds of positive law—legislative (statutory), administrative (regulatory), and case (judicial) law.

In the next two sub-sections of the chapter, we briefly explore these general ideas in more detail by briefly considering the writings of one particularly influential legal scholar (Lawrence Lessig), then four prominent policy scholars (Deborah Stone, Carol Bacchi, Sandra Braman, and Emery Roe), and in the course of doing so mention a few other important researchers.

Lessons from an Early Legal Scholar of the Internet

In perhaps the most influential early piece of scholarship on the Internet by legal scholars, *Code Is Law*, Lawrence Lessig points to two aspects of the Internet—*anonymity* and *jurisdiction*—that made it difficult to regulate behavior in its use.[16] The way in which the Internet was constructed made it possible for users to remain relatively anonymous online, or as a *New Yorker* cartoon famously said, it is hard to know you are a dog on the Internet.[17] If there is some antisocial behavior online, e.g., unauthorized use of copyrighted material, what part of the legal system has the right and responsibility to regulate this misbehavior: is it the legal system at the location where the user misbehaved, the location of their Internet service provider who provides them with access to the Internet, the location where the harmed party is located, or somewhere else? This question is particularly problematic when these locations are in different countries, where there are jurisdictional boundaries in the reach of the law, or where there might be different cultural beliefs about what constitutes an infraction and how seriously it should be punished. While early cyberlibertarians celebrated this lawless, Wild West–like frontier attitude, national governments needed to protect their national security and the way of life of their citizens, companies needed to protect their reputations and property, and individuals needed to protect themselves from identity theft and online bullying.

How, Lessig asked, could one regulate online behavior? Drawing parallels from the wider world, he identified four regulatory mechanisms. One was official *law*, including legislation, regulation by government agencies, and case law. For example, online pornography was partially regulated by legislation, and copyright was protected by case law such as in the Napster trial.[18] One can regulate by *economics*, much in the way that governments post high excise taxes on cigarette purchases to discourage smoking. A third approach is to use *social norms*, which were more effective in the early days of the Internet, when commercial activity online was widely scorned and the set of users had a high degree of homogeneity in their personal profiles and interests (but much less so in the time of the commercial Internet). Government can play a role in setting those social norms by using its bully pulpit. Finally, Lessig drew on a famous study in which Robert Moses protected segregation on New York beaches by building bridges that were too low to allow buses to pass, thus eliminating most Black bathers from some beaches, since they had to rely on public transportation to get there.[19] Lessig argued that this was an instance of regulation by *architecture*, and he likened it to the writing of particular online code that built an infrastructure that limited specific kinds of activity. An important example is the use of digital rights management (DRM) by the film and record industries, which was protected by government regulation and court decisions.

Lessig's argument was studied by every Internet legal scholar in the first decade after its publication and is still today an accepted way of thinking by these legal scholars. While many information and computer historians are familiar with Lessig's argument, there remain considerable opportunities for these historians to use Lessig's analysis as a lens on their work. We now move to a more considered discussion of how the integration of law and public policy can enhance and facilitate the history of computing and information.

How Discourse, Rhetoric, and Public Policy Matter to Historians of Information and Computing[20]

In the introduction to her widely cited and influential book *Policy Paradox: The Art of Political Decision Making*,[21] political scientist Deborah Stone famously said:

> [T]he essence of policy making in political communities [is] the struggle over ideas. Ideas are a medium of exchange and a mode of influence even more powerful than money and votes and guns. Shared meanings motivate people to action and meld individual striving into collective action. Policy making, in turn, is a constant struggle over the criteria for classification, the boundaries of categories, and the definition of ideas that guide the way people behave.

This discourse-analytic, rhetorical focus during the last several decades of public policy and legal research, including research in information policy, is particularly valuable to historians investigating computing and information history. Wayne Parsons, another well-known public policy scholar, underscores that policy research (we include the work of policy-informed historians of computing and information in this category of scholarship) is "really about persuasion and argument,"[22] what Schön and Rein call "symbolic contexts that revolve around a policy object-in-the-making."[23] This rhetorical perspective can provide historians with ways to uncover and thereby explain how "policy discourse, names and labels" have been used in computing and information initiatives "to create associations that lend legitimacy and attract support to a course of action."[24] Those looking for such legitimacy and support include governments, private sector actors, non-profit organizations, citizens groups, and others with interest in supporting, for example, some initiatives in computing and telecommunications while not supporting others. Historians can and need to tell this story.

Stone is particularly good in helping us understand that public policy's essential thrust is "classification and differentiation, about how we do and should categorize in a world where categories are not given."[25] So, how can this insight about classification and differentiation be most useful to historians of computing and information? Consider how it is that, as a polity, we distinguish areas of action that are of public concern and those that are not; how we distinguish between areas of action where political action can make progress from the current state of affairs to some more desirable state; how we distinguish between criteria for evidence that help us decide about judging states as more or less desirable; how we distinguish between courses of action that may or may not lead to the desired state of affairs; how we identify acceptable and unacceptable levels of cooperation between public and private sector actors; and how we identify means by which we can judge the effectiveness of various courses of governmental action. These are only some of the ways that classification and differentiation help historians gain insight into, and describe, how it is that governments and other actors frame and make policy decisions about which information and computing initiatives to support and which to ignore or actively oppose.

Key is how it is that governmental policy makers frame what is commonly if more than a bit erroneously called a "policy problem," i.e., what is it that policy makers regard as causing public difficulties such that these policy makers feel compelled to act and feel compelled to demonstrate that they are acting? As Australian Carol Bacchi, another influential and widely cited public policy scholar, asserts: policy makers' actions "by their nature imply a certain understanding of what needs to change (the 'problem')."[26] In other words, as her work as well as that of Deborah Stone, Emery Roe (1994),[27] Mark Considine (1994),[28] and many others in critical and discourse-sensitive public policy research show, *public policy and legal problems are humanly constructed not given.*

This fact is particularly important to recognize and examine in areas of research involved with information and communication technologies (ICTs) because these technologies play such wide and important roles in contemporary social, political, cultural, and economic developments. What policy researchers, and other scholars, including historians of computing and information, must do is show how policy "problems" are constructed—by whom, for what reasons, according to what criteria of knowledge-power relations,[29] which actors are privileged and which are undercut in representations of "problems," which political actors' values are valorized and which are not, and so on. As Schön and Rein remind us, different framings of policy concerns and courses of action are often incommensurable because different frames lead us to notice and value different facts, actors, and relationships, and to respond to different kinds and forms of arguments and evidence.[30] Thus, such scholarship can problematize and interrogate "the problematisations [*sic*] on offer" in current policy making and policy instruments.[31]

In doing such work, computing and information historians can help policy makers and others work against what some policy researchers have called the "endless opportunities of error and self-deception" that policymaking and evaluation produce in all political actors.[32] Such self-conscious and reflective research may help these historians in their own struggle against errors and self-deception. Especially important is how public policy once made is not finished; rather, it is "made and re-made endlessly [in] a process of successive approximations to some desired objective in which what is desired itself continues to change under consideration."[33] Clearly, historians can contribute deep insight into the making and remaking of public policy related to computing and information, but there is further value in considering Lindblom's assertion. What he says about public policy, of course, also applies to other fields of human endeavor, including computing and information broadly construed as well as the doing of history, as we practice endless loops of self-reflection and re-evaluation of our work and its temporal character.

Sandra Braman, the well-known information and telecommunications policy scholar, makes an observation that seems obvious, particularly in our current cultural and social moment, but that nevertheless helps us understand the salience of information policy for historians generally as well as those who work in computing and information history fields. She notes that, because information policy involves policy and legal initiatives governing the creation, sharing, and use of information, it "fundamentally shapes the conditions within which we undertake all other political, social, cultural, and economic activity."[34] Importantly, historians of all kinds, by education and inclination as well as by the norms of professional practice, are especially sensitive to the contingency of history, how things might have been different.

Emery Roe (1994) is particularly good on the contingency of public decision-making, calling it, among other things, a "hodgepodge of muddling through [as per Lindblom], groping along opportunistically, incrementally, hopefully by rules of thumb and bounded rationality."[35] Information and computing historians are quite familiar with what Roe, among others, terms the "certainty of uncertainty" in understanding the historical record and other traces of human effort and making claims about them, whether causal or otherwise.[36] Areas of public policy and legal research robustly exhibit the need for such humility.

Historians of computing and information are well-equipped for doing work in these kinds of conditions. Moreover, historical work in public policy and legal areas can contribute to researchers' understanding of the contexts in which technical and other modes of historical change take place. Just as political reasoning and public policy research are "part of a struggle to control which images of the world govern policy,"[37] public policy and legal investigations can help historians of computing and information understand which policy frames, terminologies, actors, and relationships help determine the fields of computing and information. Historians can gain insight likely not otherwise available into how the *agon*, i.e., the competition, the struggle between political actors, is important to writing and rewriting the history of information and computing.

A Deeper Dive into Privacy and Surveillance—Four Examples from Policy and Legal Studies

In this section, we look at specific examples of the approaches being used in legal and policy studies focused on computing, with the hope that they offer useful ideas about topics and approaches to computer and information historians. It is beyond our scope to survey the full range of methods used by legal and policy scholars, or the full set of policy areas in which they are applied. Instead, we illustrate how policy scholars work by taking examples from the study of privacy. It is the area of law and policy relating to ICTs that has received the most academic attention, and it is also the policy area that we two authors know best.[38]

Some remarks about the relationship of privacy studies and surveillance studies help prepare the way. As we will see just below, the privacy scholar Daniel Solove treats surveillance as one type of privacy harm. However, many scholars of the past two decades have begun to think that privacy study is subsumed under surveillance studies.[39] The relationship between the two is too complex to discuss in detail here.[40] Several observations about surveillance studies, however, suggest its intimate connections to computer and information history: (1) Surveillance has been practiced for millennia using documentation of many kinds and has been the subject of some important work in information history.[41] (2) Many actors besides the modern state practice surveillance, particularly as information about individuals and groups is increasingly commodified and monetized, e.g., for-profit corporations, information brokers, financial institutions, private security entities, Big Tech especially the Big Five (Alphabet [Google], Amazon, Apple, Meta [Facebook], and Microsoft), and non-profit organizations. (3) Surveillance studies of the past several decades has largely thrown off the formerly ubiquitous metaphors of Jeremy Bentham's and Michel Foucault's panopticon and George Orwell's Big Brother to include theories such as the culture of surveillance, the surveillant assemblage, algorithmic bias especially related to racialized and other-ed persons, contextual integrity, data feminism, self-surveillance and quantification of self (QoS), surveillance and platform capitalism, and problematization of the questionable primacy of risk assessment for making decisions. (4) Contemporary surveillance practices and regimes are *not* intended solely to observe. Rather the goal of most current modes of surveillance is the abstraction of human beings into data objects (sometimes termed "datafication") to rate and classify them and thereby to influence or monetize them. Much of surveillance studies is about the collection of information to establish power relationships of various types, often with the use of computing technologies. (5) Sociologists are prominent researchers in surveillance studies, as are Canadian, European, and other non-American scholars, giving surveillance studies a wide and heterogeneous purview.

In that context, we briefly review four different approaches to privacy: the attempts to categorize harms to privacy by Daniel Solove, in light of the failures by legal scholars to find a precise definition of privacy that is neither too narrow nor too broad; the communitarian writings of Amitai Etzioni, which focus on the good of the community rather than the good of the individual in privacy disputes; the writings on gender, race, and intersectionality by Anita Allen, who questions whether, at times, privacy protections should be held in check to attain other pro-social benefits such as accountability; and the literary studies approach of Deborah Nelson, which compares the writings of the confessional poets such as Anne Sexton and Sylvia Plath with the writings of Supreme Court justices on privacy.

Daniel Solove (A Legal Scholar)

From the 1970s, when legal scholars began to take a keen interest in privacy, until the end of the twentieth century, it was common practice for these scholars to proffer new definitions of privacy, in the hope (influenced by analytic philosophy's belief that if one can define a concept, one can study it) that if they can find just the right, nuanced definition, scholarship on this subject matter will fall neatly into place. Unfortunately for them, privacy is an elusive concept (much like other important but evasive concepts such as liberty, justice, or the good life) that has many different meanings in different settings. Thus, efforts to give a precise definition of privacy foundered. Some definitions were too broad and admitted phenomena that were not really about privacy; some were too narrow; and most were both too broad in some respects and too narrow in others.

The well-known legal privacy scholar Daniel Solove observed these definitional problems and applied an approach used by the philosopher Ludwig Wittgenstein in his later work, on the topic of family resemblance. In his consideration of what a game is, Wittgenstein argued that a game does not have a single essential defining characteristic but is instead characterized by a set of overlapping similarities that are held by some members of the family but not by all. Every member of the family shares at least some characteristics with another member of the family, but there is variation as well as similarity in the family class. Using this insight, Solove began to examine harms involving privacy and was not able to identify one singular type of harm that defined a privacy violation, but instead developed a classification scheme that included sixteen types of privacy violations, as described in Table 9.3.[42] We believe that computer and information historians can learn from Solove's and Wittgenstein's approach, not only when they study privacy in its many manifestations but also more generally in their analyses of complex or elusive concepts.

Amitai Etzioni (A Sociologist)

Overwhelmingly, privacy studies focus on harm to individuals if their privacy is abused. In the 1970s, the scholarship primarily concerned protection from the government, especially the federal government. In more recent years, privacy studies have shifted more to concerns about protection of individuals from businesses. Especially in our time of the Internet and social media, there has also been concern about protection of individuals from other individuals, e.g., criminals wanting to steal one's identity or bullies who want to extend their intimidation online through harassment, doxing, stalking, dissing, or trolling. One important early scholar, Amitai Etzioni, focused instead on the relationship of privacy to the well-being of a community, even if it means that individual privacy rights might need to be compromised.

As Etzioni himself explains communitarianism:

Table 9.3. Solove's Taxonomy of Privacy Harms

Information Collection	Surveillance	**the watching, listening to, or recording of an individual's activities**
	Interrogation	various forms of questioning or probing for information
Information Processing	Aggregation	the combination of various pieces of data about a person
	Identification	linking information to particular individuals
	Insecurity	carelessness in protecting stored information from leaks and improper access
	Secondary Use	use of information collected for one purpose for a different purpose without the data subject's consent
	Exclusion	the failure to allow the data subject to know about the data that others have about her and participate in its handling and use
Information Dissemination	Breach of Confidentiality	breaking a promise to keep a person's information confidential
	Disclosure	the revelation of truthful information about a person that impacts the way others judge her character
	Exposure	revealing another's nudity, grief, or bodily functions
	Increased Accessibility	amplifying the accessibility of information
	Blackmail	the threat to disclose personal information
	Appropriation	the use of the data subject's identity to serve the aims and interests of another
	Distortion	the dissemination of false or misleading information about individuals
Invasion	Intrusion	invasive acts that disturb one's tranquility or solitude
	Decisional Interference	the government's incursion into the data subject's decisions regarding her private affairs

[The definitions of each of the privacy harms are taken verbatim from Solove (2006, pp. 490–491).]

> Communitarianism is a social philosophy that maintains that society should articulate what is good—that such articulations are both needed and legitimate. Communitarianism is often contrasted with classical liberalism, a philosophical position that holds each individual should formulate the good on his or her own. Communitarians examine the ways shared conceptions of the good (values) are formed, transmitted, justified, and enforced. Hence their interest in communities (and moral dialogues within them), historically transmitted values and mores, and the societal units that transmit and enforce values such as the family, schools, and voluntary associations (social clubs, churches, and so forth), which are all parts of communities.[43]

Many of Etzioni's examples of a communitarian perspective on privacy do not concern information and communication technologies, although they often concern data or information. For example, in his book *The Limits of Privacy*,[44] he is a strong advocate of mandatory HIV testing for infants to protect public health and of publicly revealing the geographic locations of sex offenders to protect children in one's community. Communitarianism might support diminished individual privacy rights online, e.g., of anonymity online, to preserve some other public good such as national security or the ability to more readily enforce criminal law in cases where email is used to abet criminal actions. Etzioni's position is radically opposite to the cyberlibertarian view that, for example, applauds Apple for resisting giving access or support to law enforcement when an individual's cell phone has been involved in a criminal activity.

Etzioni's critics cite some of his more extreme ideas, such as mandatory expulsion of lepers to isolated communities in order to protect the health of the general population. Nonetheless, Etzioni provides a refreshing lens through which to understand the interactions of individuals and institutions as mediated by computing and electronic networks. His position asserts the import of community-oriented pro-social values when they are in tension with privacy interests of individuals.

Anita Allen (A Lawyer and Philosopher)

A third approach to privacy scholarship, by the feminist legal scholar and philosopher Anita Allen, is like Etzioni's communitarian critique in that it points to the importance of other pro-social values that might be oppositional to individual privacy. In Allen's book *Why Privacy Isn't Everything* (2003), she examines the tensions between two particular social values, privacy and accountability.[45]

Much as Solove identified the complexity of the concept of privacy, Allen writes about the complexity of accountability. In 1859 British philosopher John Stuart Mill wrote in his famous essay *On Liberty*, "The individual is not accountable to society for his actions, in so far as these concern the interests of no person but himself."[46] This essay led many philosophers and policy scholars of the 19th and 20th centuries to posit and naturalize a public-private divide, in which accountability was in force in the public realm but was secondary in consideration in the private realm.[47] Despite having vigorously defended privacy rights in her work as a legal and privacy scholar, Allen argued against Mill's position, claiming that accountability was equally at play in the private realm. After some analysis of various kinds of accountability—such as to herself, her friends, her family, or her race—she proffers the following definition:

> Individuals are accountable for aspects of their lives considered personal, to the extent of being obligated or impelled, whether by internal or external norms, (1) to provide information to others; (2) to explain themselves; (3) to justify themselves; (4) to face positive or negative sanctions; and/or (5) to maintain a predictable pattern of conduct on which others can reasonably rely.[48]

Allen elaborates, explaining that "we are accountable for private matters when we are bound by obligation, duty, or responsibility—or motivated by a sense of obligation, duty, or responsibility—to reckon with others. We reckon by supplying them with personal information, explanations, justifications, sanctions, or reliability."[49]

Using deeply personal, and sometimes painful, stories from her own life, Allen talks about accountability and privacy in the personal sphere as it concerns family, race, health, and sex. It is beyond the scope of this essay to explain how Allen comes to reconcile privacy and accountability in the personal sphere, but there are three lessons to note from this work of Allen for the information and computer historian.

First, privacy and accountability are two social goods that at times are in tension. It is common in public policy and law for social goods to be in contest with one another. To give an example of particular interest to the computer historian, the freedom to pursue technological innovation and distribution is sometimes in contest with national security, ecological protection, equal treatment of people, or other social goods; and the computer or information historian can learn from Allen how to examine these kinds of contests.

Second, while much of the early writing in computing history was primarily about the public sphere, for example, about the building of computers for business, scientific research, or national security, much of the more recent research about computing history is about individuals and how they carry out their daily personal lives—so, very much about the private rather than the public sphere. Thus, Allen offers suggestions about how to approach this personal sphere.

Third, Allen provides guidance to the computer and information historian about how to use stories from the author's personal life to illuminate abstractions, that is, concepts that are hard to grasp or relate to. The trick is to be able to build from the personal story to the general account, and Allen is masterful at this task.

Deborah Nelson (A Literary Scholar)

For the past few decades, ever since Jacques Derrida, many human endeavors have been able to be read as texts by scholars and thus subject to rhetorical examination.[50] Our fourth example of an approach to privacy scholarship is from a book by English literature professor Deborah Nelson, *Pursuing Privacy in Cold War America*.[51] Nelson juxtaposes two apparently disparate genres of American expression that engaged and helped define privacy, both emerging in late 1950s America: the lyric poetry of the confessional poets (especially Anne Sexton, Sylvia Plath, Robert Lowell, and W. D. Snodgrass) and a set of the constitutional law rulings of U.S. Supreme Court judges (including *Olmstead v. U.S.*,[52] *Katz v. U.S.*,[53] *Loving v. Virginia*,[54] *Griswold v. Connecticut*,[55] *Eisenstadt v. Baird*,[56] and *Roe v. Wade*[57]). Nelson employs *formal analysis*, that is, the examination of the expressive forms of these two rhetorical modes, embedded in a deep historical contextualization of the Cold War era that included "the growth of suburbia, the advent of television, the popularity of psychoanalysis, the arrival of computer databases, and the spectacles of confession associated with McCarthyism."[58]

Nelson offers several things of value to the computer or information historian who wants to write a policy-rich account. She demonstrates the value of poetry and judicial writings as primary source materials, which had heretofore been largely absent from these historical accounts. She offers a method that combines the deep reading of critical analysis and insights of literary theory with a strong historical sensitivity. Finally, she offers some major findings that can shape the perspective in our historical writings. Chief among these findings are:

- The shift from the abstract and personal tort vision of privacy (e.g., as expressed by Warren & Brandeis's foundational essay in 1890 on "The Right to Privacy"[59]) to the gendered, specific, and highly domesticated vision of privacy as enacted in the Supreme Court decisions during the second half of the twentieth century mentioned above; with the newfound implication that privacy clearly fits into the constitutional domain, emphasizing privacy's deeply political and rhetorical character.
- A recognition that the concept of privacy is tied to people rather than to places or things—as was commonly assumed in the earlier jurisprudence that emphasized papers, possessions, and the home.
- A recognition of the value of confessional poetry to explore this relational nature of privacy and its link to the voluntary confessional mode of mid- and late-20th-century lyric poetry, particularly as tied to the gender, class, and other identities of women.
- A recognition of the value of privacy to both autonomy and community action. The abstract, classically liberal tort of privacy, with its reliance on property, emphasized the white male adult of financial means, the upper- or middle-class home, and the attendant domestic relations. Instead, the growing gendered understanding of privacy that Nelson describes expands the purview of privacy

explicitly to women, reproductive behavior, and the autonomy of actors who may choose to expose or keep hidden what we might term the intimate characteristics of life. This life is one lived in community with others since privacy is relational and is a shared interest, key elements of this newer vision.

- A problematization of the assumption of a supposed public/private dichotomy (the doctrine of "separate spheres") that undergirded the earlier liberal, atomistic vision of privacy and which has been largely overthrown, particularly by power-sensitive critiques, legal feminist scholarship, and gendered, historical analyses of American politics.[60]
- A warning about our long-lived fantasy of some strong version of privacy in the past that has now been "lost." Such self-delusion not only misrepresents the past, but it also limits our imagination of how we came to our current concerns about the presumed "death of privacy," fed by the "curiously amnesiac rhetoric of the privacy crisis."[61]

Instead, Nelson calls us to a full-hearted embrace of a historically and formally informed understanding of the "intolerable ambivalence of privacy,"[62] a clear call to historians of information and computing.

Conclusion

We have emphasized three points in this chapter. First, law and public policy are powerful actors in shaping the development and use of information technologies, especially considered in its many forms. Second, computer and information historians have not entirely neglected law and public policy issues, but they have addressed them only selectively. Deeper and broader engagement with public policy and the law would enhance their historical accounts. Third, there are many examples of legal and policy scholarship that computer and information historians can learn from; we have identified some here, from a variety of academic fields and from a variety of policy and legal domains.

There are many matters, however, that we leave unaddressed. For example, one of our chapter's reviewers noted that putting more law and public policy dimensions into the history of computing and information technologies gives rise to two forms of idealism. The first type springs from what the reviewer called the "libertarian-utopian bent" among some computer scientists, as well as policy makers, business interests, and the general public. This ideological commitment leads to often exaggerated and unrealistic views of politics and competing interests. The second form of idealism comes from the fact that these actors and others in the policy realm have "aspirational views of policy," emphasizing that their own work is so important that it should drive political decisions at the expense of other points of view. In parallel to this kind of aspirational view is a wider and commonly shared form of aspiration that asserts that, if we could only get policy "right," the complex conflicts, conundrums, and contradictions that plague all polities would disappear. While all of us can understand the temptations of these kinds of idealism and aspirations, political realism and scholarly rigor demand more. Beyond this brief discussion, however, our chapter cannot explore this particular line of inquiry further.

There are two other major limitations to this chapter. First, although we have spoken of information history and computer history together, our focus has primarily been on computer history. If space permitted, we could draw out more fully similar, important points about information history. Second, this account is mostly about the United States. Although there are differences in policy and legal environments from country to country, much of what we say would apply when we consider the histories of information and computing across the globe. Even without addressing these and other limitations, we strongly maintain the value of law and public policy to writing history, particularly to computing and information historians, and we look forward to the outstanding work that will emerge with this emphasis.

Notes

1. There are important exceptions, where policy considerations are significant factors in telling a story about computer history. Examples include an older, general history of computing in America by Kenneth Flamm, *Creating the Computer: Government, Industry,*

and High Technology (Washington, DC: Brookings Institution Press, 1988); Flamm's companion volume on Japan and western Europe as well as the U.S., *Targeting the Computer: Government Support and International* Competition (Washington, DC: Brookings Institution Press, 1987); books about computing in the Cold War, notably including Atsushi Akera, *Calculating a Natural World: Scientists, Engineers, and Computers During the Rise of U.S. Cold War Research* (Cambridge, MA: MIT Press, 2008); and Paul N. Edwards, *The Closed World: Computers and the Politics of Discourse in Cold War America* (Cambridge, MA: MIT Press, 1996). Also see an important book on computers as a tool of government bureaucracy by Jon Agar, *The Government Machine: A Revolutionary History of the Computer* (reprint; Cambridge, MA: MIT Press, 2016); and several books on the history of the Internet, notably including Andrew Russell, *Open Standards and the Digital Age* (Cambridge, England: Cambridge University Press, 2014); Shane Greenstein, *How the Internet Became Commercial* (Princeton, NJ: Princeton University Press, 2015); and Peter Cowhey and Jonathan Aronson, *Transforming Global Information and Communications Markets: The Political Economy of Innovation* (Cambridge, MA: MIT Press, 2009). There is also a large article literature on this topic, well beyond our scope to list it.

2. For example, one of our chapter's reviewers helpfully noted policy actions and instruments such as non-competition clauses for high-tech workers, Bell Labs' licensing the transistor as per U.S. Department of Defense wishes, Section 230 of the Communications Decency Act of 1996 (Title V of the Telecommunications Act of 1996, PL 104–104), and various governmental actions and inactions related to artificial intelligence, among others. As we say, however, our intention is *not* to provide a list of what we regard as the most important policy instruments and historical developments related to ICTs. Even a highly curated and edited list would be immense, but, more importantly, such a list would appear prescriptive and directive in a misleading way. It might tend to inhibit rather than encourage research by not reflecting the research interests and scholarly commitments of historians wishing to explore the policy dimensions of computing and information from their own perspectives. Hence, this chapter takes a more introductory and reflective, as well as less guiding path.

3. Thomas Haigh and Paul Ceruzzi, *A New History of Modern Computing* (Cambridge, MA: MIT Press, 2021); and James W. Cortada, *All the Facts: A History of Information in the United States since 1870* (Oxford, UK: Oxford University Press, 2016).

4. Harvard Mark I (1944), ENIAC at University of Pennsylvania (1946), Project MAC at MIT (1963), Project GENIE at Berkeley (1964), PLATO IV at the University of Illinois (1972), and more distributed support for the creation of COBOL (1959) and reliable software for the Apollo space mission.

5. IBM tabulating machinery by the Internal Revenue Service (1950); IBM 701 (1953), IBM Stretch (1961), and Cray 1 (1976) by Los Alamos National Laboratory (1953); IBM 650 by NASA Ames (1955); Philco SOLO by the National Security Agency (1955); and Control Data 1604 by the U.S. Navy (1960).

6. RAND Corporation (1946), Armed Forces Security Agency (1949), ARPA (1958), National Science Foundation (1960), Office of Systems Analysis (1966), Strategic Defense Initiative (1984), and NSFNet (1986).

7. Bureau of Prisons (1930), Federal Statistics Board (1931), National Recovery Administration (1933), Agricultural Adjustment Administration (1933), Security and Exchange Commission (1934), Soil Conservation Service (1935), Works Progress Administration (1935), Office of Research, Education and Statistics (1935), Occupational Outlook Service (1940), Office of Economic Welfare (early 1940s), Office of Facts and Figures (1941), Office of Strategic Services (1942), Educational Research Information Center (1979), Paperwork Reduction Act (1980), and Department of Homeland Security (2002).

8. Privacy Act (1974), Family Educational Rights and Privacy Act (1974), Right to Financial Privacy Act (1978), Computer Matching and Privacy Protection Act (1988), Employment Polygraph Protection Act (1988), Video Privacy Protection Act (1988), Driver's Privacy Protection Act (1992), Health Insurance Portability and Accountability Act (1996), Identity Theft and Assumption Deterrence Act (1998), Children's Online Privacy Protection Act (1998), Children's Online Protection Act (1998), and Financial Modernization Act (1999). With regard to privacy, also see the disclosure of National Security Agency massive data collection on Internet use (2013), and U.S. Supreme Court ruling on access to private cell phone communication.

9. The U.S. Supreme Court decision *Sony v. Universal City Studios* (464 U.S. 417, 1984), the Audio Home Recording Act (1992), and the Digital Millennium Copyright Act (1998).

10. The data collection method used in this table significantly undercounts the coverage of Cortada because he speaks often about general trends—many of which are directly relevant to this table—and often not about specific events. Our methodology identifies only specific events.

11. William Aspray and Philip Doty, Does Technology Really Outpace Policy, and Does It Matter? A Primer for Technical Experts and Others, *Journal of the Association for Information Science and Technology* 74, no. 8 (2023): 885–904. https://doi.org/10.1002/asi.24762

12. It is beyond the scope of this chapter to demonstrate how to conduct public policy analyses, but fortunately there are many textbooks that cover this topic. See, for example, Wayne Parsons, *Public Policy: An Introduction to the Theory and Practice of Policy Analysis* (Cheltenham, UK: Edward Elgar Publishing, 1999); Cristoff Knill and Jale Tosun, *Public Policy: A New Introduction* (2nd ed., London: Bloomsbury, 2020); Cosmo Howard and Robin Miller, *Social Policy, Public Policy: From Problem to Practice* (New York: Routledge, 2020); and Scott Furlong and Michael Kraft, *Public Policy: Politics, Analysis, and Alternatives* (7th ed., Thousand Oaks, CA: CQ Press, 2020). Also see handbooks such as: Michael Moran, Martin Rein, and Robert E. Goodin, eds., *The Oxford Handbook of Public Policy* (Oxford, UK: Oxford University Press, 2008); or Eduardo Araral et al., eds., *Routledge Handbook of Public Policy* (London: Routledge, 2012).

13. Jessica Litman, *Digital Copyright* (Ann Arbor, MI: Michigan Publishing, University of Michigan Library, 2017). https://quod.lib.umich.edu/m/maize/mpub9798641/1:9/—digital-copyright?rgn=div1;view=fulltext

14. Simone Browne, *Dark Matters: On the Surveillance of Blackness* (Durham, NC: Duke University Press, 2015).

15. Julie E. Cohen, *Between Truth and Power: The Legal Constructions of Informational Capitalism* (Oxford, UK: Oxford University Press, 2022).

16. In reading an earlier draft of this section, one reviewer argued that our account is misleading because the reviewer sees Lessig first and foremost as an activist, e.g., for changes in copyright policy as associated with Creative Commons. We acknowledge Lessig's important activist achievements, but we believe that Lessig further developed and popularized an analytical framework for examining the ways to shape technology. We believe it is worthwhile to focus on Lessig's analytic framework independent of his activist persona.

17. "On the Internet, nobody knows you're a dog." Peter Steiner, cartoon, *The New Yorker*, July 5, 1993.

18. *A&M Records, Inc. v. Napster, Inc.*, 239 F.3d 1004 (9th. Cir., 2001).

19. The story is told in detail in Robert A. Caro, *The Power Broker: Robert Moses and the Fall of New York* (New York: Knopf, 1974).

20. For excellent material on the rhetorical turn in many disciplines, particularly those of interest to this chapter, see, for example, John S. Nelson, Allan Megill, Deirdre N. McCloskey, and Donald N. McCloskey, eds., *The Rhetoric of the Human Sciences: Language and Argument in Scholarship and Public Affairs* (Madison, WI: University of Wisconsin Press, 1987), particularly the chapters by James Boyd White on rhetoric and law and Charles Anderson and Charles Shapiro on rhetoric and public affairs; also see journals such as *Rhetoric and Public Affairs*, among others.

21. Deborah Stone, *Policy Paradox: The Art of Political Decision Making* (rev. ed., New York: W. W. Norton, 2002), 11.

22. Parsons, 234.

23. Donald Schön and Martin Rein, *Frame Reflection: Toward the Resolution of Intractable Policy Controversies* (New York: Basic Books, 1994), 81 and 159.

24. Stone, 154.

25. Stone, 380.

26. Carol Bacchi, *Analysing Policy: What's the Problem Represented to Be?* (Frenchs Forest, NSW: Pearson Higher Education, 2009), x. This point is underscored by Ray Rist's assertion that policy actions always "imply theories. Whether stated explicitly or not policies point to a chain of causation between initial conditions and future consequences." Ray C. Rist, Influencing the Policy Process with Qualitative Research, in Norman K. Denzin and Yvonna S. Lincoln, eds., *Handbook of Qualitative Research* (pp. 545–557) (Thousand Oaks, CA: Sage, 1994), 555.

27. Emery Roe, *Narrative Policy Analysis: Theory and Practice* (Durham, NC: Duke University Press, 1994).

28. Mark Considine, *Public Policy: A Critical Approach* (Melbourne: Macmillan Education, 1994).

29. Stone, 249.

30. Schön and Rein, 175.

31. Bacchi, 30.

32. Eugene J. Meehan, Policy: Constructing a Definition, *Policy Sciences* 18, no. 4 (1985): 291–311, 304.

33. Charles E. Lindblom, The Science of "Muddling Through," in Stella Z. Theodoulou and Matthew A. Cahn, eds., *Public Policy: The Essential Readings* (pp. 113–127) (Englewood Cliffs, NJ: Prentice-Hall), 123. (Original work published 1959).

34. Sandra Braman, *Change of State: Information, Policy, and Power* (Cambridge, MA: MIT Press, 2009), xvii–xviii.

35. Roe, x.

36. Roe, 112.

37. Stone, 381.

38. Further, as one of our chapter reviewers notes, privacy is "evergreen" in public policy studies, particularly about information and computing technologies, without making totalizing claims that privacy is *the* most important or fundamental concern in such policy studies.

39. The surveillance study literature is quite large and interdisciplinary. To get a good flavor of this scholarship, see the work of researchers such as David Lyon, Oscar Gandy, Simone Browne, and Kevin D. Haggerty and Richard V. Ericson, as well as journals such as *Surveillance & Society* and *The Information Society* and edited collections such as: T. Monahan and D. M. Wood, eds., *Surveillance Studies: A Reader* (Oxford, UK: Oxford University Press, 2018); Kirstie Ball, Kevin Haggerty, and David Lyon, eds., *Routledge Handbook of Surveillance Studies* (London: Routledge, 2012); and R. E. Dubrofsky and S. A. Magnet, eds., *Feminist Surveillance Studies* (Durham, NC: Duke University Press, 2015).

40. But see Gary T. Marx, Coming to Terms: The Kaleidoscope of Privacy and Surveillance, in Beate Roessler and Dorota Mokrosinska, eds., *Social Dimensions of Privacy: Interdisciplinary Perspectives* (pp. 32–49) (Cambridge, UK: Cambridge University Press, 2015).

41. See, for example, Andreas Marklund and Laura Skouvig, eds., *Histories of Surveillance from Antiquity to the Digital Era* (London: Routledge, 2022).

42. The taxonomy is given in Daniel Solove, A Taxonomy of Privacy, *University of Pennsylvania Law Review* 154, (2006): 477–560. Solove elaborates the taxonomy in his book *Understanding Privacy* (Cambridge, MA: Harvard University Press, 2008).

43. Amitai Etzioni, Communitarianism, *Encyclopedia of Community: From the Village to the Virtual World*, Vol. 1, A–D, Karen Christensen and David Levinson, eds., Sage Publications, 224–228.

44. Amitai Etzioni, *The Limits of Privacy* (New York: Basic Books, 1999).

45. Anita L. Allen, *Why Privacy Isn't Everything: Feminist Reflections on Personal Accountability* (Lanham, MD: Rowman & Littlefield, 2003).

46. John Stuart Mill, *On Liberty* (London: J. W. Parker and Son, 1859), chapter 5.

47. When discussing the work of literary scholar Deborah Nelson below, we note the effective problematization of this public/private "divide" by many feminist scholars.

48. Allen, 20–21.

49. Allen, 21.

50. See Jacques Derrida, *Of Grammatology* (originally published in 1967, English translation 2016, Baltimore, MD: Johns Hopkins University Press). For relevant discussion, see, for example, David Schwalkwyk, What Does Derrida Mean by "the Text"?, *Language Sciences* 19, no. 4 (1997): 381–390; or Gavin P. Hendricks, Deconstruction the End of

Writing: "Everything Is a Text, There Is Nothing Outside Context," *Verbum et Eclasia, Pretoria* 37, no. 1 (2016): 1–9.

51. Deborah Nelson, *Pursuing Privacy in Cold War America* (New York: Columbia University Press, 2002).

52. *Olmstead v. United States*, 277 U.S. 438 (1928).

53. *Katz v. United States*, 389 U.S. 347 (1967).

54. *Loving v. Virginia*, 338 U.S. 1 (1967).

55. *Griswold v. Connecticut*, 381 U.S. 479 (1965).

56. *Eisenstadt v. Baird*, 405 U.S. 438 (1972).

57. *Roe v. Wade*, 410 U.S. 113 (1973).

58. Taken from the description of the book on the back cover of the paperback edition.

59. *59.*Samuel D. Warren and Louis Brandeis, The Right to Privacy, *Harvard Law Review* 4, no. 5 (1890): 193–220.

60. See, for example, Linda K. Kerber, Separate Spheres, Female Worlds, Women's Place: The Rhetoric of Women's History, *The Journal of American History* 75, no. 1 (1988): 9–39; Judith A. McGaw, No Passive Victims, No Separate Spheres: A Feminist Perspective on Technology's History, in Stephen Cutcliffe and Robert Post, eds., *In Context: History and the History of Technology* (pp. 172–191) (Lehigh, PA: Lehigh University Press, 1989); and Rosalind Rosenberg, *Beyond Separate Spheres: Intellectual Roots of Modern Feminism* (New Haven, CT: Yale University Press, 1983).

61. Nelson, xvii.

62. Nelson, xx.

Part IV

Broadening Computing and Information History

Chapter 10

Writing the History of Telecommunications and Data Networks

Valérie Schafer

John Quarterman's book *The Matrix: Computer Networks and Conferencing Systems Worldwide*,[1] published in 1990, provides valuable insights into the proliferation of data networks during the pre-Web era. At a time when the Internet was gaining prominence, there was already a myriad of technical solutions, both private and public. These ranged from academic to entrepreneurial, encompassing university networks and national and even local solutions. This diverse environment involved various stakeholders, including institutional players, telecommunications and IT companies and research centers, universities, private providers, and a wide range of uses and users. Interoperability was a real challenge, with the telecommunications and computing sectors working sometimes in concert and sometimes independently to offer connectivity and exchange solutions to the academic community and the private sector—at this stage the general public was less involved. Within this ensemble, meticulously mapped by Quarterman, the "network of networks" already occupied its own position, without eclipsing the abundance of solutions developed from the 1970s onwards. This fueled many discussions in the 1990s, especially within standardization bodies such as the ISO (International Organization for Standardization) and ITU (International Telecommunication Union).

In this context, some early articles and books stand out. Janet Abbate's *Inventing the Internet*, for example, published in 1999,[2] has become a classic for the history of the Internet, delving into the complex negotiations, arrangements, and collaborations that shaped the genesis of ARPANET and then the Internet, primarily in the United States. This research demonstrates the interplay between various stakeholders, the innovative paths taken by technology, and the complex connections that shaped early networks and relations within a US military-scientific-industrial complex. Abbate neatly weaves these strands together, while also delving into matters of governance and questions of openness.

As the digital landscape has developed towards the general public, new avenues of research have emerged, leading to further exploration and investigation in the realm of network history. They mirror the continuous evolution of digital technology, a dynamic state of the art, and the multidisciplinary nature of research. Indeed, this history can be viewed from the angles of engineering, infrastructures, standards, and governance, but it gains even more in diversity when looking at users and content. The development of data networks for the general public and the emergence of disciplines such as digital studies, science and technology studies (STS), gender studies, and maintenance studies has had a significant impact on research—with attention now turning to highly contemporary topics (such as privacy, spam, and security) from a historical angle. Issues are expanding and diversifying, and researchers are turning to previously unexplored material (for example, born-digital heritage such as Internet forums and Web archives) and sources analyzed in new ways (e.g., network maps).

The history of data networks, therefore, stands at the intersection of several trends. It is not that historians are unsure of which avenue to explore; more, that they find themselves at a nodal point, in a dialogue with other fields such as STS or sociology of innovation. Although firmly rooted in history, their research is at the junction of local and global scales, of infrastructures and uses, of media studies, digital studies, and the history of computing and (tele)communications.

This chapter is not intended to offer an exhaustive state of the art, but rather to reflect on these evolving trends in the history of data networks.

It highlights prominent topics within this history. It starts with the narratives surrounding ARPANET and explores government support, competition among companies, governance, and regulation. It looks at how the history of the Internet is linked with the history of pre-networking technologies and alternative developments in networking technologies in countries outside the United States. It examines the way in which technology has been shaped by both users and organizations, shedding light on several trends including a strong interest in digital cultures, "missing narratives," and the practices of early adopters. It also delves into significant historical approaches, incorporating reflections on methodologies and sources, including some linked to my own research.[3] The chapter acknowledges the contributions made by major publications, research communities, and various specific approaches to improving our understanding of the history of data networks.

The first three parts explore narratives that focused more on the genesis of data networks and the way in which research on ARPANET and other networks developed. It also emphasizes how these narratives have co-evolved with STS and benefited from new perspectives on governance and standards. It is in some ways an approach rooted in infrastructures and the lower layers of data networks, but one that also digs into the role of institutions, research centers and innovators, and balances of power in the areas of networks and standards. The fourth part of this chapter focuses on the "missing narratives" that were highlighted a few years ago. These gaps have progressively been filled, leading to the emergence of new research questions that are less US-centric, less focused on innovation, and more interested in different contexts and multi-layered approaches. The last section of this chapter examines the "user turn" that characterizes some current research, in which researchers draw on born-digital heritage and other sources to explore "networking in society." The chapter closes with some reflections on the still underexplored avenues that may yet be investigated by the history of data networks, as well as possible future trends for this history, which is not bound to become a field of its own but may continue to reflect the cauldron of innovation that is networks and communication.

Roots and Origins of ARPANET and the Internet

When Manuel Castells published *The Rise of the Network Society* as volume 1 of *The Information Age*[4] in 1996, the history of data networks was still in its infancy. As already mentioned, Abbate's book in the late 1990s shed important light on the genesis of ARPANET and the Internet. It soon became a standard reference and a must-read. She delved into the discussions within the IPTO (Information Processing Techniques Office) at ARPA (Advanced Research Projects Agency) and the technical developments at BBN (Bolt Beranek and Newman), a company involved in the technical realization of the network. She also explored the relationship with the universities associated with the launch of ARPANET (ARPA's contractors) and the changing views and technical solutions that led to TCP/IP and the Internet.

There had of course already been interest in the development of networks, and changes that were affecting computing. One article worth mentioning is A. L. Norberg's "Changing Computing: The Computing Community and DARPA," in *IEEE Annals of the History of Computing* in 1996.[5] Paul Ceruzzi was also an early believer in the value of integrating the history of networks and the resulting changes in technology, places of power, and human-machine relationships into a broader reflection on the history of computing and its economic aspects. The recent, extended, and updated version of *A New History of Modern Computing* by Thomas Haigh and Paul Ceruzzi further demonstrates this growing consideration of data networks.[6]

In the second half of the 1990s, as is still the case today, historians also had to deal with the huge volumes of material being produced by data network stakeholders and journalists. The latter were quick to portray the disruptive nature of Internet development and networked communications more generally, which were beginning to extend beyond the scientific and corporate communities. For example, airlines and banks were early adopters of data networks, with examples including the SABRE network in the 1960s. In addition to Quarterman's work, Carl Malamud's 1992 book *Exploring the Internet: A Technical Travelogue*, Katie Hafner and Matthew Lyon's *Where Wizards Stay Up Late: The Origins of the Internet*, Michael Hauben and Ronda Hauben's *Netizens*, and

Howard Rheingold's *The Virtual Community*[7] are also worth mentioning. The last two focused more on communicative and social uses, especially within the first virtual communities—which did not only use the Internet but also were based in spaces such as The Well, newsgroups, or Bulletin Board Systems (BBSs).

In France, *L'imaginaire d'Internet*[8] by the sociologist Patrice Flichy is also key for Internet studies. It analyzes the cooperative modes and shared values that were central to the first Internet communities, based on openness, horizontality, and peer-to-peer reviews (P2P). This approach has had a considerable impact on the history of the "network of networks," at least in France; and there is a certain convergence of views with the writings of *early adopters*, who tend to underline the ethical, horizontal, P2P, and "democratic" aspects of the Internet.

An oral memory of ARPANET's history also emerged early on at the Charles Babbage Institute, thanks in particular to Judy O'Neill, Arthur Norberg, William Aspray, and a few other researchers.[9] Norberg and Aspray collected not only testimonies but also sources related to ARPANET. In 2000, research files containing published literature, government documents, technical reports, and correspondence from the Information Processing Techniques Office (IPTO) and Bolt Beranek and Newman were deposited at the Charles Babbage Institute, in addition to the series of interviews the CBI staff conducted with pioneers in the field.[10]

As Alexandre Serres noted in his thesis when mentioning the CBI oral histories: "With this type of document, we are already leaving the realm of 'organized memory' and celebration and are finally entering the realm of historical intelligibility and beginning to see things in perspective."[11] In the early 2000s, Serres studied the genesis of ARPANET using the CBI oral interviews and drawing on actor-network theory. He emphasized the "translation" phenomena at work since the advent of time-sharing, which contributed to the emergence of ARPANET and new modes of governance. He mentioned the RFC (request for comments) system: a set of collective technical specifications, originally published in hard copy and then online, that represents an extraordinary open record of the development of ARPANET and the Internet. Packet switching, the relationships between researchers involved in ARPANET and companies—especially BBN—and telecommunications are analyzed through the notion of "translations." These enabled Serres to trace the circulation of players and ideas. In his work, we can already see the influence of what would become core approaches in science and technology studies (STS). He draws on the work of pioneering STS researchers to explore the notion of "information infrastructure," inspired by Susan Leigh Star and Geoffrey Bowker.

The Influence of STS

This influence has grown stronger over the past two decades and is now prominent in the field of innovation and networks: researchers such as Laura DeNardis, Sandra Braman, and Francesca Musiani show a particular interest in Internet infrastructures and decentralized infrastructures in a broad sense, as well as in the multiple agencies and multistakeholderism involved in co-shaping data networks. STS and the history of technology evolved in parallel for a while, but their shared interest in sociotechnical perspectives and infrastructures has led to some convergence and calls for a vision of innovation that is sensitive to complex agencies.

Some researchers have acted as a bridge, e.g., Abbate, who in addition to shedding further light on the history of ARPANET and the Internet, particularly its privatization,[12] argued in favor of this approach in "L'histoire de l'Internet au prisme des STS"[13] ["The history of the Internet through the lens of STS"]. The same trend was clearly visible in 2019 in an article by Fenwick McKelvey and Kevin Driscoll studying the IMP (Interface Message Processor) as a boundary object. Focusing on the role of IMPs—devices that form a sub-network within ARPANET and enable openness and interoperability between equipment from different manufacturers—they invite us to reconsider the relationship between telecommunications and computing through key notions developed in STS. The article by McKelvey and Driscoll addresses a central element that has aroused the interest of historians: the relationships between telecommunications, including AT&T and the "Bell System," and computing, which converge within the field of data networks with greater or lesser difficulty, depending on economic and geographical configurations:

> To investigate the relationship between ARPANET and the Bell System, we focus on the Interface Message Processor. The Interface Message Processor, or IMP, was a boundary object in both the practical and theoretical senses of the term. Positioned between the telephone network and local computers called "hosts," the IMPs carried out the defining functions of the ARPANET: addressing, routing, and switching packets between hosts, local and remote [. . .]. Within the first IMP's industrial metal cabinet, the concerns of the Bellheads and Netheads met in a swirl of acoustic signals, electrical pulses, and digital bits, and the participation of both sets of stakeholders was necessary to bring an IMP online.[14]

From the very beginning of data network history there has been a desire to document and understand the choices made in selecting protocols, whether in Abbate's book, in the thesis I wrote in the mid-2000s on two packet-switching data networks developed in France, Transpac and Cyclades,[15] or in Andrew Russell's reflections leading to his book *Open Standards and the Digital Age* in 2014.[16] The decisions that prevailed in terms of packet switching, the foundations of the ARPANET and then Internet protocols, and technical choices made by similar or competing networks reflect the interactions between telecommunications companies and computer scientists. McKelvey and Driscoll's article sheds new light on this latter aspect, drawing on historical design documents from the *Bell Technical Journal* and public records that describe the evolution of ARPANET. It is not only their sources but also their STS-inspired approach that deepens their analysis and underlines a role for controversy studies and the study of technical systems as political arrangements and negotiations.

Standards and Governance

The way the Internet developed was disruptive not only from a technical perspective, but undoubtedly also because it led to the establishment of multiple new governance bodies and procedures. This gave rise to sometimes critical views from early contemporaries, notably Milton Mueller.[17] When the Internet Corporation for Assigned Names and Numbers

(ICANN) was created in 1998, he was highly critical of its close dependence on the United States.

As Musiani and I noted, in parallel to the major debates on Internet governance, the academic field has co-shaped the concept of global governance of the Internet and played a part in these debates. Several phases can be identified, the first of which focuses on "Internet exceptionalism" and the understanding of the Internet as a space of its own:

> While the first document that comes to mind as a symbol of this phase is perhaps the 1996 flamboyant "Declaration of Independence of Cyberspace" by John Perry Barlow, from a political and academic standpoint, this phase was marked by discussions about whether the Internet should develop its specific regulation system, more decentralized and multi-centered, and not primarily based on state-centered control [. . .], as well as by debates on "cyberspace sovereignty."[18]

We identified the second phase as starting in the late 1990s and involving the search for a novel framework and structure for Internet governance. These issues became particularly salient with the creation of ICANN.

Two challenges were central to the governance debate. The first was the openness of Internet standards. David Clark's words on the 24th anniversary of the Internet Engineering Task Force (IETF) in 1992 have become famous: "We reject kings, presidents and voting. We believe in rough consensus and running code." But there were nonetheless internal divisions and debates, as well as conflicting memories and testimonies, such as those of Alex McKenzie, who shared his experience of discussions on Internet debates in the Nethead community. He reflected on the consequences of the choices pushed by Robert Kahn within ARPANET:

> History is not an experiment that can be run multiple times from different initial conditions, but I have often speculated about what might have been different if DARPA had chosen to adopt the INWG 96 proposal in early 1976. I once thought that if the research community were united behind one End-to-End Protocol, we might have avoided most or all of the incredibly wasteful ISO Open Systems Interconnection activity. However, the OSI effort was probably an

> inevitable clash between computer manufacturers and PTTs, and of both against IBM.[19]

Historians are also interested in a second aspect, namely the often-heated debates within standardization bodies. The discussions around multi-layered networks were far from consensual, as Russell showed in "'Rough Consensus and Running Code' and the Internet-OSI Standards War"[20] before proposing a longer-term analysis in his book *Digital Standards*. I followed a largely parallel line of investigation based on the French case that is closely intertwined with these international issues, and we then wrote a joint article on the history of Cyclades[21] following Russell's visit to Paris to conduct oral interviews with French researchers in 2012.

In addition to the question of layered architectures within the Internet and the OSI model, the history of protocols calls into question the power of some traditional players, in particular the ITU (International Telecommunication Union), and more generally the role of telecommunications in defining or shaping Internet governance. This issue played out at several levels, from the national picture, illustrated, for example, by the development of Cyclades and Transpac in France, to the international level I explored in a book edited by Gabriele Balbi and Andreas Fickers on the history of the ITU. The policy of promoting open "de facto standards" within the Internet has changed the process of standardization, which was traditionally negotiated in well-recognized bodies like the International Organization for Standardization (ISO), the ITU and—in particular for data networks—the Consultative Committee for International Telephony and Telegraphy (CCITT). By analyzing organizational archives, ITU reports, press materials, and Web archives, I demonstrated how the CCITT struggled to maintain real power while the Internet was developing in parallel through dedicated bodies like the IAB (Internet Architecture Board), IETF, and ICANN. The ITU sought to restore its techno-diplomatic role through multistakeholderism, with the World Summit on Information Society (2003 and 2005) marking a crucial step in this process.[22]

My PhD (defended in 2007) on the Cyclades and Transpac networks drew on oral histories and the rich archives of Inria (the French National Institute for Research in Digital Science and Technology). The latter collections contain abundant documentation on the French case, as well as providing reports relating to strategies deployed internationally and records of discussions with other networks and teams. In particular, this material gave a precise picture of discussions with the British, who played a role in packet switching. They were connected to ARPANET at an early stage; and, as in France, different technical solutions for packet-switching networks were proposed, for example, by the NPL (National Physical Laboratory) and the British Post Office.[23] It should be noted, however, that the tensions between proponents of a packet-switching approach akin to ARPANET (i.e., the Cyclades network[24]) and the virtual circuits adopted by telecom operators were particularly prevalent in France. There is a wealth of French archives on these "two schools" of packet switching, although the archives of telecom operators were not always easily accessible since they were spread across several locations that were not always well classified or open. I had to deal with a certain asymmetry of sources when studying these two networks. This asymmetry and the issue of access to sources may also partly explain the level of interest shown by researchers in ARPANET and the Internet compared to some other networks. The Internet's players have made its abundant documentation widely available via RFCs, network maps, numerous oral testimonials, and equally numerous digital traces. In this respect, we should mention the remarkable work by Morgan Currie and Bradley Fidler on Internet maps.[25] It testifies to the possibility of a fruitful internal and external critique of this type of document, further enriching the history of ARPANET. The initial focus on its overall genesis now seems to have shifted to a critical analysis of more specific aspects.[26]

We can see a real battle of protocols when unfolding the complex thread of national and international issues around packet switching and the many technical solutions at work in the 1980s and 1990s. There are the protocols of the Internet, those deployed around layered architectures within ISO, and those developed within telecommunications standardization bodies. In particular, the X25 protocol, adopted in 1976, held out for a long time and was deployed in networks such as Transpac.[27] It supported the Minitel in France, and X25 was also largely used in South America. This is documented by two players at the European level, Howard Davies and Beatrice Bressan, in *A History of International Research Networking*.[28] It is a fascinating story in terms of engineering cultures and economic, technical, and political issues. It is one that is difficult to trace without oversimplification. I remember submitting a book manuscript to a French publisher in 2011, only to be asked to remove some acronyms and players to make it easier to read and to change the title of the book, "Des réseaux et des hommes" [Networks and Men], in such a way as to make us wonder whether France invented or missed the Internet. I ultimately chose another much less renowned publisher who kept the complexity of the narrative intact. The same level of complexity can also be seen in the book by James Pelkey, Andrew Russell, and Loring Robbins.[29] I thus avoided the tendency to interpret all network histories from the angle of the ultimate success of the Internet. Indeed, the latter trend is palpable in the French media and even in the French literature, e.g., Eric Reinhardt's recent novel *Comédies françaises*.[30] The author offers a fictionalized version of the story of Cyclades, allowing itself some flagrant departures from historical reality to highlight the role of the French in the Internet, while holding the telecommunications sector fully responsible for a missed opportunity.

There are disputes over authorship and conflicting memories, which have grown in importance in recent years. They are a classic feature of the history of technology and innovation, which is full of innovators—as perfectly exemplified in Walter Isaacson's book *The Innovators: How a Group of Hackers, Geniuses, and Geeks Created the Digital Revolution*.[31] This book sparked critical reactions among historians, prompting alternative perspectives on the author's portrayal of the creators of the digital revolution and his vision of innovation and computing. There are also debates on the respective roles of Paul Baran, Leonard Kleinrock, and Donald Davies in the creation of packets in networks, as well as debates on the invention of email following Shiva Ayyadurai's claim to be its creator, trying to erase the role of Ray Tomlinson at BBN.[32] In the 2000s, when I was writing my thesis, rather than focus on quarrels over authorship, the general predominant view was that France had missed an opportunity with Cyclades and had taken a step backwards with the Minitel. It can be difficult to invite readers to think about innovation in context, without the inevitable outcome of the Internet "victory," and to explore other choices and visions. This led to strong and sometimes angry reactions to some of my research among those from the telecommunications or computing sides who had hoped for a "historical rehabilitation" or a clear stance in favor of a technical solution.

Yet not all paths led to the Internet, and beyond the case of France, other countries were influenced by the debates that unfolded in the 1990s and the complex choices that had to be made between the Internet and solutions developed by the telecommunications sector, such as X25. The case of Costa Rica and South America, for which Ignacio Silès provides a very detailed analysis of the issues at stake, is particularly interesting.[33] Focusing exclusively on the relationship with the Internet does not allow us to fully consider paths of influence that do not always pass through the United States, as in the case of the success in South Korea of the translation of the famous French 1978 Nora-Minc Report on *The Computerization of Society*,[34] which I discovered through Jo Dongwon's research on digitization in South Korea. These national histories are gradually filling some gaps in the history of data networks.

Missing Narratives

While our discussion so far has largely focused on ARPANET, the Internet, and packet-switching networks, we should return to the criticism voiced in 2013 by Martin Campbell-Kelly and Daniel Garcia-Swartz, who regret that "most histories of the Internet, both scholarly and popular, focus on its origins in the ARPANET network built by the U.S. Advanced Projects Research Agency in the late 1960s" and suggest looking at other solutions such as "the development of computerized networks for banks and the rise of the Electronic Data Interchange movement [. . .], the first online information providers, such as Lexis and Dialog," or "the first consumer networks such as CompuServe and Prodigy, the first commercial email services, and Videotext services that became the basis for national information infrastructures in many countries."[35]

In these various fields, we can cite multiple works and researchers who have filled these gaps in the last decade, e.g., related to a precise national space, specific sectors (e.g., banking, with the work of Lana Swartz, Martin Schmitt, and Sebastian Giessmann[36]) or videotext, in particular the French Minitel.[37]

Of course, these approaches were not available in 2002, when I began my doctoral research on the genesis of packet-switching networks in France in the 1970s. Works such as *L'imaginaire d'Internet* offered specific perspectives and imaginaries. But it became obvious that the French context was markedly different from the one in the United States: there was none of the hustle and bustle of the American campuses; telecommunications was based on a monopoly on lines and the development of solutions for key accounts; computing research at Iria (later, renamed Inria) was not organized in the same way as the research environment at MIT or UCLA; and the young French company CII (Compagnie Internationale pour l'Informatique) had little in common with key IT players in the United States. Common threads may be identified, such as the interest in time-sharing by scientists who would develop network solutions in France, trips to the United States by some of the key protagonists, a consulting contract between BBN and Iria, links between Cyclades and ARPANET, the participation of French telecommunications researchers and computer scientists in the first public demonstration of ARPANET, and discussions at joint conferences. However, the development of French data networks was based more on industrial and research policies. My study relied heavily on internal sources from research centers, such as the Cyclades archives at Inria, or the files tracing the role of the DGT (Direction Générale des télécommunications) or CNET (Centre national d'étude des télécommunications) in the development of the Transpac network. What we see is the unfolding of a French scientific and industrial policy firmly rooted in the objectives of the "Plan Calcul" launched by de Gaulle to develop French computing, and in the desire of the telecommunications sector to offer pragmatic, affordable solutions to companies seeking access to connectivity solutions. This history is part of a vibrant international context, while remaining relatively far removed from the US imaginary described later by Fred Turner in his exciting book about Stewart Brand.[38]

The importance of placing these narratives in national contexts—because in the 1970s and 1980s, visions of what computing and networks should be were also largely political[39]—is also palpable in the first part of the book co-edited by Gerard Goggin and Mark McLelland, *The Routledge Companion to Global Internet Histories.*[40] Other scholars, including Paolo Bory for Italy and Miglé Bareikyté for Lithuania, are also enriching our knowledge of new geographical areas,[41] taking into account different cultural, economic, and societal contexts.

Another very valuable input is the effort to position the history of networks within the wider history of computing, making it part of a broader story of development. This is the case in the books mentioned above by Pelkey, Russell, and Robbins,[42] and by Haigh and Ceruzzi.[43] It can also be seen in the book by Peter Freeman, W. Richards Adrion, and William Aspray on the history of computing at the National Science Foundation (NSF), albeit in a different way as it focuses on a more targeted history.[44] They trace the role played by the NSF in the development of modern computing over 60 years, referring to NSFNET and the way the NSF supported the development of the Internet for the general public.

Other issues that run through the history of data networks are considered in innovative ways. These include spam, for which Finn Brunton's work on media archeology remains unsurpassed,[45] and cookies and privacy issues,[46] which highlight links with earlier networks. Contemporary and past issues also come together in Julien Mailland's approach to net neutrality, which he starts from the Minitel.[47] The launch of the journal *Internet Histories* in 2017 paved the way for contributions at the intersection of Internet and Web history, as well as for more hybrid approaches exploring technology and digital cultures. Historians interested in the history of networks and those more interested in digital communications or Web content did not previously co-exist in the same publication spaces: the former tended to publish in journals related to the history of computing such as *IEEE Annals of Computing*, while the latter targeted journals such as *New Media & Society* (although there may of course be some exceptions).

Finally, another trend that is helping to break away from an overly Internet-centric vision is the emergence of more specific studies of other networks. Following on from my work on Cyclades and Transpac, I traced the origins of the French research and education network RENATER developed in the 1990s.[48] This history involves Unix communities, and it is also worth noting the importance of Usenet and the work by Camille Paloque-Berges on Unix communities and the adoption of computer-mediated communications in Europe.[49] The history of Usenet and the study by Steve Jones and Guillaume Latzko-Toth on PLATO[50] bring another dimension to data network analysis: that of early adopters, whether user-designers or users of educational networks. In the former case, Paloque-Berges explores not only protocol debates but also the development of codes of conduct and communication in computer-mediated communications. She looks at the start of the first forums and the transnational links in computer communities that developed around Unix, particularly at French and European levels. In the second case, Jones and Latzko-Toth analyze a pioneering educational computer platform developed at the Computer-Based Education Research Laboratory at the University of Illinois in the 1960s and 1970s. They describe the development of social interactions (message boards, etc.) and online gaming and look at the "social" features of online communities.

Networks and Networking in Society

While more local or national approaches to the history of data networks enable us to decentralize our gaze from the Internet and look to other contexts and specific regulations, the history of the period prior to the development of the Web initially focused on scientists, engineers, designers, and technical aspects rather than uses. As Paloque-Berges and Driscoll note:

> How can one history contain the experiences of ARPA-funded researchers at Stanford in 1976 and an elderly cell phone user living in Beijing in 2017? In the face of the essential plurality of the Internet,

> we propose that researchers adopt an alternative approach to internet historiography rooted in the experience of users.[51]

Users are of course not totally absent from previous accounts, but they are sometimes limited to engineers and user-testers, for example, at universities involved in Cyclades or ARPANET. Users also played a key role in telecommunications, for example, with the development of Transpac and the use of packet switching to build systems that were previously based on point-to-point lines between companies. However, the data network uses analyzed in early histories of data networks remain modest, often scientific, and under-analyzed. In my case the interest in users became stronger and more noticeable in the Minitel study that brought the French general public into the era of electronic exchanges on a massive scale. Yet, from the outset, networks have always clearly had a communicative dimension over and above questions of file transfer, shared computing power, etc. This topic has recently been addressed by a new generation of researchers and approaches.

As an introduction to the special issue of *Internet Histories* on the ARPANET, appearing in 2019, Paloque-Berges and I quoted Stephanie Dick's presentation at SHOT St. Louis 2018 on the MIT Macsyma (Mac Symbolic Manipulator) project. Dick pointed out that the PDP 10 was one of the most popular nodes in 1972 within ARPANET but also that it was very difficult to become a user of the Macsyma system. As she explained, the community formed around the Macsyma project did not exist before but developed as the project progressed.

This willingness to take an interest in early users is also present in Fred Turner's book, which highlights Stewart Brand's important role as an intermediary; and in Julie Momméja's thesis, which takes a more cultural look at the communities that accompanied the first steps of networked communications in the San Francisco Bay Area.[52]

While debates surrounding the history of networks are rare and relatively peaceful—at least between historians—stakeholders are often far more entrenched in their views when it comes to questions of fatherhood. There seems to be a general trend to emphasize the central role of intermediaries and users and the importance of the socio-cultural context. For example, Dominique Cardon in his preface to the French translation of Turner's book, and other authors from the fields of sociology and information and communication sciences place particular emphasis on the 1960s context—hippies, LSD, and so on. In general, the aim is to support arguments that focus more on uses, the openness of the Internet, and its values, and to question "Internet democracy."[53] The idea is to respond to contemporary questions about freedom and the potential impacts of digital technology by drawing in part on its past, especially the values attributed to the first communities. The focus shifts from institutional issues to innovation from below. This approach is undoubtedly interesting, inasmuch as it allows us to move away from the historiography of institutions, but it sometimes draws heavily on the writings of early users. While it allows us to take the undeniable creativity of early uses seriously, it runs the risk of detaching itself from an important technical, institutional, and economic background and missing out on an entire line of history.

These approaches therefore need to be balanced so as not to create a divide between a research and institutional analysis to the genesis of the Internet and one that focuses on a more bottom-up approach. This is the case in the PhD thesis by Momméja, who clearly tries to hold the two approaches together. Technology is never neglected: while the author is clearly more interested in people than components, she incorporates the latter when it is relevant, from the IBM 80 to the development of microcomputing, the Community Memory system, the Altair, and Apple. She also creates a continuity with contemporary movements such as Revive & Restore, the bio-hacker movement, the Counter Culture Labs project, and Maker Faire, weaving well-thought-out links between past, present, and future.

This can also be seen in the 2018 book by Joy Lisi Rankin, *A People's History of Computing in the United States*, which strikes a balance between a technical and institutional approach and a more socio-cultural perspective. She repositions the history of PLATO in a largely user-centered context and in a more general history of computing, for example, when she analyzes the emergence of the BASIC language. She explores the use of new forms of computer-mediated communication through Bulletin

Board Systems (BBSs) and emails in PLATO. She also takes a gendered approach, devoting several chapters of her book to gender issues ("Making a Macho Computer Culture" and "How Computing Went from Male to Masculine") while emphasizing intersectionality. By suggesting, "Let us overwrite the Silicon Valley mythology. Let us look beyond the narrow inspiration of digital Founding Fathers. Let us recover the many people's histories of computing,"[54] Rankin joins Russell's criticism of the focus on pioneers and innovators.[55] Above all, Rankin's book is part of a trend in the history of computing to investigate new fields, highlighting hidden challenges and players and paying more attention to inclusiveness. This is clearly illustrated by recent books such as *Your Computer Is on Fire*, edited by Mar Hicks, Benjamin Peters, Thomas Mullaney, and Kavita Philip.[56] It is surely in this trend towards histories that are closer to users and current challenges, for which Nathan Ensmenger also argued in an article in 2012,[57] that the history of data networks is becoming increasingly integrated into the contemporary history of computing. In *The Modem World: A Prehistory of Social Media* and in his thesis "Hobbyist inter-networking and the popular Internet imaginary: Forgotten histories of networked personal computing, 1978-1998,"[58] Driscoll traces the roots of the "modem world" back to ham and citizens band radio shacks, phone phreaking, and hobbyist microcomputer clubs, while also shedding light on the history of Fido Net and BBS boards and referring to broader practices and digital cultures.

When it comes to Internet practices and locations, another approach is to look at connection locations, the materiality of the Internet, ISP choices, Internet advertising, and practices in cyber cafés, for example, as I did in 2017 or as Stefan Udelhofen suggested the same year in "Über Computerkästen in Cafés. Räumlich-materielle Anordnungen und symbolische Ordnungen von Internetcafés zwischen 1991 und 2003."[59]

Finally, it is worth noting that the desire to view the history of data networks from a usage-based perspective is accompanied by attempts to consider continuities and changes in analog and digital technologies as part of a long history of media. One of the first books to establish certain forms of continuity was Tom Standage's *The Victorian Internet: The Remarkable Story of the Telegraph and the Nineteenth Century's*

On-line Pioneers.[60] The book I recently edited with Gabriele Balbi, Nelson Ribeiro, and Christian Schwarzenegger, *Digital Roots: Historicizing Media and Communication Concepts of the Digital Age*,[61] is an example of this attempt to cut across the long history of media while highlighting notions such as interactivity, amateurism, echo chambers, and fake news. This long-term approach is rather different, however, and establishes genealogies that reach different audiences.

New Avenues

Have digital developments diluted or renewed the history of data networks? While they have undoubtedly blurred the boundaries between history, STS, sociology, and media history, they have also led to conceptual enrichment and new forms of dialog. The history of data networks continues to be written in the pages of journals such as *IEEE Annals of the History of Computing* and *Technology & Culture*, and by members of SIGCIS (the Special Interest Group for Computing, Information, and Society), but it is also being explored by other audiences in other spaces, such as ECREA conferences on the history of communication and the Tensions of Europe conferences on the wider history of technology and infrastructures. The investigation of born-digital materials, oral histories, infrastructures and uses, and digital communities makes this a broad, flexible field, one that is of interest beyond the history of computing and telecommunications alone. While the adoption of a user-centered approach undoubtedly reflects contemporary trends, it may also prove "easier" in logistical terms for more recent periods, as documents and sources related to strategies pursued by today's Internet giants and big companies are often well guarded and not easily accessible.

The history of data networks is constantly being forged in the in-between, demonstrating a certain inventiveness; and in just twenty years it has carved out a space for itself in the research landscape. There is still much to study, particularly with the potential for new contributions from digital humanities. For instance, text mining could uncover trends and imaginaries and offer a fresh perspective on some documents. Performing distant reading and topic modeling on the Charles Babbage Institute's interviews could yield fascinating insights. The same is true for some newsgroups and early Internet fora. Newsgroups remain largely under-explored and hold great potential for investigation. This is demonstrated, for example, by the study Hanna Zeavin conducted on archived posts from two Usenet newsgroups in "'This Is Womenspace': USENET and the Fight for a Digital Backroom, 1983–86," which intertwined communication history with women's history.[62] In the field of born-digital heritage, early Web archives in various areas such as servers, security, infrastructure, advertising, and domain name providers still have a lot to reveal.

Several networks and communities also remain relatively untapped, like the community of physicists, which played a significant role in Europe through networks like HEPnet and at CERN (the European Organization for Nuclear Research) in the early days of the Web.

The focus when it comes to users is often early adopters, for example hobbyists. There are several reasons for this: first, the development of digital technologies has strongly diluted and broadened user categories, making it difficult to grasp perspectives and uses and to document practices at scale. Second, hobbyists, amateurs, and early adopters are often talkative, especially on BBSs and other fora, and this can help us to reconstruct imaginaries and practices. Finally, they often also create spaces (whether "IRL" or virtual) to discuss their practices, difficulties, and challenges, or to network; and this can provide interesting sources for historians. However, focusing on early adopters, hobbyists, hackers, and first users is of course not enough. It creates bias, whether related to gender, socio-professional status, or technical skills; and it may fuel a new kind of heroic narrative that focuses on "firsts" (first uses, adopters, etc.). As noted by Russell, "We must resist the urge to sanctify even the kindest of individuals who feature in our histories."[63] While Russell suggested delving into "histories of networking" as a way of more directly analyzing accounts that do not specifically deal with computer networks, "networking in society" could also likely serve as a robust starting point that encompasses a diversity of users and practices. There are multiple genealogies and histories related to network uses that are yet to be traced. As an example, online video game practices must be contextualized within the broader field of the history of the gaming industry and gaming at large, while the history of games also needs to be seen as part of the larger field of history of computing, as recently demonstrated by Laine Nooney.[64] The same is true for digitalization. One example is CD-ROMs, which influenced design practices and digital uses, while also becoming an early entry point to the Internet for the public.

Investigating infrastructures, materialities, and maintenance[65] would also bring us back to telecommunications and other players in the history of networks. There is still a huge amount to explore—for example, the history of Bluetooth that we aim to trace together with Giessmann and colleagues at the University of Siegen, or the Internet of Things—and such research may provide a fascinating complement to the history of engineering, standardization, and human-machine interactions.

Furthermore, there are some recurring themes that regularly come to the forefront, progressively gaining in complexity. For instance, the tension between decentralization and centralization, openness and closure, is constantly replayed—particularly in the context of issues such as blockchain, the acquisition of Twitter, and others. The dissolution of networks into ambient computing and the development of the Internet of Things, and now digital twins, is likely to prompt a re-evaluation of this history, with cybernetics, AI, human-computer interactions, bots, and the imaginaries of networked communities being reinterpreted and further explored. However, while these issues may reveal new origins, challenges, and historical narratives, this should not diminish the significance of previously examined approaches, such as the role of regulation, standardization bodies, and institutional history. The imaginaries and regulatory concerns surrounding AI, as well as its potential uses, for example in education, align with previous historical trends, prompting a re-evaluation of these perspectives.

These possibilities represent a far-from-exhaustive list. However, they give an idea of the challenges that may lie ahead and how we might continue to approach the history of networks in a decentralized, peer-to-peer way, between centers and peripheries, mainstream and marginal tendencies, rooted in historical context while also taking account of new digital developments.

Notes

1. John Quarterman, *The Matrix: Computer Networks and Conferencing Systems Worldwide* (Bedford: Digital Press, 1990).

2. Janet Abbate, *Inventing the Internet* (Cambridge, MA: MIT Press, 1999).

3. It started from 2002, with a PhD on the history of two packet-switching networks developed in France during the 1970s, Cyclades and Transpac.

4. Manuel Castells, *The Rise of the Network Society* (Malden, MA: Blackwell Publishers, 1996).

5. Arthur L. Norberg, Changing Computing: The Computing Community and DARPA, *IEEE Annals of the History of Computing*, vol. 18, no. 2 (1996): 40–53.

6. Thomas Haigh and Paul Ceruzzi, *A New History of Modern Computing* (Cambridge, MA: MIT Press, 2021).

7. Carl Malamud, *Exploring the Internet: A Technical Travelogue* (Hoboken: Prentice Hall, 1992); Katie Hafner and Matthew Lyon, *Where Wizards Stay Up Late: The Origins of the Internet* (New York: Simon & Schuster, 1998); Michael Hauben and Ronda Hauben, *Netizens: On the History and Impact of Usenet and the Internet* (Los Alamitos: IEEE Computer Society, 1997); Howard Rheingold, *The Virtual Community* (Reading: Addison-Wesley, 1993).

8. Patrice Flichy, *L'imaginaire d'Internet* (Paris: La Découverte, 2001).

9. Judy O'Neill and Arthur Norberg, *Transforming Computer Technology: Information Processing for the Pentagon, 1962–1986* (Baltimore: Johns Hopkins University Press, 1996).

10. https://archives.lib.umn.edu/repositories/3/resources/264#

11. Alexandre Serres, Aux sources d'Internet: l'émergence d'ARPANET. Exploration du processus d'émergence d'une infrastructure informationnelle. Description des trajectoires des acteurs et actants, des filières et des réseaux constitutifs de la naissance d'ARPANET. Problèmes critiques et épistémologiques posés par l'histoire des innovations, PhD thesis (Rennes: Université Rennes 2, 2000), 46.

12. Janet Abbate, Privatizing the Internet: Competing Visions and Chaotic Events, 1987–1995, *IEEE Annals of the History of Computing*, vol. 32, no. 1 (2010): 10–22.

13. Janet Abbate, L'histoire de l'Internet au prisme des STS, *Le Temps des médias*, vol. 18, no. 1 (2012): 170–180.

14. Fenwick McKelvey and Kevin Driscoll, ARPANET and its boundary devices: Modems, IMPs, and the inter-structuralism of infrastructures, *Internet Histories*, 3:1 (2019): 31–50.

15. The former was developed in the 1970s in the telecommunications sector by the French public telecommunications operator, which had a monopoly position on telecommunication lines; and the latter within a computer research institute, Iria (later, Inria). Valérie Schafer, Des réseaux et des hommes. Les réseaux à commutation de paquets un enjeu pour le monde des télécommunications et de l'informatique françaises (des années 1960 au début des années 1980), PhD thesis under the supervision of Prof. Pascal Griset (Paris: Sorbonne University, 2007).

16. Andrew Russell, *Open Standards in the Digital Age. History, Ideology and Networks* (Cambridge: Cambridge University Press, 2014).

17. Milton Mueller, ICANN and Internet Governance: Sorting Through the Debris of "Self-regulation," Info, vol. 1, no. 6 (1999).

18. Francesca Musiani and Valérie Schafer, Global Governance, in Gabriele Balbi, Nelson Ribeiro, Valérie Schafer, and Christian Schwarzenegger (eds.), *Digital Roots: Historicizing Media and Communication Concepts of the Digital Age* (Berlin: De Gruyter, 2021): 117–136.

19. Alex McKenzie, INWG and the Conception of the Internet: An Eyewitness Account, *IEEE Annals of the History of Computing*, vol. 33, no. 1 (2011): 66–71.

http://alexmckenzie.weebly.com/inwg-and-the-conception-of-the-internet-an-eyewitness-account.html

20. Andrew Russell, "Rough Consensus and Running Code" and the Internet-OSI Standards War, *IEEE Annals of the History of Computing*, vol. 28, no. 3 (2006): 48–61; and OSI: The Internet That Wasn't, *IEEE Spectrum* 50 (August 2013).

21. Andrew L. Russell and Valérie Schafer, In the Shadow of ARPANET and Internet: Louis Pouzin and the Cyclades Network in the 1970s, *Technology & Culture* 55, no. 4 (2014): 880–907.

22. Valérie Schafer, The ITU Facing the Emergence of the Internet, 1960s–Early 2000s, in Andreas Fickers and Gabriele Balbi (eds.), *History of the International Telecommunication Union (ITU)* (Berlin: De Gruyter, 2020): 321–344.

23. A complementary presentation of this history has been pieced together by Jacob Ward in his book on British telecommunications: Jacob Ward, *Visions of a Digital Nation: Market and Monopoly in British Telecommunications* (Cambridge, MA: MIT Press, forthcoming 2024).

24. Cyclades relied on datagrams and contributed to international reflections on the issue.

25. Bradley Fidler and Morgan Currie, The Production and Interpretation of ARPANET Maps, *IEEE Annals of the History of Computing*, vol. 37, no. 1 (2015): 44–55.

26. See, for example, Quinn DuPont and Bradley Fidler's work on cryptography and cybersecurity: Quinn DuPont and Bradley Fidler, Edge Cryptography and the Codevelopment of Computer Networks and Cybersecurity, *IEEE Annals of the History of Computing* vol. 38, no. 4 (2016): 55–73.

27. Rémi Després, X.25 Virtual Circuits—TRANSPAC IN France—Pre-Internet Data Networking [History of communications], *IEEE Communications Magazine*, vol. 48, no. 11 (2010): 40–46.

28. Howard Davies and Beatrice Bressan (eds.), *A History of International Research Networking: The People Who Made It Happen* (Hoboken, NJ: Wiley-Blackwell, 2010).

29. James Pelkey, Andrew Russell, and Loring Robins, *Circuits, Packets, and Protocols: Entrepreneurs and Computer Communications, 1968–1988* (New York: ACM Book Series, 2022).

30. Eric Reinhardt, *Comédies françaises* (Paris: Gallimard, 2020).

31. Walter Isaacson, *The Innovators: How a Group of Hackers, Geniuses, and Geeks Created the Digital Revolution* (New York: Simon & Schuster, 2015).

32. Thomas Haigh, Seven Lessons From Bad History, *Communications of the ACM*, vol. 55, no. 9 (2012): 26–29.

33. Ignacio Silès, *A Transnational History of the Internet in Central America, 1985–2000: Networks, Integration, and Development* (London: Palgrave Macmillan, 2020).

34. Simon Nora and Alain Minc, *L'informatisation de la société: rapport à M. le Président de la République* (Paris: Seuil, 1978).

35. Martin Campbell-Kelly and Daniel Garcia-Swartz, The History of the Internet: The Missing Narratives, *Journal of Information Technology* 28 (1) (2013): 18–33.

36. Lana Swartz, *New money: How payment became social media* (New Haven: Yale University Press, 2020); Martin Schmitt and Vernetzte Bankenwelt, Computerisierung in der Kreditwirtschaft der Bundesrepublik und der DDR, in Frank Bösch (ed.), *Wege in die digitale Gesellschaft: Computernutzung in der Bundesrepublik 1955–1990* (Göttingen: Wallstein Verlag, 2018): 126–147; Sebastian Giessmann, Money, Credit, and Digital Payment 1971/2014: From the Credit Card to Apple Pay, *Administration and Society* 50/9 (2018): 1259–1279.

37. Valérie Schafer and Benjamin Thierry, *Le Minitel, l'enfance numérique de la France* (Paris: Nuvis, 2012); Julien Mailland and Kevin Driscoll, *Minitel. Welcome to the Internet* (Cambridge, MA: MIT Press, 2017).

38. Fred Turner, *From Counterculture to Cyberculture: Stewart Brand, the Whole Earth Network, and the Rise of Digital Utopianism* (Chicago: University of Chicago Press, 2006).

39. See the recent book by Dick Van Lente (ed.), *Prophets of Computing: Visions of Society Transformed by Computing* (Kentfield: Morgan & Claypool Publishers, 2023).

40. Gerard Goggin and Mark McLelland, *The Routledge Companion to Global Internet Histories* (New York: Routledge, 2017).

41. Paolo Bory, The Italian Network Hopes: Rise and Fall of the Socrate and Iperbole Projects in the Mid-1990s, *Internet Histories* 3:2 (2019): 105–122; Miglé Bareykité, *The Post-Socialist Internet: How Labor, Geopolitics and Critique Produce the Internet in Lithuania* ([Transcript] Digital Society, 2022).

42. James Pelkey, Andrew Russell, and Loring Robins, *Circuits, Packets, and Protocols: Entrepreneurs and Computer Communications, 1968–1988* (New York: ACM Book Series, 2022).

43. Thomas Haigh and Paul Ceruzzi, *A New History of Modern Computing* (Cambridge, MA: MIT Press, 2021).

44. Peter Freeman, W. Richards Adrion, and William Aspray, *Computing and the National Science Foundation, 1950–2016: Building a Foundation for Modern Computing* (New York: ACM Books, 2019).

45. Finn Brunton, *Spam: A Shadow History of the Internet* (Cambridge, MA: MIT Press, 2013).

46. Meg Leta Jones, Practicing privacy on other networks: Network structures, economic arrangements, and identity strategies before cookies, *Internet Histories*, vol. 4 (2020): 142–160.

47. Julien Mailland, Building Internet policy on history: Lessons of the forgotten 1981 network neutrality debate, *Internet Histories*, vol. 2 (2018): 1–19.

48. Valérie Schafer, "Part of a Whole: RENATER, a Twenty-Year-Old Network within the Internet," *Information & Culture*, vol. 50, no. 2 (2015): 217–235.

49. See, for example, Camille Paloque-Berges, Mapping A French Internet Experience: A Decade Of Unix Networks Cooperation (1983–1993), in Gerard Goggin and Mark McLelland (eds.), *The Routledge Companion to Global Internet Histories* (New York: Routledge, 2017): 153–170.

50. Steve Jones and Guillaume Latzko-Toth, Out from the PLATO Cave: Uncovering the Pre-Internet History of Social Computing, *Internet Histories*, 1:1–2 (2017): 60–69.

51. Kevin Driscoll and Camille Paloque-Berges, Searching for Missing "Net Histories," *Internet Histories*, 1:1–2 (2017): 47.

52. Julie Momméja, Du Whole Earth Catalog à la Long Now Foundation dans la Baie de San Francisco: co-évolution sur la "Frontière" créative (1955–2020), PhD thesis (Paris: Université Paris 3, 2021).

53. Dominique Cardon, *La démocratie Internet. Promesses et limites* (Paris: La République des Idées/Le Seuil, 2010).

54. Joy Lisi Rankin, *A People's History of Computing in the United States* (Cambridge: Harvard University Press, 2018): 242.

55. Andrew Russell, Hagiography, Revisionism & Blasphemy in Internet Histories, *Internet Histories*, 1:1–2 (2017): 15–25.

56. Mar Hicks, Benjamin Peters, Thomas S. Mullaney, and Kavita Philip (eds.), *Your Computer Is on Fire* (Cambridge, MA: MIT Press, 2021).

57. Nathan Ensmenger, The Digital Construction of Technology: Rethinking the History of Computers in Society, *Technology and Culture* 53 (4) (2012): 753–776.

58. Kevin Driscoll, *The Modem World: A Prehistory of Social Media* (London: Yale University Press, 2022); Kevin Driscoll, Hobbyist Inter-Networking and the Popular Internet Imaginary: Forgotten Histories of Networked Personal Computing, 1978–1998, PhD thesis (Los Angeles: University of Southern California, 2014).

59. Valérie Schafer, Les cybercafés dans la France des années 1990, espaces médiatiques et médiatisés, *Le Temps des médias*, vol. 28, no. 1 (2017): 159–172; Stefan Udelhofen, Über Computerkästen in Cafés. Räumlich-materielle Anordnungen und symbolische Ordnungen von Internetcafés zwischen 1991 und 2003, in Christina Bartz et al. (eds.), Gehäuse. Mediale Einkapselungen (Paderborn: Fink, 2017): 53–69.

60. Tom Standage, *The Victorian Internet: The Remarkable Story of the Telegraph and the Nineteenth Century's On-line Pioneers* (London: Walker & Company, 1998).

61. Gabriele Balbi, Nelson Ribeiro, Valérie Schafer, and Christian Schwarzenegger (eds.), *Digital Roots: Historicizing Media and Communication Concepts of the Digital Age* (Berlin: De Gruyter, 2021).

62. Hannah Zeavin, "This is Womenspace": USENET and the Fight for a Digital Backroom, 1983–86, *Technology and Culture* 63, no. 3 (2022): 634–664.

63. Andrew Russell, Hagiography, Revisionism & Blasphemy in Internet Histories, *Internet Histories*, 1:1–2 (2017): 25.

64. Laine Nooney, The Apple II Age: How the Computer Became Personal (Chicago: University of Chicago Press, 2023).

65. Andrew Russell and Lee Vinsel, After Innovation, Turn to Maintenance, *Technology and Culture* 59, no. 1 (2018): 1–25.

Chapter 11

Writing Data into the Histories of Computing and Information

Matthew L. Jones

"A multiplicity of large, remote-access computer systems, if interconnected," Paul Baran told a committee of the US Congress in 1966, "can pose the danger of loss of the individual's right to privacy—as we know it today." Worries about one giant database to know us all were misplaced. "The composite information data base may be so large and so easily accessible that it would permit unscrupulous individuals to use this information for unlawful means."[1] Decades before surveillance advertising ruled the Internet he had helped imagine long before, the RAND analyst Baran sought to educate legislators about the privacy implications of the technical capacities of interconnected networks. Far from being misguided, their worries about the ethics and politics of data collection and analysis were not—to his mind—amplified enough, given the realities of security around data and the growing power of analysis using it.

In his comments, Baran connected new material infrastructures of storage and communication, mathematical procedures of analysis, and fundamental legal, ethical, and security questions. Using data as an optic within the history of information and computing locates us squarely at the intersection of these phenomena. In them, we see the history of computing deeply implicated in political and ethical effects and debates. And we see the history of power fundamentally involves the capacities and possibilities of digital information systems. The history of data braids the history of information and computing tightly with the history of statistics and applied mathematics. All these histories come together to enable models of producing corporate profits and state intelligence through data collection and analysis, and thus raise pressing issues around privacy and security.

Truths and Falsehoods

No sharp line divides the history of data from the history of information. So, what does a focus on "data" over "information" afford the historian? I take the answer to be fairly pragmatic: in working with data as our central category—and subject—of analysis, we attend to different things with historiographic and analytic traditions complementary to those more prominent in the history of information, well-illustrated in the chapters in this volume.

So, what even is the subject of a history of "data"? Data is a form of knowledge; it is likewise a category of knowledge. Thus, we might study:

1. History of things considered to be "data" at any time in the past within some relatively modern definition, for example, Babylonian astronomical observations;[2]
2. History of the term itself through its dramatic shift in meaning in the eighteenth century, from axiomatic givens to that which is given in nature through observation, discrete observations of information.[3]

My focus here will primarily on the first, with some attention to claims about the nature of data. Understanding data today requires accepting one fundamental truth and one false, though powerful belief. The fundamental truth is that there is no "raw" data.[4] The false belief is that there is just such "raw data." Writing the history of data requires showing the artifice of data; but it also requires documenting the power and continuing causal impact of the ideal that data is an unmediated source of truth for reasoning, government, making decisions, and producing scientific knowledge. In his well-known fusillade about the dangers of "innovation" within the history of technology, David Edgerton noted, "Innovation-centred and knowledge-centred accounts of technology are central to twentieth century culture." So too with accounts of the transparency and neutrality of data. However much historians, sociologists, and philosophy may deplore it, this false belief is central to 21st-century culture.

Data is artificial, for our purposes, in at least two ways: first, data is something made by humans and instruments, not something simply revealed in nature; second, data is stored, maintained, duplicated, evaluated, and investigated through concrete technologies, whether of paper, metal, or digital machines, dependent on human labor, conjoined to forms of classification, implicit and not.[5] The history of data presented here goes beyond the history of data sets to the history of their analysis—scientific and not—and use—commercial, military, and beyond, the effects of those uses, and the legal and ethical questions around collection and use of data.[6]

The first time I presented my initial findings on the history of the data sciences to a large group of computer scientists at my then institution, a colleague came up to introduce himself and to thank me for giving the subspecialty of database studies visibility. While computer science departments were expanding quickly with new positions in machine learning and data science, often looked upon with envy by other departments in engineering schools, the subfields that made all that data analysis now possible were too often forgotten. That forgetting is one of the slippages that has rebranded automatic decision-making systems, grounded in data infrastructures, into "artificial intelligence." It is only one of the major forgettings that obscures our understanding of the conditions of possibility of our data-mediated world.

INFRASTRUCTURES OF STORAGE AND CLASSIFICATION

In April 1959, Solomon Kullback opened a symposium on "Data Storage and Data Retrieval" in Fort Meade, Maryland. The statistics and the problems raised during the two-day event suggest the unanswered questions around data infrastructures at the time. The participants freely intermingled questions of storage, optimizing retrieval, and undertaking analysis of data—digital and analog alike.

It was no accident that the US National Security Agency should hold such a gathering in the late 1950s. If much scientific computing of the day focused on high-speed computation, NSA needed to contend with the torrent of communications it was sweeping up: "The Agency became as much or more a data processing center than a 'cryptanalytic center." And so, it sought "high speed substitutes for the best data processors of the era, tabulating equipment."[7] In focusing "on the manipulation of large volumes of data and great flexibility and variety in non-numerical logical processes," NSA required infrastructure more like large corporations than like physicists of the day.[8]

At the 1959 meeting, NSA's intelligence officers detailed the sheer scale of their problems to a select group of renowned scientific experts. They had, for example, files in a relatively simple ontology encompassing:

> PERSONS: 3 million people—half Russians, half from the rest of the world. Besides names, this is the place we record residence, occupations, affiliations, etc. and keep in effect a dossier of all we can get on these people.
>
> THINGS: 100,000 commodity subject items.
>
> PLACES: 500,000 place names and associated information such as organizations and people in charge, manufactures and stockpiles, transportation, tec [*sic*].
>
> PROCESSES: 100,000 items.[9]

(The civil libertarians of the 1970s and 2000s were not, let us say, wholly wrong about secret government databases.)

The NSA staffers likewise described a veritable onslaught of data produced through daily collection: in one example, "Magnetic Tape is being generated . . . at a rate of 14,000 tapes (45,000 intercept hours) monthly, and within a year will rise to 16,000 tapes (55,000 hours)." The various divisions of NSA sought to digitize all this information, to sort it, and to be able to query it. They even hoped for automatic processing to extract entities and, *grosso modo*, to discern the unknown from data:

> Mechanical extraction from texts would support a library. The library would be capable of generating complete lists (people, commodities, etc.). . . . It should yield up original material on request. It should make associations of data, and analysts would be particularly interested in associations of which they were previously unaware.[10]

Even if questions of converting documents into machine-readable form were to be solved, William Weaver, one of the invited august scientific experts, opined, it "may not make much impact if the machine must approach problem in a very formal way, without the associative power of the human mind. Some basic research needs to be done to understand how people get trained in logic, linguistics, etc. We have to know the relationships of words and ideas."[11]

The NSA workshop illustrates the irresolution around 1959 about how to store and analyze data within digital computers with large-scale storage, even as it reveals the continuity of data-focused computation with punch card processes. One NSA official offered different possibilities for storage and retrieval:

There are three classes of ways in which we could handle our information:

1. Keep only one copy, and use a classification system with cross references. [. . .]
2. Make multiple files, anticipating future use. [. . .]
3. Keep only one copy, but have a multiplicity of ways to get at it.[12]

Such questions remained open in the development of database software through the 1960s.

The NSA case is revealing because its early problems and goals, and the fondest wishes of its staff, would become the problems, goals, and wishes of many more organizations in the next half century. Computer historian Thomas Haigh has documented the development of early database management systems from the 1950s until the 1970s, up through the crucial standardization around the "relational model," in which multiple facets of data could be placed into distinct tables.[13] In the relational model, users could use common "keys" to combine the data across different tables in diverse ways. With this approach to databases, users could focus on relationships in and among complex tables, rather than having to know the order of relations in internal storage: "Most users should interact with a relational model of the data consisting of a collection of time-varying relationships (rather than relations). Each user need not know more about any relationship than its name together with the names of its domains (role qualified whenever necessary)."[14] The relational model abstracted the forms of data as experienced by the user from forms of representation on the machine—something all the more important as the data increased in size. Database infrastructures thus essentially enabled abstraction from physical representation to enable greater freedom in querying.

A generation of subsequent work brought database technologies, based primarily on the relational model, to the robustness and security necessary for global business throughout the end of the twentieth century. In the 1970s and 1980s, database researchers had endeavored to improve the security, reliability, and resiliency of databases.[15] Central to the developments of database practice was an insistence that changes to the database be "atomic": any set of connected changes to a database must be entirely completed or not completed at all. If you transfer $100 from one account to the other, the database system must either alter both accounts or neither. Researchers had to solve the problem for large-scale systems, where hardware, software, and power were presumed to fail with great regularity: "We must guarantee that all the effects of a transaction are installed in the database, or that none of them are, and this guarantee must be kept even when a system crash loses the contents of main memory."[16]

In 1973, the business information visionary Charles Bachman proclaimed that data storage was undergoing a "Copernican Revolution." This revolution "is changing the programmer from a stationary viewer of objects passing before him in core into a mobile navigator who is able to probe and traverse a database at will."[17] Rather than envisioning data as something examined serially, the advent of new random-access storage technologies such as disk drives meant that digital computers could escape the traditions of serial information storage and retrieval inherited from early technologies of punch cards. Older technologies of retrieval, such as the index, could find analogues in digital technologies, if only programmers could shift their worldview. From this perspective, the task was to organize and store data for diverse programs and programmers to use. However much relational databases had opened avenues for investigation, much remained to be done to create knowledge—and value—from those databases.

Yet, as the hardware and software developed, storage prevailed over analysis and retrieval. By the 1990s, many observers were announcing a crisis: programmers had figured out how to securely store data, but they had not done nearly enough to provide tools for investigating that data. In 1995 "data warehousing" expert Ralph Kimball argued:

> For the last dozen years we have had to put off getting the data out in order to concentrate on getting the data in. We needed systems that could absorb at least 1,000 transactions per second to enable us to store our basic enterprise data. Grafting transaction processing onto relational databases nearly put the lights out.[18]

In other words, many of the concerns of the NSA database workshop had been answered—but questions of querying, retrieval, and analysis, not so much.

Too Little—and Then Too Much—Data

In 1966, Paul Baran testified around the time that data was moving from expensive and challenging to getting plentiful—perhaps too much so. For all the long lineages of complaints of too much to know, scientists including Francis Galton and Karl Pearson deplored just how arduous obtaining data on people was. In many domains, complaints about the paucity of data remained central well into the 1980s and beyond. Large databases required infrastructural capacities and the financial support—and often coercion—to acquire.[19]

No story of data and information would be complete without concern about overload. "Every shrill insistence" about the horrors of information overload, Chad Wellmon writes, "brings with it a particular solution." These solutions are not just technical; they are "fundamentally normative," he explains. "They made particular claims not only about what was good or bad about print, for example, but about what constituted 'true' knowledge."[20] The same holds true for digital or other forms of information reproduction. The normativity is both epistemic and ethical, for it concerns the preferred qualities of the knower. Discussing the seeming eternal recurrence of the sense of novel information overload, the anthropologist Nick Seaver argues,

> What can we make of the long history of overload's unrelenting novelty? If we treat these stories as instances of a myth and look for their shared structure, we might note that they are not just historical accounts of the state of media; they are stories about a scalar relationship between archives and individuals. Overload requires more than an overwhelming amount of stuff; it requires someone to feel overwhelmed.[21]

Solution to problems of overload are solutions to what we value as knowledge and whom we value as knowledgeable.

In the 1990s, the growing perception of a crisis in analyzing corporate and scientific data propelled the emergence of a movement called Knowledge Discovery in Databases, often referred to simply as "data mining." The movement sought to bring together large-scale corporate database infrastructures with new forms of analysis drawn from a variety of engineering and scientific disciplines—from signals processing to machine learning. Unlike most of those fields, data miners insisted on the need to work within computational constraints and upon "dirty" "real world" data collected in everyday business. For database researchers, data mining was viewed as a superior form of querying.[22]

Google search emerged in this context. As graduate students, Sergey Brin and Lawrence Page were involved in a group named MIDAS—**MI**ning **D**ata **A**t **S**tanford, with the support of several faculty members, each a pioneer in database management.[23] Before taking on the entire World Wide Web, Brin and Page, along with their other collaborators, argued that the scale of the web made it so challenging but simultaneously made it deeply promising:

> We take advantage of one central idea: the Web provides its own metadata through its link structure, anchor text [the visible name of the link], and partially redundant content. This is because a substantial portion *of* the Web is *about* the Web.[24]

Many in the data mining community focused on turning their foremost problem, the vastness of data, into a crucial solution. Scale might be the resource for dealing with scale itself. In particular, the World Wide Web involves tremendous work to classify and categorize itself. No artificial intelligence was needed: rather a process of taking advantage of all this human classification was needed—what they called PageRank. They had to achieve this with existing commodity hardware. "Google's data structures," they explained, "are optimized so that a large document collection can be crawled, indexed, and searched with little cost."[25]

Google's efforts from the start rested on the labor of everyone creating webpages linking to other webpages. Labor of many kinds, often even less acknowledged, structures the creation and analysis of data, both in the digital realm and its prehistory.[26] As the infrastructure for data collection and analysis grew, so did the environmental impact of computing at a hitherto unknown scale. Inasmuch as NSA often led the way, concerns about the impact of NSA's large computational facility presaged worries about data centers far more globally.[27]

While data about the web on the web provided its initial focus, Google soon found other data it had begun collecting to be far more valuable: the data collected about every single person using it, which could be used not just to serve better search results, but to serve ads.[28] Collections of personal data had become among the most valuable of corporate properties.

Data-Intensive Science

A *Wired* article from 2008 celebrated "a world where massive amounts of data and applied mathematics replace every other tool that might be brought to bear." Theory be gone, the prophets appeared to be saying! "Out with every theory of human behavior, from linguistics to sociology. Forget taxonomy, ontology, and psychology. Who knows why people do what they do? The point is, they do it, and we can track and measure it with unprecedented fidelity. With enough data, the numbers speak for themselves."[29] In a marginally more measured 2009 piece, "The Unreasonable Effectiveness of Data," three Google affiliates premised the power of huge data sets on the dispersed human curation of the data available "in the wild." The Google researchers praise the discovery of the multitude of rules found, say, in understanding translation.[30] Memorization, not first principles, have allowed the advances in machine translation:

> Instead of assuming that general patterns are more effective than memorizing specific phrases, today's translation models introduce general rules only when they improve translation over just memorizing particular phrases (for instance, in rules for dates and numbers). Similar observations have been made in every other application of machine learning to Web data . . .[31]

Data-focused machine learning, the authors contend, has supplanted the dominant models of academic linguistics with its traditional scientific goals of simplicity and brevity. As in other fields where data scientists dared to tread—and they dared, these claims hardly went unopposed—but often such dissent found fewer ears, and fewer online clicks.

Transformation of domains into data-intensive computational sciences is neither trivial nor pregiven.[32] For one, making large-scale data sets involves vast work, as Paul Edwards has documented in the case of climate research, where he documents the distinct process of "making global data"—accumulating weather data from around the world—and of "making data global"—"building complete, coherent, and consistent global data sets from incomplete, inconsistent, and heterogeneous data sources."[33] But even with the creation of large-scale data sets, in many quarters of the postwar period, data-driven empiricism paled in prestige next to more socially and philosophical prestigious forms of mathematization, as historian Alma Steingart has argued. "It was not measurement and quantification that characterized the mathematization of the social sciences after the war, but axiomatization."[34] Rather than accepting a deterministic narrative that ever-larger stores of data led inexorably to an overturning of the nature of science, we should look to those who turned worries about too much data into opportunities for new kinds of scientific work.

From the 1970s, a slowly growing number of statisticians and computer scientists argued that increasing computational power alongside ever-greater sets of data ought to be shifting the possibilities of knowledge dramatically. In a 1983 article in *Scientific American*, Stanford researchers Bradley Efron and Persi Diaconis argued:

> Most statistical methods in common use today were developed between 1800 and 1930, when computation was slow and expensive. Now computation is fast and cheap; the difference is measured in multiples of a million. In the past few years there has been a surge in the development of new statistical theories and methods that take advantage of the high-speed digital computer. [. . .] The payoff for such intensive computation is freedom from two limiting factors that have dominated statistical theory since its beginnings: the assumption that the data conform to a bell-shaped curve and the need to focus on statistical measures whose theoretical properties can be analyzed mathematically.[35]

Working alternatively with military data and environment data as a subcontractor to the US government, the statistician Leo Breiman celebrated that expanding computing capacity meant a greater ability to deal with rich granular data. For all the celebration of the increased power of computers, Diaconis, Efron, and Breiman were clear that new algorithms for real-world data on fast—but not infinite—machines had to be created. "With large data sets involving many variables," Breiman and colleagues argued, "more structure can be discerned and a variety of different approaches tried." The complexity of the data is key, and may include:

High dimensionality

A mixture of data types

> Nonstandard data structure
>
> and, perhaps most challenging, nonhomogeneity; that is, different relationships hold between variables in different parts of the measurement space.[36]

The existing methods of statistics were simply inadequate. "To analyze and understate complex data sets, methods are needed which in some sense select salient features of the data, discard the background noise, and feed back to the analyst understandable summaries of the information."[37] The tools Breiman and others in computational statistics, machine learning, and pattern recognition developed, in turn, were picked up by the corporate data miners in the 1990s.

Through the Cold War and beyond, the US and Soviet military and intelligence establishments funded empirically focused computational statistics in a variety of loosely defined fields, such as pattern recognition and machine learning. In the United States, this funding supported corporate labs such as Bell Labs and IBM, organizations such as the Stanford Research Institute, and academic labs across the nation. Other US government agencies, such as NASA and later the Environmental Protection Agency (EPA), funded data-focused statistical work, including Breiman's; the NSA long supported applied mathematical and statistical work, still mostly classified, to contend with exactly the issues discussed above. In the Soviet Union, the Institute for Control Studies promoted a remarkable strand of machine learning heavily focused on an instrumentalist approach to classifying data for industrial purposes.[38] Much as vast state spending nurtured the microelectronics industries of California and New England that made the personal computer revolution possible, long-term government support of computational methods of contending with data helped make the explosion of large-scale machine learning possible. The booming innovations of our day rest on an intellectual and computational infrastructure paid for during the Cold War and its immediate aftermath.

In the 2000s, the explosion of new businesses focused on corporate data and then the influx of streaming data from the Internet leveraged these generations of work. The wealth subsequently generated in turn amplified these research efforts in an extraordinary feedback loop. By the early 2010s, Internet giants in China such as Baidu competed with the American heavyweights for the stars in machine learning. Scholars who had been scraping by on government grants now find themselves flush with millions from the dominant platforms; long discounted areas of study, such as neural nets, find themselves heralded—and funded—as the bringers of the next industrial revolution.

Data and Classification

At the 1959 NSA symposium, the legendary analyst Juanita Moody "pointed out the need for a common heading format which could receive information as it develops and store it for future use."[39] The history of data must include histories and sociologies of classification and the material infrastructures for preforming and using classifications. More than generations of philosophical debates over realism of classification, the approach Geoffrey Bowker termed "infrastructural inversion" provides crucial tools for thinking through classification. "Infrastructural inversion," Bowker and Star explain, "means recognizing the depths of interdependence of technical networks and standards, on the one hand, and the real work of politics and knowledge production on the other."[40] While this approach focuses us on the material work of classification and its infrastructures, it likewise provides a potent lens for seeing how those classifications and infrastructures affect the material worlds. Classifications and the tools that materialize them causally alter actions in the world and thus the human, animal, and material world alike. "Under the sway of cognitive idealism," Bowker and Star argue, "it is easy to see classifications as properties of mind and standards as ideal numbers or floating cultural inheritances. But they have material force in the world. They are built into and embedded in every feature of the built environment (and in many of the nature-culture borderlands, such as with engineered genetic organisms)."[41] Any deflationary view of the impact of computing on society is perhaps best countered through an understanding of how new technical informational infrastructures have transformed the very world they purport to document.

However challenging for individual scientific research groups, standardizing data is even more contested when crossing institutional, disciplinary, and national lines. In their study of model organism databases, Sabina Leonelli and Rachel Ankeny describe the development of formalized data curators responsible for "(1) the choice of terminology to classify data and (2) the selection and provision of information about experimental settings in which data are produced, including information about specimens and protocols."[42] To be useful, datasets must include metadata: information about the production of the data, essential for evaluating it in further analysis, indeed essential to future work with it. Dispute over metadata can produce what one group calls "science friction."[43]

Every scientific database depends upon a set of decisions—and compromises—about how to represent which aspects of data collection to include and how to store them on actual computer systems. Different choices enable different ranges of knowledge production. The cataloging of genomes saw the "evolution of GenBank from flat-file to relational to federated data-bases," according to Hallam Stevens. These "paralleled biologists' moves from gene-centric to alignment-centric to multielement views of biological action."[44] Leonelli and Ankeny likewise argue, "Through classification systems such as the Gene Ontology, databases foster implicit terminological consensus within model organism communities, thus strengthening communication across disciplines but also imposing epistemic agreement on how to understand and represent biological entities and processes."[45] While databases do not exclusively enable only one theory, different databases lend themselves to particular types of investigation and make others more challenging. Different ways of storing data have different investigative affordances. Like models, database can be performative.[46]

Advocates of the introduction of computation into various scientific fields draw heavily upon technological determinist narratives to justify the necessity of new epistemic practices and differently skilled practitioners. To justify the intrusion of computational statistical methods into taxonomy, for example, the biologist George Gaylor Simpson explained that they "become quite necessary as we gather observations on increasing large numbers of variables in large numbers of individuals."[47] While historians must not take such deterministic narratives at face value, their claims are crucial for understanding funding decisions and transformations of disciplines.

Ethics, Privacy, Propriety

The University of California, Irvine has for many years hosted a depository of data sets used to benchmark models in machine learning. An early addition to the collection includes data on the prevalence of diabetes among an Indigenous community living at the Gila River Indian Community Reservation, the Akimel O'odham (often called the "Pima Indians"). The machine learning community has made considerable use of this data set as a "toy" or example problem for various sorts of prediction algorithm. In 2017, the anthropologist Joanna Radin wrote about the creation and repurposing of this data set, amid a broader set of struggles by Akimel O'odham to secure autonomy of knowledge about themselves. The "history of a particular data set known as the Pima Indian Diabetes Dataset (often referred to as PIDD)," Radin argues, "makes political and economic subjectivity visible in an algorithmic age sustained by a steady diet of repurposed data."[48] The Akimel O'odham case invites us to think broadly about the range of issues often reduced to narrower categories of privacy and ownership. Which returns us to the moment of great concern about government and corporate databases around 1970, and the ways those concerns were too easily brushed aside, both legally and analytically.

In the United States, the initial wave of privacy concerns about data peaked in the wake of Watergate and the "year of intelligence" that revealed the CIA, FBI, and NSA had indeed been snooping on Americans, especially on civil rights and anti-war protestors. Legislation to provide a robust set of privacy protections about government and commercial collection of use of data found supporters on both sides of the aisle. The robust protections envisioned included:

1. There must be no personal data system whose very existence is secret.

2. There must be a way for an individual to find out that information about him is in a record and how that information is to be used.
3. There must be a way for an individual to correct information about him, if it is erroneous.
4. There must be a record of every significant access to any personal data in the system, including the identity of all persons and organizations to whom access has been given.
5. There must be a way for an individual to prevent information about him collected for one purpose from being used for other purposes, without his consent.[49]

With no small dose of corporate lobbying, the scope of the Privacy Law of 1974 was narrowed to limit only the US federal government, largely by requiring transparency—not limitations—on collection. No general principle that individuals had privacy rights was established and legally reinforced, and the perception that the government posed the greatest danger was reinforced. While legislation restricting some sectors—higher education, credit ratings, and healthcare—came into force, commercial data collection remained—and remains—very lightly regulated in the US. Decades later, the European Union pursued a dramatically different approach to privacy with the GDPR. Protecting one's privacy had increasingly become a highly individualized activity. And the lack of privacy proved crucial to the business models of the data-intensive corporations of the Internet era.

Critics rightly recognized the weakness of the new privacy legislation. Technical experts including Baran, with extensive experience with highly classified government computation, spoke in remarkably similar terms to civil libertarians. The Privacy Protection Study Commission warned in 1977: "The real danger is the gradual erosion of individual liberties through the automation, integration, and interconnection of *many small*, separate recordkeeping systems, each of which alone may seem innocuous, even benevolent, and wholly justifiable."[50] Such collections allow for the classification not just of individuals, but of entire communities, as worthy and unworthy, as deserving of credit or of suspicion. Much recent scholarship on the profound power of large automatic decision-making systems to reinforce systemic inequalities have shown just how incisive such concerns were.[51]

Contending with the Near Ubiquity of Data-Driven Models

Digital computers have long been information processors, as well as symbolic and calculative machines: today they thoroughly incorporate machine learning models trained on large data. Our interactions with quotidian computers, whether phones, "smart devices," or laptops, are modulated through such data-driven models. Input, whether by keyboard or speech, is filtered through predictive (and corrective) algorithms, with pretrained models onboard. Not simply computerized, contemporary automobiles brim with data-based models that "aid" driving and collect information on their drivers and their conduct. Booting up and using a contemporary operating system—whether on a laptop, phone, or automobile—involves in most cases numerous transmissions reporting on locations and facets of your usage. Yet the job of historians of computing centrally involves resisting teleological narratives that all this—this constellation of collection, analysis, and ethical and political choice—had to come into being in these ways, whether because of the relentless march of technology or of neoliberalism or your monocausality of choice. The same holds of historians of science or engineering in the face of the explosion of data-centric practice and theorizing. Rather than continuing older debates about whether computerization largely continues or breaks from previous practice, historians are and should continue documenting how different forms of data production, metadata creation, and algorithmic analysis play out in different scientific disciplines. Historians of computing, of science, and of technology have a crucial role in transmuting smooth teleologies into accounts reflecting of the disputes, challenges, and limits of these approaches. Given the centrality of data-driven computation in politics, engineering, communication in the last fifty years, focusing on these phenomena places historians of computing and information into an ever-greater role in informing the political, cultural, and economic history itself of the last half decade: their distinctive skills are needed to understand not just the technical but fundamental political and cultural change.

How does focusing on data affect the history of computing and of information? Working through the history of data, information, and computation in the last hundred years requires a diverse array of historians, many outside the typical streams of history of technology, history of computing, and history of information. It absolutely defies an easy divide between material and ideational histories of computing and information technologies, and indeed involves considerable scholarship that does not necessarily place computers or information at the center of its historical enterprise. Yet, it also requires the disciplinary competencies of historians of computing and of information to replace easy teleologies with empirical precision, to provide grounded accounts of historical contingency, needed as much for superior policymaking as for better scholarly monographs.

"Every time someone proposes a system that handles embarrassing information," Paul Baran told the US Congress in 1966, "we are commonly reassured by the words, 'Only those having a legitimate need to know will have access to the information.' While the statement rings seductively of safety and is granted in good faith, its validity is sometimes overstated, particularly when such systems are interconnected."[52] Baran was right. The history of data must concern infrastructures, collection, algorithms. And it equally must contend with powerful representations, true and false, about what data enables, the knowledge it produces, what it obscures, who it empowers, and whom it takes from. Historians of computing and of information are primed to treat all these questions with care. We should do so.

Notes

1. U.S. Government Printing Office, *The Computer and Invasion Of Privacy*, 1966, 125, http://archive.org/details/U.S.House1966TheComputerAndInvasionOfPrivacy.

2. For a careful example, see Francesca Rochberg, "Observing and Describing The World Through Divination and Astronomy," in *The Oxford Handbook of Cuneiform Culture*, ed. Karen Radner and Eleanor Robson, 1st ed. (Oxford University Press, 2012), 622, https://doi.org/10.1093/oxfordhb/9780199557301.013.0029.

3. For the history of the category, see Daniel Rosenberg, "Data as Word," *Historical Studies in the Natural Sciences* 48, no. 5 (November 1, 2018): 557–67, https://doi.org/10.1525/hsns.2018.48.5.557; Daniel Rosenberg, "Data Before the Fact," in *Raw Data Is an Oxymoron*, ed. Lisa Gitelman (Cambridge, MA: The MIT Press, 2013), 15–40.

4. "Raw data," the title of a prominent edited collection from 2013 proclaimed, "is an oxymoron." Lisa Gitelman, ed., *Raw Data Is an Oxymoron*, Infrastructures Series (Cambridge, MA: The MIT Press, 2013).

5. Compare Sabina Leonelli, *Data-Centric Biology: A Philosophical Study* (Chicago: University of Chicago Press, 2016), 70–77.

6. I have been fortunate to be a co-organizer of the Mellon-Sawyer seminar on the History of Artificial Intelligence, based in Cambridge, England, the last two years, aimed precisely at this effort. The volume of papers to appear soon as an issue of BJHS Issues will dramatically augment our knowledge of the global history of AI. https://www.hps.cam.ac.uk/research/projects/histories-of-ai.

7. Colin B. Burke, *It Wasn't All Magic: The Early Struggle to Automate Cryptanalysis, 1930s-1960s* (Fort Meade, MD: Center for Cryptological History, NSA, 2002), 264, http://archive.org/details/NSA-WasntAllMagic_2002.

8. Samuel S. Snyder, "Computer Advances Pioneered by Cryptologic Organizations," *Annals of the History of Computing* 2, no. 1 (1980): 66.

9. "Data Storage and Data Retrieval Symposium" (Ft. Meade, MD: NSA, April 16, 1959), 6, https://archive.org/details/doc-id-6815270.

10. "Data Storage and Data Retrieval Symposium," 3.

11. "Data Storage and Data Retrieval Symposium," 7.

12. "Data Storage and Data Retrieval Symposium," 6.

13. Thomas Haigh, "How Data Got Its Base: Information Storage Software in the 1950s and 1960s," *Annals of the History of Computing, IEEE* 31, no. 4 (2009): 6–25; Thomas Haigh, "'A Veritable Bucket of Facts' Origins of the Data Base Management System," *ACM SIGMOD Record* 35, no. 2 (2006): 33–49.

14. Edgar F. Codd, "A Relational Model of Data for Large Shared Data Banks," *Communications of the ACM* 13, no. 6 (1970): 380.

15. The main academic histories of database systems are Thomas J. Bergin and Thomas Haigh, "The Commercialization of Database Management Systems, 1969–1983," *Annals of the History of Computing* 31, no. 4 (2009): 26–41; Avi Silberschatz, Michael Stonebraker, and Jeffrey D. Ullman, "Database Systems: Achievements and Opportunities," *ACM SIGMOD Record* 19, no. 4 (1990): 6–22.

16. Silberschatz, Stonebraker, and Ullman, "Database Systems," 11.

17. Charles W. Bachman, "The Programmer as Navigator," *Commun. ACM* 16, no. 11 (November 1973): 654, https://doi.org/10.1145/355611.362534.

18. Ralph Kimball, "The Database Market Splits," accessed June 18, 2013, http://www.kimballgroup.com/1995/09/01/the-database-market-splits/.

19. Paul Edwards, *A Vast Machine: Computer Models, Climate Data, and the Politics of Global Warming* (Cambridge: MIT Press, 2010) offers one well-known study of this transition.

20. Chad Wellmon, *Organizing Enlightenment* (Baltimore: Johns Hopkins University Press, 2015), 9.

21. Nick Seaver, *Computing Taste: Algorithms and the Makers of Music Recommendation* (Chicago: University of Chicago Press, 2022), 36–37.

22. I've treated this movement briefly in Matthew L. Jones, "Querying the Archive: Database Mining from Apriori to Page-Rank," in *Science in the Archives: Pasts, Presents, Futures*, ed. Lorraine Daston (Chicago: University of Chicago Press, 2016), 311–28.

23. A printed version of this material appeared as Sergey Brin, Rajeev Motwani, and Terry Winograd, "What Can You Do with a Web in Your Pocket?," *Data Engineering Bulletin* 21 (1998): 37–47.

24. Brin, Motwani, and Winograd, 2.

25. S. Brin and L. Page, "The Anatomy of a Large-Scale Hypertextual Web Search Engine," in *Seventh International World-Wide Web Conference (WWW 1998)*, 1998, http://ilpubs.stanford.edu:8090/361/, §4.2.

26. Sarah T. Roberts, *Behind the Screen: Content Moderation in the Shadows of Social Media* (New Haven: Yale University Press, 2019); Lilly Irani, "Justice for 'Data Janitors,'" *Public Books* (blog), January 15, 2015, http://www.publicbooks.org/justice-for-data-janitors/; Mary L. Gray and Siddharth Suri, *Ghost Work: How to Stop Silicon Valley from Building a New Global Underclass* (Boston: Houghton Mifflin Harcourt, 2019). These complement major works in the history of computing, especially J. Abbate, *Recoding Gender: Women's Changing Participation in Computing* (MIT Press, 2012); Jennifer S. Light, "When Computers Were Women," *Technology and Culture* 40, no. 3 (1999): 455–83; Mar Hicks, *Programmed Inequality: How Britain Discarded Women Technologists and Lost Its Edge in Computing* (Cambridge, MA: MIT Press, 2017).

27. Mél Hogan, "Data Flows and Water Woes: The Utah Data Center," *Big Data & Society* 2, no. 2 (December 1, 2015): 2053951715592429, https://doi.org/10.1177/2053951715592429; Nathan Ensmenger, "The Environmental History of Computing," *Technology and Culture* 59, no. 4 (2018): S7–S33, https://doi.org/10.1353/tech.2018.0148; Zane Griffin Talley Cooper, "Of Dog Kennels, Magnets, and Hard Drives: Dealing with Big Data Peripheries," *Big Data & Society* 8, no. 2 (July 1, 2021): 20539517211015430, https://doi.org/10.1177/20539517211015430.

28. Matthew Crain, *Profit Over Privacy: How Surveillance Advertising Conquered the Internet* (Minneapolis: University of Minnesota Press, 2021); Tim Hwang, *Subprime Attention Crisis: Advertising and the Time Bomb at the Heart of the Internet* (Farrar, Straus & Giroux, 2020); Chris Wiggins and Matthew L. Jones, *How Data Happened: A History from the Age of Reason to the Age of Algorithms* (W. W. Norton & Company, Inc., 2023), ch. 12.

29. Chris Anderson, "The End of Theory: The Data Deluge Makes the Scientific Method Obsolete," Wired Magazine On-Line, 2008, http://archive.wired.com/science/discoveries/magazine/16-07/pb_theory.

30. See Xiaochang Li, "Divination Engines: A Media History of Text Prediction" (NYU, 2017) for the best current history around approaches to speech recognition and translation.

31. A. Halevy, P. Norvig, and F. Pereira, "The Unreasonable Effectiveness of Data," *Intelligent Systems, IEEE* 24, no. 2 (April 2009): 12, https://doi.org/10.1109/MIS.2009.36.

32. Charles N. Yood, *Hybrid Zone: Computers and Science at Argonne National Laboratory, 1946–1992* (Docent Press, 2013). For the notion of domain in the data world, see David Ribes et al., "The Logic of Domains," *Social Studies of Science* 49, no. 3 (June 1, 2019): 281–309, https://doi.org/10.1177/0306312719849709.

33. Edwards, *A Vast Machine*, 251.

34. Alma Steingart, *Axiomatics: Mathematical Thought and High Modernism* (Chicago, IL: University of Chicago Press, 2023), 57.

35. Persi Diaconis and Bradley Efron, "Computer-Intensive Methods in Statistics," *Scientific American* 248, no. 5 (1983): 116–31; see the earlier Bradley Efron, "Computers and the Theory of Statistics: Thinking the Unthinkable," *SIAM Review* 21, no. 4 (1979): 460–80.

36. Leo Breiman et al., *Classification and Regression Trees* (Chapman & Hall, 1984), 7. For these developments, see Matthew L. Jones, "Decision Trees, Random Forests, and the Genealogy of the Black Box," in *Algorithmic Modernity: Mechanizing Thought and Action, 1500–2000*, ed. Morgan Ames and Massimo Mazzotti (New York: Oxford University Press, 2023), 190–215. For a crucial parallel story, see Xiaochang Li, "'There's No Data Like More Data': Automatic Speech Recognition and the Making of Algorithmic Culture," *Osiris* 38 (July 2023): 165–82, https://doi.org/10.1086/725132.

37. Breiman et al., *Classification and Regression Trees*, 8.

38. See "History of the Institute," Institute of Control Sciences, accessed July 7, 2017, http://www.ipu.ru/en/node/12744.

39. "Data Storage and Data Retrieval Symposium," 3.

40. Geoffrey C. Bowker and Susan Leigh Star, *Sorting Things Out* (MIT Press, 1999), 34. Internal citation omitted.

41. Bowker and Star, 39.

42. Sabina Leonelli and Rachel A. Ankeny, "Re-Thinking Organisms: The Impact of Databases on Model Organism Biology," *Studies in History and Philosophy of Science Part C: Studies in History and Philosophy of Biological and Biomedical Sciences* 43, no. 1 (March 2012): 31, https://doi.org/10.1016/j.shpsc.2011.10.003.

43. Paul Edwards et al., "Science Friction: Data, Metadata, and Collaboration," *Social Studies of Science* 41, no. 5 (2011): 667–90.

44. Hallam Stevens, *Life Out of Sequence: A Data-Driven History of Bioinformatics* (Chicago: University of Chicago Press, 2013), 168.

45. Leonelli and Ankeny, "Re-Thinking Organisms," 32.

46. Geoffrey C. Bowker, "Biodiversity Datadiversity," *Social Studies of Science* 30, no. 5 (October 1, 2000): 675–76.

47. George Gaylord Simpson, "Primate Taxonomy and Recent Studies of Nonhuman Primates," *Annals of the New York Academy of Sciences* 102, no. 2 (1962): 504, https://doi.org/10.1111/j.1749-6632.1962.tb13656.x.

48. Joanna Radin, "'Digital Natives': How Medical and Indigenous Histories Matter for Big Data," *Osiris* 32, no. 1 (September 1, 2017): 46, https://doi.org/10.1086/693853;

for other key examples, see Amandalynne Paullada et al., "Data and Its (Dis)Contents: A Survey of Dataset Development and Use in Machine Learning Research," *Patterns* 2, no. 11 (November 12, 2021): 100336, https://doi.org/10.1016/j.patter.2021.100336.

49. United States Congress Senate Committee on Government Operations Ad Hoc Subcommittee on Privacy and Information, *Privacy: The Collection, Use, and Computerization of Personal Data: Joint Hearings Before the Ad Hoc Subcommittee on Privacy and Information Systems of the Committee on Government Operations and the Subcommittee on Constitutional Rights of the Committee on the Judiciary, United States Senate, Ninety-Third Congress, Second Session . . . June 18, 19, and 20, 1974* (Washington: US Govt. Print. Off, 1974), I:53.

50. United States and Privacy Protection Study Commission, *Personal Privacy in an Information Society*, 533 (original all in italics).

51. See the now very well-known danah boyd and Kate Crawford, "Critical Questions for Big Data," *Information, Communication & Society* 15, no. 5 (June 1, 2012): 662–79, https://doi.org/10.1080/1369118X.2012.678878; Cathy O'Neil, *Weapons of Math Destruction: How Big Data Increases Inequality and Threatens Democracy*, 1st edition (New York: Crown, 2016); Virginia Eubanks, *Automating Inequality: How High-Tech Tools Profile, Police, and Punish the Poor* (New York, NY: St. Martin's Press, 2017); Safiya Umoja Noble, *Algorithms of Oppression: How Search Engines Reinforce Racism* (New York: New York University Press, 2018); Ruha Benjamin, *Race after Technology: Abolitionist Tools for the New Jim Code* (Cambridge, UK; Medford, MA: Polity Press, 2019); Emily M. Bender et al., "On the Dangers of Stochastic Parrots: Can Language Models Be Too Big?," *FAccT '21*, 2021, 14; Kate Crawford, *Atlas of AI: Power, Politics, and the Planetary Costs of Artificial Intelligence* (New Haven, CT: Yale University Press, 2021), https://doi.org/10.12987/9780300252392.

52. U.S. Government Printing Office, *The Computer and Invasion of Privacy*, 120.

Chapter 12

Writing about Writing about Social Informatics

Opportunities and Challenges

Howard Rosenbaum

A Meditation on Social Informatics

Social informatics (SI) is entering its fifth decade of researching and critically analyzing the complex, dynamic, and varied relationships among people, information, and computing technologies and the contexts of their design, implementation, management, and use. This chapter is a meditation on writing about SI. It discusses issues that colleagues, co-authors, and I have faced as we have written and continue to write about SI and use the approach in our work. It describes my journey through the discipline of SI by providing some background on its origins, a story that has been told in greater detail elsewhere,[1] its development over time, and my involvement in research and writing about SI. The theory of scientific and intellectual movements (SIMs), proposed by sociologists Frickel and Gross,[2] will be used as an organizing frame to describe the ways in which writing about SI has changed. It is an apposite framework to use here because SI is a classic example of a SIM, defined as a movement based on "collective efforts to pursue research programs or projects for thought in the face of resistance from others in the scientific or intellectual community."[3] To provide a common basis for understanding SI, the working definition here is that it is "an approach to studying the social aspects of computing that takes into account the complex interplay between information and communication technologies (ICT), the people who design, implement, manage, and use them, and the contexts of their design and use."[4]

I have been involved in and writing about social informatics since 1996. Over the last 25 or so years, the opportunities and challenges in doing this work have shifted and evolved through what I see as three main stages. At the beginning, the challenge was to spread the word and define the approach for a wider audience in ways that would both attract attention and new converts. This involved making the argument in multiple venues about the importance of SI for the study of people, computing, and information. The second stage involved writing that attempted to solidify its place in the academic landscape by publishing and presenting theoretical and empirical work grounded in SI. During this stage, researchers have been engaged in writing intended to contribute to the core body of knowledge that constitutes SI. The third stage involves writing intended to contribute to this body of knowledge and that critically analyzes the state of the approach in the face of what Fichman, Sanfilippo, and Rosenbaum[5] have called its transformation stage. These three stages are not discrete and sequential; although they have emerged at different times, they actually now overlap with and flow into each other. This chapter is therefore an extended reflection on the current state of SI with Frickel and Gross'[6] framework providing a context that offers a glimpse into what the future may hold.

In the Beginning: The Importance of Social Informatics for the Study of Information and Computing

Rob Kling, one of the scholars most closely associated with social informatics, described it in 2000 as " . . . research [that] pertains to information technology developments and uses in any social setting, not just organizations."[7] Several years later, Lamb and Sawyer claimed that it is "emerging as a transdiscipline that draws researchers who focus on the inter-relationships among people, their institutional and cultural contexts, and their uses of ICTs."[8] Day adds that it arose out of " . . .

disjunctions between popular and professional claims about the social values and uses of information and communication technologies and the empirical reality of such during the last two decades of the 20th and the beginning of the 21st centuries."[9] So how do we describe the emergence and development of SI?

Vehovar, Smutny, and Bartol[10] find that SI activities, measured by the number of publications, research and educational activities, and conferences and symposia, increased between 2005 and 2014, then began a "slight" decline. After nearly half a century of work, SI as a form of inquiry has not converged around a core theoretical or methodological focus.[11] Smutny and Vehovar[12] describe the current state of SI as "expanding, even as community fragmentation, topical dispersion, and methodological diversity continue to increase." This is attributed to the variegated emergence of SI which began in Norway and Slovenia in the 1980s[13] and quickly and apparently independently emerged in other parts of the world, arriving finally in the United States in the mid-1990s. They identify seven schools of SI that they label by country of origin: German, Japanese, Norwegian, Russian, Slovenian, United Kingdom, and United States.[14] The implication of their analysis is that "SI is not a single path; instead, each school brings its own specific perspective, which is developed under specific circumstances."[15] However, one theme that seems to cut across these versions of SI is a shared concern for understanding and unpacking the interactions between people and ICT/computing in group, organizational, and societal settings.[16]

This concern was instrumental in the development of SI in the United States, which was motivated by a response from a small community of researchers studying the social impacts of computing to the prevailing ideology which they labeled "technological determinism." Led by the late Rob Kling and colleagues at the University of California at Irvine in the 1970s, this community began to articulate an alternative approach to studying computing in society that drew upon different theoretical and conceptual frameworks, mainly from the social sciences and more specifically, in Kling's case, sociology.[17] With this work, they laid the foundation for what would become, in time, what has been described above as the "United States" school of SI.[18] The initial critical and oppositional impulse in their research led them to integrate social science theory and concepts into their analyses of computing, which they now saw as being embedded in social contexts; originally, these contexts were primarily organizational, but over time, the scope of the research widened to include larger swaths of the social world. For about twenty-five years, this flavor of SI research has made use of a range of social theories to unpack the complexities at the intersection of ICT, people, and society.

As far back as the early 1970s, Kling[19] and colleagues had been reflecting upon the prevailing intellectual approaches to the study of computerization and arguing for what he called then a "person centered" approach to computing. Considering that at this time, he was in a computer science department, this was a bold move. By 1977, this approach had become explicitly social with Kling and Gerson,[20] publishing articles in the inaugural and second issues of the journal *Symbolic Interaction*. In this sense, the drive towards SI in the United States began in the late 1970s with the work of the "Irvine School" where Kraemer, Kling, and Danziger and their colleagues and students were among "the first scholars to begin careful empirical study of the effect of computerization in government agencies, and soon started a research program on IT and local government at UCI's Public Policy Research Organization."[21] The network of scholars who passed through Irvine during this time left with a continued interest in the social aspects of computing that eventually became formalized as SI.

In 1982, Kling, working with colleagues Danziger, Dutton, Kraemer, Star, and King, provided empirical evidence of the social impacts of computing, "arguing that computing reinforced bias and structures of control in the organizations they studied."[22] By the mid-1980s, this approach to the social impacts of computing had evolved to the point at which Kling[23] could offer a reading list on computing and organizations that features several sociologists' work and could assert that "[i]n addition to their functional capabilities as an information processing tool, computer-based technologies are also social objects which may be highly charged with meaning."[24] Further, computerization was cast as a complex phenomenon composed of "technical, social and political processes."[25] Computer systems, at this point, were clearly seen to involve social choices with a mandate that a "wide variety of noneconomic factors must be considered to fully understand the process of rapid computerization."[26] In an influential article about the social contexts of computing, Kling and Iacono[27] drew upon sociological theories of social movements which became the foundation of their analogous concept of "computerization movements." At the beginning of the 1990s, Kling argued that "questions of the social consequences of computerization [are] central for the information systems field" and that research on the social impacts of computing should "examine what kinds of social transformations have and have not happened because of computerization and why. And it is important for us to articulate the real social choices that remain."[28] This stream of research was clearly motivated by the "dissatisfaction with what (were) perceived to be the dominant intellectual practices or expectations in a given field or set of fields," an important early step in the eventual emergence of SI as a scientific and intellectual movement.[29]

Drawing on earlier work by Kling and Scacchi,[30] Kling advocated for the analytic use of "web models" which include a "focal technology, its immediate users, and a rich ecology of social relationships with other social groups and organizations in which the technology is developed, adopted and used."[31] This approach was intended to turn researchers' attention to the social contexts of computing technologies in order to better understand the complex impacts of computing in organizations. With this move, the study of the impacts of computerization was explicitly characterized as being rooted in social science and information systems were likewise characterized as a "form of social organization."[32] By explicitly focusing on the recognition and consequences of embedding computer systems in their social and organizational contexts,[33] the foundations of the United States school of SI were beginning to come into view. As well, the operationalization of the web model made clear another important feature of Kling's approach to the social study of computing—that "web analysts have a more difficult time in defining crisp boundaries and identifying a precise computing infrastructure without examining concrete organizational arrangements."[34] As opposed to the approach he and his like-minded colleagues criticized for simplifying the analysis of computing, they argued that the social worlds in which computer systems were embedded were complex, dynamic, messy, and a primary focus of investigation.

From this time forward, there was no longer as pressing a need to explicitly challenge technological determinism, although systems rationalism remained a critical target because of the ways in which it, in their view, over-simplified both organizational settings and computer systems. The question instead became one of the ways in which computer-based systems transformed society, or phrased somewhat differently, of the social consequences of technology design and use. Computer systems, Kling explained, "can play key roles in restructuring major social relationships—interpersonal, intergroup, and institutional"[35]—for better or for worse. Speaking directly to sociologists, he asserted that "ideas about social change so permeate the discourse and images of computerization that they should be at the center of attention for sociological inquiry."[36] In this article, Kling describes organizational informatics as a "field that studies the development and use of computerized information systems and communication systems in organizations."[37] This disciplinary label was useful because it made clear that computerization was impacting a range of organizations, settings, and people besides businesses including health care, military operations, air traffic control, libraries, and home uses, to name a few. It was also useful because it foregrounded social factors and relationships and used them to explain the impacts of computerization. In this context, Kling made clear that his understanding of organizational informatics was rooted in sociological constructivism, stating that "[p]eople construct their worlds, including the meanings and uses of information technologies, through their social interactions."[38]

This position, however, was not without its critics,[39] and in 1991 and 1992, Kling engaged in a debate with Woolgar and Grint[40] in the pages of *Science, Technology, and Human Values* which allowed them to clarify their differing approaches to understanding the social and organizational impact of computing. Woolgar and Grint argued that Kling's quest to recover the social dimension in the analysis of computing was well-intentioned but insufficient because it was "insufficiently sensitive to interpretivism."[41] In their view, there were two versions, restricted, which assumes that social construction had its greatest effects during the design and development of an artifact after which its meaning stabilized, and thoroughgoing, which assumes that the nature of the artifact remains fluid beyond the point of use.[42] In response, Kling described his approach as a reconstructive interpretivism which he saw as "encouraging analysts to construct new narratives about technological and social systems after they have examined the ways that important ambiguities are reified or taken for granted."[43] He argued that Woolgar and Grint's approach relegated the analyst to a "reactive role of a constant critic, who is rarely the narrator"; the analyst should, in his view, "be able legitimately to construct compelling narratives about technology."[44]

As is detailed in several places,[45] Kling and colleagues gathered at UCLA in early 1996 for an invitational workshop on the "Social Aspects of Digital Libraries." Although the primary purpose of this gathering was to develop a definition of and research agenda for the study of digital libraries that could provide common ground for researchers interested in the topic from widely varying disciplines, the authors noted that it "is not by accident that a term for this community, 'social informatics,' originated at the UCLA workshop."[46] As described by Kling, Rosenbaum, and Sawyer this came about because

> some participants in this research community agreed that the scattering of related research in a wide array of journals and the use of different nomenclatures was impeding both the research and the abilities of "research consumers" to find important work. They decided that a common name for the field would be helpful. After significant deliberation, they selected "social informatics."[47]

At this point, SI was emerging as a scientific and intellectual movement (SIM) following a pattern described by Frickel and Gross in which a group of high-status academics engages in "collective efforts to pursue research programs or projects for thought in the face of resistance from others in the scientific or intellectual community."[48] The groundwork had been laid in the two decades preceding when Kling and colleagues challenged the prevailing conventional wisdom about computing in society and it came to a head at the 1996 workshop, when the SIM was given a name, social informatics. However, as described above, this version of SI did not burst forth *sui generis*; after gathering momentum since the 1970s, it was given a name and its champions began a process intended to allow it to take its place in the intellectual landscape of cognate forms of inquiry that also focused on technology and society. Some of these approaches included science and technology studies (STS), actor-network theory (ANT), the social construction of technology (SCOT), and a tradition of sociotechnical studies in the United Kingdom going back to the 1950s. Perhaps with the exception of ANT, these approaches "rely on a priori assumptions, either stated or unstated, that either social considerations or technical considerations are of primary importance."[49] What followed was a concerted and coordinated attempt to grow SI as a new approach to be used in the study of computing in society.

My involvement with this version of social informatics began in 1996, soon after Kling joined the faculty of what was then the School of Library and Information Science at Indiana University Bloomington. Early on, he was quite generous with his time and as he made the rounds of the faculty in his new school, and over the course of several discussions, we quickly found common ground in sociological theory, particularly in the work of the symbolic interactionists, the ethnomethodologists, and the sociologists of knowledge. From the first two traditions came the growing interest in the importance of discourse in SI and from the third, the insight that knowledge and information are grounded in sociocultural and historical contexts. In what was an early strategic move to bring SI to the attention of the academy, Kling sought and received a grant from the National Science Foundation in 1996 to hold a workshop, "Advances in Social Informatics," at Indiana University in 1997, the goal of which was to articulate a research agenda for social and organizational informatics. Kling enrolled me in this project, and I worked with him to organize and run the invitation-only workshop in Bloomington that year. This workshop was the next step following the UCLA workshop and was a successful meeting. It is also an important step in the evolution of SI as a SIM because, according to Frickel and Gross, "[o]pportunity structures that afford SIM participants broad license to use available resources with relative impunity are better, from a SIM emergence perspective."[50] Twenty-five scholars from information science, sociology, communications, information systems, social psychology, computer science, and anthropology came to Indiana University Bloomington to clarify systematically the domains of social informatics by exploring the state of knowledge about the integration of computerization and networked information into social and organizational life, and the roles of information and communication technology (ICT) in social and organizational change.[51]

The work that followed the workshop marked the first stage in writing about SI. The primary task was to work out a definition of the approach, describe the research agenda that had emerged from the workshop, and begin the effort to raise awareness of SI in information science, information systems, computer science, and other cognate disciplines. What was challenging at this point was to write about SI in a way that was welcoming and that indicated that it was a rigorous, useful, and high-quality form of intellectual inquiry that addressed fundamental questions, issues, and problems in the study of people, computing, and society in ways that could lead to compelling research findings, useful theory, and positive change. There was an immediate deliverable that had to be written—the report on the workshop for the NSF—that led to an additional challenge. A group of five people who had attended the workshop had to synthesize the diversity of opinions and insights from the twenty-five participants into a single voice that would serve the purposes outlined above. Led by Kling, the work was completed and the workshop report[52] submitted to the NSF. Although the report's official audience was the funders at the National Science Foundation, Kling decided to make the report more widely available by posting it on the website of the Center for Social Informatics, located at Indiana University; the document currently lives at ResearchGate.net.

This document then became the basis for a manuscript that Kling intended to be the next step—a declaration of intent to establish social informatics as a viable and valuable form of social science inquiry into computing and society. Drawing on report material, Kling[53] published a brief article about social informatics in *D-Lib Magazine*. While working on the manuscript and in consultation with Kling, Sawyer and

Rosenbaum[54] also published what was an early attempt to offer a definition of SI and provide a brief state-of-the-art review of its development in and relevance for information science. This paper was written within a year of Kling's article and was targeted at a different audience; Kling published in *D-Lib Magazine* in 1999, the primary audience for which was the digital library research community, while Sawyer and Rosenbaum published in *Informing Science* in 2000, an online publication for the larger information science community. Sawyer and I were working on the manuscript that was eventually published in 2005 and wrote this paper to provide an early overview of SI. Kling's[55] definition of social informatics was used in this paper and one contribution was to introduce three orientations that could characterize SI research; normative SI research suggests "alternatives for professionals who design, implement, use, or develop policy about ICTs"; analytic research develops theoretical approaches about "ICTs in institutional and cultural contexts or to empirical studies that are organized to contribute to such theorizing"; critical research involves "examining ICTs from perspectives that do not automatically (uncritically) adopt the goals and beliefs of the groups that commission, design, or implement specific ICTs."[56] As well, five centers were identified as loci for SI research: Indiana University, Napier University, the University of California, Irvine, Syracuse University, and the University of Oslo. Table 1 in this paper is a summary of seven main findings from organizational and social informatics research.[57]

Around this time, I was invited to submit an entry on SI for the *Encyclopedia of Human-Computer Interaction*. This was an opportunity to explain this form of inquiry to a new and different audience and, in a small way, contribute to the attempt to circulate the approach more widely in the intellectual community.[58] Kling was an early and careful reader of drafts of this entry, which was published after his death. Most useful were his comments about framing SI as a rigorous form of inquiry that could be easily adapted to what is referred to below as the "social turn" in HCI. The goal in this exercise in writing about social informatics was to provide an overview of its relevance for HCI researchers. The argument was that

> social informatics is relevant for human-computer interaction because of a theoretical and methodological trend that has been noticeable in HCI since the late 1990s. This trend, spurred by (among other factors) a technical development, the rise of ubiquitous, pervasive, and mobile computing, is leading many HCI researchers to take a social turn in their thinking.[59]

Between 1997 and 2003, Kling, Sawyer, and I expanded and deepened the themes described in the workshop report into a book published in 2005, two years after Kling's untimely death. The book was intended to be a definitive statement that summarized the current state-of-the-art thinking about and research motivated by SI. It was to be an announcement that SI was primed and ready to take its place as the most important approach to study the social aspects of computing. This step could not have been taken without the assistance and support of many friends of SI such as the scholars at the 1997 workshop, those who worked on the workshop report, and those who helped from a distance, such as Kling's colleague and frequent co-author, Suzi Iacono, who was a supporter of our efforts from within the National Science Foundation, where, at the time, she was Program Director for Digital Society and Technologies in the Directorate for Computer and Information Science and Engineering and later headed up the Information Technology Research (ITR) Program.

In the book, we offered a definition of SI based on Kling,[60] a discussion of the consequences of ICTs for organizations and social life, and a set of guidelines intended to demonstrate the utility and value of the approach for designers, developers, and implementers of information systems. There was a chapter devoted to explaining how SI could contribute to policy analysis by recasting a set of policy issues from a SI perspective. There was an argument intended to persuade educators in information science, computer science, and other cognate social science disciplines to integrate SI research and approaches into their graduate curricula. Finally, there was a lengthy discussion of ways in which researchers could communicate the findings and insights of SI research to ICT professionals and to the academic and research communities. A challenge faced in writing this book was to describe SI in such a way as to persuade these different stakeholders to take it seriously. The hope was that researchers in different disciplines would adopt SI as a descriptor of their research, that faculty in these disciplines would teach SI courses and encourage a new generation of graduate students to become SI researchers, and that policy makers and practitioners would be motivated to draw upon SI research as they made important decisions. As we optimistically concluded:

> We have laid out the case for Social Informatics moving from being a useful but alternative approach to being one that is the core knowledge basis for understanding computing. The rationale for taking organizational and social informatics seriously is that by doing this we will improve the professional practices of designing, developing, deploying, and using ICT and ICT-based systems. Systems designs informed by Social Informatics will be more useful. ICT policy will be more effective, and future researchers will engage more substantial questions and provide more useful insights than those who ignore the interconnections among the social and technical aspects of computing.[61]

The publication of this book marked the midway point of the first phase of my involvement in writing about SI and was, in many ways, a novel experience. As an early career academic, I found that working on this manuscript involved a different type of writing than that in which I had previously been engaged. Rather than describing the results of research or carefully working through a conceptual argument, the goal was to provide a state-of-the-art description of a new form of inquiry and to do so in the form of an extended persuasive argument announcing to academic and practitioner communities the arrival of SI, a new, exciting, insightful, and rigorous[62] approach to the study of the social aspects of computing of which they should take notice and advantage. In addition, through extensive discussion and reading during the preparation of this manuscript, Kling shared his version of the pre-history of social informatics, described above. This knowledge helped to put the work we were doing into an historical and intellectual context which has proved useful in subsequent writing about SI. This type of writing was very different compared to writing about a SI research project, and Kling's guidance in the year before his passing was an important factor in the completion of the manuscript.

Upon reflection, it is not clear that the book has had the impact that we had hoped for; in 18 years, it has been cited 285 times. The other publications about SI published during this time, however, may have had greater impact. Kling's[63] article has been cited once (according to Google Scholar) although when posthumously reprinted in *The Information Society*,[64] it has been cited 972 times. Sawyer and Rosenbaum (2000)[65] has been cited 173 times, and the NSF Report[66] has been cited 84 times. Meyer agrees, stating that "we must admit as part of any stock-taking that social informatics as an academic pursuit has not achieved the scholarly traction we might have hoped for."[67] Compared to the uptake of similar approaches such as science and technology studies and actor-network theory, these foundational SI publications have not had equivalent impact; however, this work was important in the early establishment of SI as a SIM because

> however conceptualized and implemented, these programs involve the transformation of thoughts or research findings into ideas and knowledge that are circulated widely within the intellectual community, subjected to scrutiny and contestation, embraced by some, and rejected by others, and that may emerge from the process deemed credible or true.[68]

Once the book was done, the opportunities and challenges involved in the ongoing attempt to establish SI as a respectable form of inquiry shifted. The next challenge that arose in writing about SI involved serious engagement in historical revisionism. Soon after the book was published, I received a personal communication from a Norwegian scholar, Dr. Ingar Roggen, who had noticed a brief section in the book that implied that SI had sprung, *sui generis*, from the discussion at the 1996 digital libraries workshop. There was also text he singled out that described Kling as the "inventor" of SI. He explained that the history of SI was richer, deeper, and more complex than the book had set out. Further correspondence revealed that there was another more interesting origin story that had its roots in Scandinavian scholarship that brought a social science perspective to the study of people, technology, and society in the early 1980s. In fact, according to Roggen,[69] Kling had visited the Norwegian SI research group at the end of the 1980s at which time they explained to him that they had been studying the complex interactions among people, technology, and social and organizational contexts from sociological and social-psychological perspectives. The challenge, then, was to rewrite the history of SI in a more expansive and inclusive way. The chance to do so came several years later.

In an entry on the topic of SI in the third edition of the *Encyclopedia of Library and Information Science*,[70] I had the opportunity to update and expand upon Kling's[71] entry in the second edition of the publication. This relatively brief essay included a definition of SI, a description of the research domain, a brief history, its key assumptions and findings, and a discussion of its relevance for information science. This was an early attempt to summarize and synthesize the historical information that I had gathered; it began with the first mention of SI in Norway is 1982, although the term "informatics" has been in the academic lexicon for many years, and, in the early 1980s

> . . . the term acquired the modifier "social" when it was used by Norwegian social scientists whose work focused on the intersections among technology, organizations, and work. Roggen claims that "social informatics, the interdisciplinary field of informatics, . . . [was] established as a science in 1982 by Stein Bråten" at the University of Oslo.[72]

This entry was another example of writing about SI that was intended to increase awareness and interest in the information science research community, concluding that there was considerable overlap between the concerns of SI and library and information science.

A panel on SI at a Social Informatics Special Interest Group Research symposium held at the 2013 Association for Information Science and Technology annual conference led to an invitation from a publisher to Pnina Fichman to edit a collection of papers on the past, present, and future of social informatics (which had been the theme of the panel), which she accepted and then invited me to co-edit the volume. This was another opportunity to provide a clearer and more complete recounting of the history and development of SI than had been possible in the encyclopedia entry. The retelling of the story involved responsibly internationalizing SI, describing its emergence and development in what turned out to be six different locations around the world, beginning in Norway in 1982. During the course of the research and writing, it became clear that there were sometimes subtle and nuanced differences among these versions of SI (including the label used, such as socio-informatics [Japan]), with the chapter attempting to explain why the version that developed in the United States became the most dominant at the end of the 20th century. The origins and early development of social informatics were placed in a theoretical context based on Frickel and Gross' framework,[73] introduced above. The revision of the history of SI was, I think, an important contribution and a needed corrective to the previous US-centric version I had published and I am grateful to Dr. Roggen for helping me to understand this. This book was published in 2014. Recently, Smutny and Vehovar[74] have explored the global emergence of SI in more extensive and useful detail, carrying this version of the history of SI much further.

The most recent occasion in which I was involved in writing about social informatics is Fichman, Sanfilippo, and Rosenbaum.[75] This book is part of Morgan and Claypool's Synthesis Lectures on Information Concepts, Retrieval, and Services and is the (current) culmination of the work in which colleagues and I have been involved that has been intended to ensure that SI remains a viable SIM. The synthesis makes three main contributions. First, it provides a summary of the international development and emergence of SI leading to a focus on the United States school. Second, it invokes an ecological metaphor to explain the evolution of SI in the United States in four stages—foundational work (1980s), a period of expansion (1990s), coherence (early 2000s), and transformation (2005 to the present). Third, it systematically reviews and organizes the literature with an original and useful framework of SI according to its principles, approaches and methods, concepts, topics, and findings as these attributes emerged in the four stages. The book was our best effort to write about SI in a way that is accessible to an information science audience of researchers, teachers, and students. In a sense, it is an update, albeit in more concise form, of Kling, Rosenbaum, and Sawyer.[76] It is aimed at a similar academic audience although not the audience of practitioners and policy makers. The publication of this book in 2015 marked the end of the first stage of writing about social informatics.

In the Middle: Developing a Core Body of Knowledge for Social Informatics as a Scientific and Intellectual Movement

If a SIM is to succeed, there must be a "knowledge core" towards which "participants are consciously oriented, regardless of their understanding of it."[77] This core is composed of a slowly growing collection of empirical findings, and theoretical developments, and "consists of intellectual practices that are contentious relative to normative expectations within a given scientific or intellectual domain."[78] The second stage in the writing about social informatics has involved an accretion of contributions to the core body of SI knowledge. As mentioned above, Fichman, Sanfilippo, and Rosenbaum[79] compiled and systematically and critically reviewed the core body of SI knowledge as it was in 2015 using a framework that organized the literature chronologically through four stages and conceptually, in terms of its principles, approaches and methods, concepts, topics, and findings. Contributing to the knowledge core is essential to a SIM and in this section describes my involvement in this type of more conventional writing about SI.

For example, Rosenbaum and Joung[80] used the concept of the "sociotechnical interaction network" (STIN) to analyze the American Memory Project (AMP), a Library of Congress digital library program. Coming online in 1994,

> American Memory is the online resource compiled by the Library of Congress National Digital Library Program. With the participation of other libraries and archives, the program provides a gateway to rich primary source materials relating to the history and culture of the United States.[81]

Kling, McKim, Fortuna, and King[82] introduced the concept of a STIN in an arvix.org preprint paper, explaining that it is based on insights from actor network theory and "web models,"[83] although the concept remains somewhat vague in this paper and is described through examples. However, in Kling, McKim, and King (2003)[84] a STIN is clearly defined as

> . . . a network that includes people (including organizations), equipment, data, diverse resources (money, skill, status), documents and messages, legal arrangements and enforcement mechanisms, and resource flows. . . . The network relationships between these elements include: social, economic, and political interactions.

A key element of a STIN is "that technology-in-use and a social world are not separate entities—they co-constitute each other."[85] In their work, the STIN was used to model what they called "electronic scientific communication forums." Rosenbaum and Joung made use of and extended the concept by incorporating "enrollment," from actor network theory, which "refers to the moment when another actor accepts the interests defined by the focal actor, which in this case is the main actor in the STIN."[86] Primary, secondary, and tertiary enrollment strategies were described to demonstrate that the AMP was, in fact, a successful STIN.

Meyer, Hara, and Rosenbaum[87] made use of the concept of computerization movements, introduced by Kling and Iacono,[88] refined in Iacono and Kling,[89] and taken up by Hara and Rosenbaum.[90] Computerization movements are based on the sociological concept of social movements and describe the ways in which "momentum gathers around a core ICT leading to a socially constructed process of societal mobilization."[91] In this research, photo sharing sites, such as Flickr.com and photoblogs.org were studied from a SI perspective to determine whether these sites represented an emerging computerization movement or were just a passing fad. Analysis of the data indicated that photobloggers had the potential to become a computerization movement, but such a movement would likely be constrained in in its scope and reach. This work demonstrates that the concept is useful outside of the organizational settings to which it had typically been applied, thereby contributing to the core body of SI knowledge.

Rosenbaum and Davenport[92] took on the problem of interdependence, also called the macro-micro problem, in information systems research and proposed a SI approach as a strategy for dealing with it. This paper argues that

> as an unintended consequence of the social turn in information systems research, the problem can be expected to arise and without useful conceptual tools, much intellectual capital will be spent thinking it through. It offers a concept from social informatics, Kling et al.'s (2003) "sociotechnical interaction network" (STIN), as a way to resolve the problem by offering a link between the micro-order of information and communication technology use and the larger macro-order of organizations and institutions.[93]

One interesting contribution of this paper was to augment the STIN with insights from Dourish's[94] interactional view of context and Sheff's[95] interpretation of Goffman's frames.[96] This type of integrative contribution that draws upon concepts from cognate disciplines is one way to grow the core body of knowledge.

Hara and Rosenbaum[97] contributed to the core body of SI knowledge by refining and expanding the concept of the computerization movements and in doing so, broadened the remit of SI. This paper is an expanded version of Hara and Rosenbaum[98] that made use of conceptual modeling for data analysis. The paper was motivated by the insight that while for Iacono and Kling[99] the primary settings in which these movements developed were in organizations, the world of computing had changed drastically. Their approach made sense given the fact that at the time of their writing, information systems were being mainly implemented and used in private and public sector organizations. The spread and uses of smartphones seemed to be becoming more widespread and we wondered if it was an example of a computerization movement; this motivated our earlier paper.[100] Using conceptual modeling in the later paper, Hara and Rosenbaum[101] showed that computerization movements were also taking place in social (non-organizational) settings as could be seen with, for example, the rise of computer gaming, the rise in ubiquity of online social networks, the use of personal computing devices, and the increasing power and sophistication of smartphones, all of which take place outside of organizational settings. These and related phenomena were supported and enabled by the pervasive spread of broadband and high-speed internet to the home. We suggested that the domain was ripe for SI researchers.

Shachaf and Rosenbaum[102] extended the reach of SI to online social reference sites and proposed a research agenda based on a STIN framework. After describing social reference as a phenomenon characterized by people interacting with each other through digitally mediated platforms, they argued that a STIN framework was appropriate because it could be used to understand "complex networks of people, ICTs, organizations, and their structured interrelationships."[103] The paper concluded by posing seven research questions that could be investigated using the STIN framework.

Rosenbaum and Shachaf[104] was motivated by our attempt to investigate some of these questions. When we studied online question and answer sites, we noticed that much of the research on these communities focused on issues of information retrieval, information seeking and use, and information intermediation. The community itself was assumed to play an important but unspecified role but despite "the frequent use of the term community in these studies, Q&A sites have rarely been examined as online communities."[105] With the objective of the research being to understand the social dynamics of these sites, it seemed appropriate to make use of social theory. The question was which theories would be the most useful. We developed an approach on structuration theory,[106] the duality of technology,[107] and communities of practice.[108] This approach foregrounded the social, technical, and contextual factors that shaped the trajectories of online Q&A sites over time and led to our main theoretical contribution, a structurationally informed concept of communities of practice. Our emphasis on social theory in this work was to provide a corrective to research on online Q&A sites in which the online "community was assumed to exist and provide the context for the various studies"[109] to information retrieval research that typically did not review literature on online communities, and to information behavior studies of these communities that tended to minimize the role of technology. Making use of social theory in this way allowed us to "address the gap in the literature on Q&A sites, and online communities, a sociotechnical approach that integrates structuration and CoPs to analyze and understand these communities is a useful way to proceed."[110]

Yang, Zhu, and Fichman[111] is an edited collection addressing the ways in which the use of ICTs shaped and were shaped by the COVID-19 pandemic and represents an important addition to the core body of SI knowledge. The common theme that connects the nine chapters of this book is reliance on a SI approach. As they explain,

> The concept of "social informatics" in this book is used in two ways. First, it serves as an underlying theoretical perceptive across all the chapters. . . . Second, social informatics serves as a theoretical foundation for developing new theories in context because of its flexibility.[112]

My contribution to this book[113] addresses the second use of SI. This chapter is a conceptual analysis of the impacts that algorithms are having in the daily lives of the online and connected public and argues for the importance of social theory as a foundational approach for SI. Based in critical algorithm studies, the chapter makes use of the social phenomenology of Schutz[114] and Rosenberger and Verbeek's[115] postphenomenology, particularly the concepts of the lifeworld and the natural attitude, to explain the hold that algorithmic assemblages have on the ways in which people interact with and understand their social worlds. The main argument is that "algorithmic assemblages are shaping the lifeworld and the natural attitude and, in doing so, take part in the constitution of social worlds."[116] The contribution to SI made here is the demonstration that

> . . . the conceptual foundations of social informatics can be productively informed by the introduction of a perspective based on social and postphenomenology. This approach provides a social ontology, the lifeworld, and a social epistemology, the natural attitude, that can be employed to understand how "technologies can influence the 'worlds of

> direct experience, of contemporaries, of predecessors and of successors.'" (Coeckelbergh & Reijers, 2016, 344)[117]

The core body of SI knowledge continues to grow. Fichman and Hara are assembling another set of papers about SI research and theory for an edited volume, tentatively titled *Social Informatics in the Brave New World*, to appear in 2024. They intend that the contributions will use SI to empirically and theoretically understand emerging issues related to ICTs that are pervasive in almost every aspect of our daily lives. I am working on a contribution that is a reflection on the grounds of SI. I argue that there is a need to critically examine the foundations of SI if it is to take its rightful place in the study of the complex interrelationships between society and technology. The North American version of SI has been around for more than twenty-five years and, in that time, has established itself as a viable and useful way to study the complex interrelationships that take place between people and technologies in their many and varied contexts of use. It has been used in a variety of disciplines although, as mentioned above, it has not cohered around a theoretical or conceptual core; instead, it has been continually defined as a "problem-centered" approach and "[r]ecent SI scholarship has begun to redefine and reinvent the approach to studying computerization and society."[118] Achieving this type of coherence, however, may have never been the point—in 2005, Kling, Rosenbaum, and Sawyer[119] explained that social informatics was a problem-centered form of inquiry that was theoretically and methodologically agnostic. Although this was part of a strategy to expand the scope of social informatics to cover work on the social aspects of technology across multiple disciplines, fifteen years on it seems that this may have been a useful short-term move with a long-term cost; without a coherent intellectual foundation SI has become fragmented and diluted. This has not gone unnoticed; a decade ago, Meyer commented that " . . . one would be hard-pressed to point to a canon of social informatics publications that have coalesced into anything resembling a field or discipline or, in fact, anything more than a loosely-knit professional community of like-minded scholars."[120]

What happens to a SIM if there is a core body of knowledge and scholars are not making use of it? Could an articulation of the grounds of SI begin to address this question? Upon what foundations does SI rest? What is the research philosophy that animates it? How is the world constituted when viewed through the lens of SI? This is the question of its ontological grounds. Given this, how can we best come to know this world? This is the question of its epistemological grounds. As Alvesson and Skoldberg argue, "ontology and epistemology determine good social science."[121] The former is important because it "helps researchers recognize how certain they can be about the nature and existence of objects they are researching" and the latter because it specifies "what constitutes a knowledge claim" and how it can be acquired.[122] This chapter, if persuasive, will have two main contributions. First, it will open the discourse on the philosophical grounds of social informatics, a topic whose time has come. Second, SI as originally conceived was intended to have a clear critical lens which has receded somewhat over the years. This chapter will contribute to the recovery and foregrounding of SI's critical impulse by integrating insights from structuration, sociomateriality, and postphenomenology.

Now: Critical Reflection on the Current State of Social Informatics as a Scientific and Intellectual Movement

The third stage of writing about social informatics involves reflection on the current state of the approach. There are three issues that, in my opinion, need critical attention, all of which, I think, have a bearing on its future. The first is the question of the best way to represent its history and development when writing about SI. This is the question of the selection of the appropriate metaphor. With colleagues, I have used two main metaphors to describe the emergence and development of SI. In our books,[123] an ecological metaphor was used to portray SI as having moved through five stages: emergence, a foundational stage, a period of expansion, one of coherence, and the current period, transformation. The second metaphor is that of social movements, borrowed from sociology. This metaphor is at the core of the concept of the scientific and intellectual movement, used in this chapter and in Rosenbaum.[124] Each has its advantages. The ecological metaphor, with its implication of sociotechnical natural selection, is a useful way to organize the SI literature and show how its principles, concepts, approaches, topics, and findings have evolved over time. It highlights the complexity, diversity, and interconnectedness of the theories, methods, and findings that constitute the core body of SI knowledge. The metaphor also has its limitations, one of which is that it can be an evolutionary instantiation of the free marketplace of ideas, implying that through competition, the best ideas rise to the top and the rest fade away. In addition, this approach does not give sufficient attention to the social, cultural, and political embeddedness of SI.

One of the advantages of the SIM approach is that it addresses this limitation directly, which makes it useful in this context because of the focus on the importance of the structural and cultural environment in the success and failure of this type of social movement and the argument that its trajectory is as political as it is based in the intellectual work of the participants in the movement. Frickel and Gross[125] single out as factors critical to the success of a scientific and intellectual movement the need for early entrants to be high-status intellectual actors, the importance of access to resources such as grants and other funding, the availability of publication and presentation venues, and the ability to attract participants to the movement including training the next generation of researchers and teachers. By emphasizing aspects of SI using insights from the study of social movements, it becomes possible to draw insights that can help explain the challenges of writing about SI over time. This metaphor is also not without some limitations. In some ways it has limited scope and may not apply easily to other paths by which ideas become widely known; it is possible that some influential ideas may emerge and spread without fitting the social movement model. This has not been the case when considering SI, which seems to be a clear example of a SIM.

A second issue in writing about SI is boundary maintenance: Can a scholar be considered a part of the United States (or any) school of SI if they do not identify as such? Beginning during the writing of the NSF workshop report[126] and continuing during the preparation of Kling, Rosenbaum, and Sawyer,[127] there was a lively discussion with Kling about who was to be considered a SI scholar. Kling's position was that if they studied people, technology, and their contexts, they were in; my position was that if they did not identify as SI scholars, they were out. It is certainly possible to argue that a person may be considered a SI scholar if their work aligns with the core concepts and objectives of the field, regardless of whether they actively use that label to describe themselves. The recognition of their contributions by peers, the relevance of their research to the field's concerns, and their impact on advancing the understanding of the relationships between technology and society may be more important indicators of their status as SI scholars. This was Kling's position at the time and his position prevailed; his vision was that SI should be a large tent.

Now, I believe that this large-tent issue remains a concern. In my view, determining the boundaries of SI involves emphasizing the significance of scholarly self-identification, shared values, and involvement in the SI community. Identifying as a SI scholar indicates a commitment to the values, methodologies, and principles of the field. By choosing to label oneself in this way, an individual is signaling a recognition, and hopefully an acceptance, of the field's core concepts, theories, and research agendas. If the person does not label themselves as such, to do so without their consent seems to be an imposition with ethical implications. For example, if an individual does not identify as a SI scholar, it may indicate a lack of intention to align their work with the field's specific goals and principles, and, in this case, identification should not be externally imposed.

If a paper is focuses on some or all of the SI triad (people, technology, context of use) but does not use the phrase in the title and it does not appear in the abstract or body of the text, should it be considered a part of the corpus of SI? It is possible to argue that if the content of the paper addresses the core SI concerns, such as the interplay between technology and society, it can be classified as SI research. The absence of the term itself does not negate the relevance of the research to the field. Especially since SI research frequently involves collaborations between researchers from various disciplines, such as sociology, information science, communication studies, and computer science, a paper that engages with insights from these disciplines to explore technology's societal implications can be considered SI informatics research.

It is also possible to argue that the absence of the term "social informatics" in a paper's title, abstract, or body might hinder clarity and effective communication within the academic community. If research is relevant to SI but avoids the terminology, it could hinder the growth and establishment of the field as distinct from other forms of inquiry. Also, failing to use the term "social informatics" might result in the paper being overlooked or not taken seriously by scholars who are specifically seeking SI research. Explicitly using SI as a descriptor or keyword contributes to the growth, development, and recognition of the field. I do not think it is appropriate to claim such a paper as part of the core body of social informatics knowledge nor to identify the author(s) as SI scholars. However, this is an issue that is clearly debatable and one with which the SI discourse should engage.

A third issue in writing about SI concerns the extent to which social theory is a necessary component of SI research. This might seem to be an easy issue to resolve—if there is no social theory in the research, then it is not SI. However, it is not so simple since there are versions of SI such as that offered by Sadiku, Eze, and Musa, who explain that "(s)ocial informatics is a multidisciplinary field drawing on theories and methods from sociology, computer science, information science, business, and social sciences."[128] This leaves the door open for a paper to claim a SI heritage but only use theories, concepts, and methods from, say, computer science and business. In fact, it may be argued that social theory is not necessary for SI research. Research can focus on empirical analyses, case studies, technological design and implementation, and other aspects without reliance on explicit social theory. Research that focuses on designing and implementing technology with a deep understanding of its societal contexts may make valuable contributions. This type of research is often practical and solution-oriented, and it does not make use of social theory. Computational methods grounded in theoretical computer science and mathematics can be used to explore concerns central to SI with social theory nowhere to be found. All of these may be considered examples of social informatics research.

On the other hand, there is also an argument, towards which I am inclined, that social theory is necessary if research is to be considered as belonging to the core body of SI knowledge. Social theory provides a conceptual foundation for analyzing and interpreting the complex interplay between people, technology, and their contexts of use. It offers insights into the cultural, social, and historical contexts that shape technology adoption, use, and outcomes within a framework that can organize concepts and variables related to the main concerns of social informatics. The "social" in SI means that there are ongoing concerns for the consequences of such phenomena as power dynamics, social change, inequality, and ethics. Social theory provides tools to critically examine these issues, and a paper without such theoretical engagement might not fully address the broader implications of the interplay among people, technology, and society. Without any engagement with social theory, a paper may lack the depth and rigor required to qualify as SI research. This is another issue that should be part of the discourse about the future of SI.

Conclusion

Writing about writing about social informatics has involved a look back at the history and development of SI. This chapter has been written in the first person because I was there soon after the beginning of what Smutny and Vehovar[129] have called the United States school, working alongside Kling and colleagues for almost a decade to make the case for the importance of SI for the study of the social consequences of computing. This work resulted in several papers, a report, and a book, all of which were writings about SI. This was an effort to raise the profile of SI by demonstrating that it has "a more or less coherent program for scientific or intellectual change or advance."[130] In the intervening years, I have been fortunate to collaborate with Fichman[131] and Fichman and Sanfilippo[132] on more recent publications that have had the same objective.

This was followed by almost two decades of periodic contributions intended to deepen and enrich the core body of SI knowledge. This is a necessary activity to ensure (but not guarantee) the success of SI as a SIM and involves "the transformation of thoughts or research findings into ideas and knowledge that are circulated widely within the intellectual community, subjected to scrutiny and contestation, embraced by some and rejected by others, and that may emerge from the process deemed credible or true."[133] Finally, in what Fichman, Sanfilippo, and Rosenbaum[134] have called a period of transformation and Smutny and Vehovar[135] have called a period of fragmentation, we are in the third stage, which calls for critical reflection in the current state and future possibilities of SI as a SIM.

One way is to consider the fate of SIMs. SI is a relatively young form of inquiry; the oldest version having been in existence for five decades (Norway) and one of the youngest (United States) for two and a half decades. After a flurry of activity between the late 1990s and 2014, activity as measured by publications and conference presentations seems to be plateauing.[136] This begs the question of the future of SI. As a SIM, SI could face one of four main outcomes. In two of these, the movement essentially disappears. The positive outcome occurs when the SIM becomes institutionalized, taking its place alongside other accepted forms of intellectual inquiry. In this outcome, SI becomes a viable alternative that can partner with, for example, science and technology studies. Departments may be formed around SI, degree programs established, suites of courses offered, and funding calls written incorporating SI themes and research problems. In this sense, it disappears into the intellectual landscape. Sawyer and Tapia[137] discussed the ways in which SI could achieve this level of institutionalization. The negative outcome is that it simply disappears. The high-status scholars who were its early champions lose interest and move on or retire without any replacements. Research funding and publication and conference venues dry up. The next generation of SI scholars never appears. A third outcome is that the SIM simply "putters along" at a low level of activity, not quite disappearing, not quite catching fire. There are scattered pockets of activity where research is being done without drawing much attention outside the small community of like-minded SI researchers. Frickel and Gross[138] describe this type of SIM as one "not likely to reverberate in the lives of most people." The fourth outcome, which they believe does not happen often is that the "social and scientific/intellectual changes" that are at the core of the SIM "become revolutionary."[139]

Which Outcome Best Describes the Future of Social Informatics?

I am cautiously optimistic about the future of SI as a SIM. This form of inquiry seems to have taken root in information science, where there are many of the characteristics singled out by Frickel and Gross[140] as being critical for the success of such a movement. In the United States and elsewhere there are high-status academics who publish SI work, SI courses and research centers in iSchools, conferences devoted to the topics, journals that publish SI research and the occasional special issue, special interest groups in professional organizations, and a small cohort of students and early career faculty who seem willing to become the next generation of SI researchers and teachers. There are also increasing opportunities to engage in relevant SI research. As new technologies such as artificial intelligence, large language models, and algorithmic management become more deeply integrated into daily life, the study of technology, people, and contexts of use becomes even more crucial. SI researchers can provide insights into how technology influences human behavior, communication, and societal structures, how humans shape these technologies through use, and how emerging technologies such as artificial intelligence, blockchain, and virtual reality interact with people and social and organizational contexts.

In time, these developments may lead to the positive institutionalization of SI as described above. Recalling that there are seven schools of social informatics,[141] my hope is that all will flourish. Finally, this is the current challenge I see in writing about social informatics: How can we write about social informatics in ways that help it catch fire and succeed in the long-run as a scientific and intellectual movement?[142]

Notes

1. Fichman, P., Sanfilippo, M. R., & Rosenbaum, H. *Social Informatics Evolving*. (Springer Nature, 2015); Vehovar, V., Smutny, Z., & Bartol, J. Evolution of Social Informatics: Publications, Research, and Educational Activities. *The Information Society*, 38, 5 (2022), 307–333; and Smutny, Z., & Vehovar, V. Social Informatics Research: Schools of Thought, Methodological Basis, and Thematic Conceptualization. *Journal of the Association for Information Science and Technology*, 71, 5 (2020), 529–539. Fichman, P. & Rosenbaum, H. (eds.) (2014). *Social Informatics, Past, Present and Future*. Cambridge Scholars Publishing.

2. Frickel, S., & Gross, N. A General Theory of Scientific/Intellectual Movements. *American Sociological Review*, 70, 2 (2005), 204. Smutny, Social Informatics Research, p. 232.

3. Frickel, Scientific/Intellectual Movements, p. 206.

4. Fichman, *Social Informatics Evolving*, p. 4.

5. Fichman, *Social Informatics Evolving*.

6. Frickel, Scientific/Intellectual Movements.

7. Kling, R. Learning About Information Technologies and Social Change: The Contribution of Social Informatics. *The Information Society* 16, 3 (2000), p. 228.

8. Lamb, R., & Sawyer, S. On Extending Social Informatics from a Rich Legacy of Networks and Conceptual Resources. *Information Technology & People*, 18, 1 (2005), p. 10.

9. Day, R. E. Kling and the "critical": Social Informatics and Critical Informatics. *Journal of the American Society for Information Science and Technology* 58, 4 (2007), p. 575.

10. Vehovar, Evolution of Social Informatics, p. 307.

11. Fichman, *Social Informatics Evolving*.

12. Smutny, Social Informatics Research, p. 529.

13. Vehovar, V. Social Informatics: An Emerging Discipline? In *Social Informatics: An Information Society for All? In Remembrance of Rob Kling: Proceedings of the Seventh International Conference on Human Choice and Computers* (HCC7), IFIP TC 9, Maribor, Slovenia, September 21–23, 2006 (pp. 73–85). Springer US.

14. Smutny, Social Informatics Research, p. 531.

15. Smutny, Social Informatics Research, p. 537.

16. Vehovar, Evolution of Social Informatics, p. 22.

17. Robbin, A. Rob Kling In Search of One Good Theory: The Origins of Computerization Movements. *The Information Society*, 23, 4 (2007); and Wellman, B., & Hiltz, S. R. Sociological Rob: How Rob Kling Brought Computing and Sociology Together. *The Information Society*, 20, 2 (2004), 91–95.

18. Vehovar, Evolution of Social Informatics, p. 532.

19. Kling, R. Toward a Person-Centered Computer Technology. In *Proceedings of the ACM Annual Conference*, Atlanta, GA, Aug. 1973, pp. 387–391.

20. Kling, R., & Gerson, E. The Social Dynamics of Technical Innovation in the Computing World. *Symbolic Interaction*, 1, 1 (Fall 1977), pp.132–146; and Kling, R., & Gerson, E. "Patterns of Segmentation and Intersection in the Computing World." *Symbolic Interaction* 1, 2 (Spring 1978), pp. 24–43.

21. King, J. L. Rob Kling and the Irvine school. *The Information Society*, 20, 2 (2004), p. 97.

22. Meyer, E. T., Shankar, K., Willis, M., Sharma, S., & Sawyer, S. The Social Informatics of Knowledge. *Journal of the Association for Information Science and Technology*, 70, 4 (2019), p. 308.

23. Kling, R. The Social Dimensions of Computerization. *ACM SIGCHI Bulletin*, 18, 4 (1986), pp. 337–339.

24. Kling, R., & Scacchi, W. The Web of Computing: Computer Technology as Social Organization. *Advances in Computers* 21 (1982), pp. 1–90.

25. Kling, Social Dimensions.

26. Kling, R., & Iacono, S. The Mobilization of Support for Computerization: The Role of Computerization Movements. *Social Problems*, 35, 3 (1988), pp. 226, 227.
27. Kling, Mobilization of Support, p. 228.
28. Kling, R., Information Systems, Social Transformations, and Quality of Life. In *Proceedings of the Conference on Computers and the Quality of Life* (August 1990), p. 84.
29. Frickel, Scientific/Intellectual Movements, pp. 209–210.
30. Kling, Web of Computing.
31. Kling, R. Behind the Terminal: The Critical Role of Computing Infrastructure in Effective Information Systems Development and Use. UC Irvine, ICS Technical Reports (1991), p. 4.
32. Kling, Behind the Terminal, p. 5.
33. Kling, Behind the Terminal, p. 37.
34. Kling, Behind the Terminal, p. 49.
35. Kling, R. Computerization and Social Transformations. *Science, Technology, & Human Values* 16, 3 (1991), p. 344.
36. Kling, Computerization and Social Transformations, p. 345.
37. Kling, R. Organizational Analysis in Computer Science. *The Information Society*, 9, 2 (1993), p. 75.
38. Kling, Organizational Analysis, p. 80.
39. Kling, R. Audiences, Narratives, and Human Values in Social Studies of Technology. *Science, Technology, & Human Values* 17, 3 (1992), 349–365.
40. Woolgar, S., & Grint, K. Computers and the Transformation of Social Analysis. *Science, Technology, & Human Values* 16, 3 (1991), 368–378.
41. Woolgar, Computers and the Transformation, p. 375.
42. Woolgar, Computers and the Transformation, pp. 369–70.
43. Kling, Audiences, p. 355.
44. Kling, Audiences, p. 355.
45. Fichman, *Social Informatics Evolving*; and Kling, R., Rosenbaum, H., & Sawyer, S. *Understanding and Communicating Social Informatics: A Framework for Studying and Teaching the Human Contexts of Information and Communication Technologies*. (Medford, NJ: Information Today, Inc., 2005).
46. Borgman, C. L., Bates, M. J., Cloonan, M. V., Efthimiadis, E. N., Gilliland-Swetland, A. J., Kafai, Y. B., . . . & Maddox, A. B. Social Aspects of Digital Libraries. Final Report to the National Science Foundation (1996), p. 4.
47. Kling, *Communicating Social Informatics*, p. 6.
48. Frickel, Scientific/Intellectual Movements, p. 206.
49. Meyer, E. T. Examining the Hyphen: The Value of Social Informatics for Research and Teaching. In Pnina Fichman and Howard Rosenbaum (Eds.), *Social Informatics: Past, Present and Future*. (Newcastle upon Tyne: Cambridge Scholars Publishing, 2014), p. 57.
50. Frickel, Scientific/Intellectual Movements, p. 214.
51. Kling, R., Crawford, H., Rosenbaum, H., Sawyer, S., & Weisbad, S. Learning from Social Iinformatics: Information and Communication Technologies in Human Contexts. Report submitted to the National Science Foundation (2000), p. 205.
52. Kling, NSF Workshop report.

53. Kling, R. What is Social Informatics and Why Does it Matter? *D-Lib Magazine*, 5, 1 (1999), ISSN 1082–9873.
54. Sawyer, S., & Rosenbaum, H. Social Informatics in the Information Sciences: Current Activities and Emerging Directions. *Informing Science*, 3 (2000), 89.
55. Kling, What is Social Informatics (*D-Lib*).
56. Sawyer, Social Informatics, p. 90.
57. Sawyer, Social Iinformatics, p. 94.
58. Frickel, Scientific/Intellectual Movements, p. 214.
59. Rosenbaum, H. Social Informatics. *Encyclopedia of Human-Computer Interaction*. (Great Barrington, MA: Berkshire Publishing, 2004), p. 633.
60. Kling, What is Social Informatics (*D-Lib*).
61. Kling, *Communicating Social Informatics*, pp. 153–154.
62. Kling, R., Crawford, H., Rosenbaum, H., Sawyer, S., & Weisband, S. Learning from Social Informatics: Information and Communication Technologies in Human Contexts (2000); and Sawyer, S., & Eschenfelder, K. R. Social informatics: Perspectives, Examples, and Trends. *Annual Review of Information Science and Technology* 36 (2002), 427–465.
63. Kling, What is Social Informatics (*D-Lib*).
64. Kling, R. What is Social Informatics and Why Does it Matter? *The Information Society* 23, 4 (2007), pp. 205–220.
65. Sawyer, Social Informatics.
66. Kling, NSF Workshop report.
67. Meyer, Examining the Hyphen, p. 56.
68. Frickel, Scientific/Intellectual Movements, p. 206.
69. Roggen, I. (2005). Personal communication, October 17.
70. Rosenbaum, H. Social Informatics. In M. Bates (ed.), *Encyclopedia of Library and Information Science*, 3rd edition. (New York: Marcel Dekker, 2010), pp. 4814–4819.
71. Kling, R. Social Informatics. In M. Drake (ed.), *Encyclopedia of Library and Information Science*, 2nd edition, Volume 1, pp. 2656–2661.
72. Rosenbaum, Social informatics, LIS, 1416.
73. Frickel, Scientific/Intellectual Movements.
74. Smutny, Social Informatics Research.
75. Fichman, Social Informatics Evolving.
76. Kling, *Communicating Social Informatics*.
77. Frickel, Scientific/Intellectual Movements, p. 206
78. Frickel, Scientific/Intellectual Movements, p. 206
79. Fichman, Social Informatics Evolving.
80. Rosenbaum, H., & Joung, K. Socio-Technical Interaction Networks as a Tool for Understanding Digital Libraries. In *Proceedings of the 67th Annual Meeting of the American Society for Information Science*, vol. 41 (2004). Medford, NJ: Information Today, Inc.
81. Library of Congress (n.d.). American Memory Project. https://memory.loc.gov/ammem/amabout.html
82. Kling, R., McKim, G., Fortuna, J., & King, A. Scientific Collaboratories as Socio-Technical Interaction Networks: A Theoretical Approach (2000), arXiv preprint cs/0005007.

83. Kling, R., & Scacchi, W. (1982). The Web of Computing: Computer Technology as Social Organization. In *Advances in Computers* 21 (1982), pp. 1–90.

84. Kling, R., McKim, G., & King, A. A Bit More To It: Scholarly Communication Forums as Socio-Technical Interaction Networks. *Journal of the American Society for Information Science and Technology*, 54, 1 (2003), p. 48.

85. Kling, A Bit More To It, p. 54.

86. Rosenbaum, Socio-Technical Interaction Networks, p. 208.

87. Meyer, E., Hara, N., & Rosenbaum, H. How Photobloggers are Framing a New Computerization Movement. Association of Internet Researchers Annual Meeting: Internet Research 6.0. Chicago, October 8 (2005).

88. Kling, Mobilization of Support.

89. Iacono, S., & Kling, R. Computerization Movements and Tales of Technological Utopianism. In R. Kling (ed.), *Computerization and Controversy: Value Conflicts and Social Change* (2nd ed., San Diego, CA: Academic Press, 1996), pp. 85–105; and Iacono, S., & Kling, R. Computerization Movements: The Rise of the Internet and Distant Forms of Work. In J. A. Yates & J. Van Maanen (Eds.), *Information Technology and Organizational Transformation: History, Rhetoric and Practice* (Thousand Oaks, CA: Sage Publications, 2001), pp. 93–136.

90. Hara, N., & Rosenbaum, H. On the Success and Failure of Computerization Movements. Proceedings of the Workshop "Extending the Contributions of Professor Rob Kling to the Analysis of Computerization Movements." CRITO, Irvine, CA, 2005.

91. Iacono, Computerization and Rise of the Internet, p. 97.

92. Rosenbaum, H., & Davenport, E. Why the Context Matters: A Social Informatics Spproach to the Problem of Interdependence in Information Systems Research. Paper 494. *Proceedings of the 12th America's Conference on Information Systems*, 2006.

93. Rosenbaum, Context Matters, p. 1.

94. Dourish, P. What We Talk about When We Talk about Context. *Personal and Ubiquitous Computing*, 8 (2004), 19–30.

95. Scheff, T. The Structure of Context: Deciphering Frame Analysis. *Sociological Theory*, 23, 4 (2005), 368–385.

96. Goffman, E. *Frame Analysis: An Essay on the Organization of Experience* (Cambridge, MA: Harvard University Press, 1974).

97. Hara, N., & Rosenbaum, H. Revising the Conceptualization of Computer Movements. *The Information Society*, 24, 4 (2008) 229–245.

98. Hara, Success and Failure.

99. Iacono, Computerization Movements; and Iacono, Computerization and Rise of the Iinternet.

100. Hara, Success and Failure.

101. Hara, Revising the Conceptualization.

102. Shachaf, P., & Rosenbaum, H. Online Social Reference: A Research Agenda through a STIN Framework. *Proceedings of the 2009 iConference*, February 9, University of North Carolina at Chapel Hill, 2009.

103. Shachaf, Online Social, 3.

104. Rosenbaum, H., and Shachaf, P. A Structuration Approach to Online Communities of Practice: The Case of Q&A Communities. *Journal of the American Society for Information Science and Technology*, 61, 9 (2010), 1933–1944.

105. Rosenbaum, Structuration, p. 1933.

106. Giddens, A. *Central Problems in Social Theory: Action, Structure, and Contradiction in Social Analysis* (Berkeley, CA: University of California Press, 1979); and Giddens, A. *The Constitution of Society* (Berkeley, CA: University of California Press, 1984).

107. Orlikowski, W. J. The Duality of Technology: Rethinking the Concept of Technology in Organizations. *Organization Science*, 3, 3 (1992), 398–427; and Orlikowski, W. J. Using Technology and Constituting Structures: A Practice Lens for Studying Technology in Organizations. *Organization Science*, 11, 4 (2000), 404–428.

108. Wenger, E. *Communities of Practice: Learning, Meaning, and Identity* (Cambridge: Cambridge University Press, 1998); Lave, J. (1991). Situating Learning in Communities of Practice. In L. B. Resnick, J. M. Levine, & S. D. Teasley (Eds.), *Perspectives on Socially Shared Cognition* (American Psychological Association), pp. 63–82; and Lave, J., & Wenger, E. *Situated Learning: Legitimate Peripheral Participation* (Cambridge: Cambridge University Press, 1991).

109. Rosenbaum, Structuration, p. 1941.

110. Rosenbaum, Structuration, p. 1942.

111. Yang, S., Zhu, A., & Fichman, P. (Eds.) *The Usage and Impact of ICTs during the Covid-19 Pandemic* (London: Taylor and Francis, 2023).

112. Yang, *Usage and Impact*, p. 3.

113. Rosenbaum, H., Algorithmic Assemblages, the Natural Attitude, and the Social Informatics of the Pandemic Lifeworld. In S. Yang, A. Zhu, & P. Fichman (Eds.), *The Usage and Impact of ICTs during the Covid-19 Pandemic* (Taylor and Francis, 2023).

114. Schutz, A. *The Phenomenology of the Social World* (Chicago: Northwestern University Press, 1967).

115. Rosenberger, R., & Verbeek, P. P. (Eds.) *Postphenomenological Investigations: Essays on Human–Technology Relations* (Lanham, MD: Lexington Books, 2015).

116. Rosenbaum, Algorithmic Assemblages, p. 249.

117. Rosenbaum, Algorithmic Assemblages, p. 250.

118. Fichman, *Social Informatics Evolving*, p. 24.

119. Kling, *Communicating Social Informatics.*

120. Meyer, Examining the Hyphen, p. 56.

121. Alvesson, M., & Skoldberg, K. *Reflexive Methodology: New Vistas for Qualitative Research* (2nd ed., London: Sage, 2009), p. 8.

122. Moon, K., & Blackman, D. A Guide to Ontology, Epistemology, and Philosophical Perspectives for Interdisciplinary Researchers, https://i2insights.org/2017/05/02/philosophy-for-interdisciplinarity/, 2017.

123. Fichman, *Social Informatics Evolving*; and Fichman, Social Informatics: Past, Present, and Future.

124. Fichman, Social Informatics: Past, Present, and Future.

125. Frickel, Scientific/Intellectual Movements, p. 221.

126. Kling, NSF Workshop report.

127. Kling, *Communicating Social Informatics*.

128. Sadiku, M. N., Eze, K. G., & Musa, S. M. Social Informatics. *International Journal of Advanced Research in Computer Science and Software Engineering*, 9, 2 (2019), p. 6

129. Smutny, Social Informatics Research.

130. Frickel, Scientific/Intellectual Movements, p. 206.

131. Fichman Social Informatics: Past, Present, and Future.

132. Fichman, *Social Informatics Evolving*.

133. Frickel, Scientific/Intellectual Movements, p. 206.

134. Fichman, *Social Informatics Evolving*.

135. Smutny, Social Informatics Research.

136. Vehovar, Evolution of Social Informatics, p. 309.

137. Sawyer, S., & Tapia, A. From Findings to Theories: Institutionalizing Social Informatics. *The Information Society* 23, 4 (2007), 263–275.

138. Frickel, Scientific/Intellectual Movements, p. 225.

139. Frickel, Scientific/Intellectual Movements, p. 225.

140. Frickel, Scientific/Intellectual Movements.

141. Smutny & Vehovar, 2020.

142. Acknowledgements: I would like to thank Bill Aspray, Greg Downey, and Geoff Bowker for their careful reading of the draft and for their insightful and helpful comments.

CHAPTER 13

Writing Geography into the History of Computing and Information

Greg Downey

An Interdisciplinary Career in Computing and Information

Like many contemporary scholars of computing and information, throughout my academic career of teaching and writing about knowledge technologies, I have enthusiastically branded myself as an "interdisciplinary" researcher.[1] Those of us who raise that flag often do so for a variety of interrelated reasons. Maybe we were trained in multiple graduate disciplines; for example, I pursued a joint doctorate in both a humanities department and an engineering department. Or maybe we found academic employment in a split position of some sort; I am currently a tenured professor shared 50 percent in a journalism school and 50 percent in an information school. Maybe, the ubiquity of our objects of study themselves—constantly-expanding computing and information infrastructures involving every aspect of our increasingly digitally-mediated lives—implies an interdisciplinary orientation by default. But in the end, any claim to interdisciplinarity has to reflect the origins of our research curiosity and questions themselves: What do we perceive as interesting, puzzling, or problematic, and thus worthy of systematic study and explanation? For me, this is why *writing* the history of computing and information must always start with an effort of *seeing* the artifacts, actors, and processes of computing and information in a particular way: through geographical lenses such as place, space, and scale.

I am drawing on the spatial metaphor of "seeing" here on purpose, to suggest a mode of perception that sounds almost automatic but yet can be frustratingly elusive—especially in the field of computing and information, where it is commonplace to think of data and technology (and the crucial human labor tied to both) as present both everywhere and nowhere. After all, "the computer" is often defined as some sort of aspatial, mathematical ideal: systems of hardware, software, and human practice geared toward universal application and generalizable processing. For more than 75 years, computer science students have been told that such Turing machines differ only by degree of physical size, algorithm speed, and storage capacity (under Moore's famous law). As an undergraduate computer science student myself in the mid-1980s—just as the batch-terminal, command-line interface was transitioning to the desktop PC, WYSIWYG environment—I originally chose this field because of the employment security offered by the (even then) universal presence of digital technology within almost all professional jobs. Computers seemed to be everywhere, or would be soon, so they were a safe career choice for a curious and indecisive student like me. (And I expect the same calculation drives many of today's booming undergraduate computer science and data science majors as well.)

But in retrospect, I now understand that my own early career as a professional software developer/analyst in the early 1990s offered a case study in the differing effects of spatial context on social activity within technological infrastructure. My first full-time tech job was with a global marketing firm headquartered in a 50-story building within the dense commercial district of a major US city, just as the microcomputer networking revolution was displacing older mainframe and dumb terminal systems. The cultural meaning of one's "place" in that corporate hierarchy literally attached to which floor one's cubicle was on—the higher up in the sky, the greater your power and privilege within the firm. Information technology services were on floor six (just above the archival media storage and cafeteria floors), where one sat with one's team in a well-defined desk floorplan mirroring the department organization chart. And the nature of the technology one used was clearly coded to the place one occupied as well: database programmers like myself in one section filled with PCs, networking professionals in another among the server racks, legacy mainframe programmers using terminals in a third area, and those tasked with assisting the "creatives" with digital design tools on their Macs congregating in a fourth space. More than anything, though, the place was one of isolation—the software we developed never left the building, electronic communication was tightly controlled, and this entire technology group was never meant to be seen or visited by the main clients of the firm.

The job I found after that could not have been more different in its social and technological construction of place: a low-rise office building within a suburban greenfield research park that was part of a local private university's efforts to attract national industrial and military funding for basic cognitive science research. My workspace was intentionally jumbled in among those of other programmers, digital artists, linguistics graduates, and cognitive science professors, shifting location dozens of times as project configurations changed. Corporate and government visitors regularly toured our workspaces, while liaisons from funding partners joined our work teams. The technology here was open and mobile: laptops and Newtons with connections to the new World Wide Web supported online discussion and open-source development projects. And our work was so purposefully visible as to be prominently featured in *Wired* magazine. Thus, these two very different places of computation could both be "seen" in terms of the different ways they represented and organized power and permanence, expertise and collaboration, and technology and labor, for different purposes and constituencies.

But simply noting the geographical aspects of a computing and information environment does not automatically lead to conceptualizing, analyzing, and exploring those geographical aspects in ways that result in a different historical narrative. For me, it took first returning to graduate school (allowing me to experience a serendipitous collision of ideas from both the history of technology and human geography) and then forging a new career in academia (applying my training in both an information school and a journalism school) to fully realize that geography matters a great deal to the history of computing and information—so much so that I believe one simply cannot write the history of computing and information without it.

Making geography matter when studying computing and information is both a systematic scholarly choice and an idiosyncratic "way of seeing." In this chapter—a personal and autobiographical tour through the intersections between history, geography, and the study of computing and information—I hope to explain why "writing geography into the history of computing and information" is a choice I have found success with in my own work, and to provide some structure, suggestions, and resources for others to do so as well.

Seeing Place and Space within the History of Technology

Reading and Seeing: Nature's Metropolis

I stumbled into this combination of history and geography by happenstance—but my first step was toward history. I credit that choice almost entirely to a book I found during some earlier evening graduate studies while still working as a software professional: *Nature's Metropolis* by William Cronon.[2] At the time I did not realize the novelty of reading a volume written by a professor employed in both a history department and a geography department. I did not appreciate the amazing accomplishment of having a deeply researched scholarly monograph end up as a best seller in popular bookstores around the nation. And I did not expect a book about nature and cities to be, at its core, about the human relationships and technological networks created over a century to share the information and knowledge that knit a frontier town (Chicago) together with its rural hinterlands to become a megalopolis. But even before I had a language for it, I knew that this was the kind of synthetic historical and conceptual work that I wanted to aspire to in my own academic career—telling a story of the "digital metropolis" in my case—so I set about searching for a university that would train me to do so.

In reviewing graduate programs where I could explore the social antecedents and consequences of the kinds of digital information systems I had worked on as an early-career IT professional, I ended up enrolled in a history of science, medicine, and technology department. As a new graduate student, I did not appreciate the differences between "science and technology studies" or "information studies" or "the history of science, medicine, and technology"—but studying the history of computing and information in parallel with other kinds of technological tools, techniques, and systems, ranging in scale from the pencil to the city, immediately made a kind of intuitive sense to me, raised as I was with a direct experience of computing that leaped from the handheld electronic game to the global information superhighway during my lifetime. And it was my great fortune to work with advisers like Stuart Leslie and Robert Kargon, whose work on topics such as Cold War science and the Silicon Valley regional economic development model considered the connections between specific sites and extensive regions with technological change—situating innovations and artifacts, organizations and movements, in the landscape as well as on the timeline.[3] I think of this strategy as both finding *the place of technology*, and seeking *the technology of place*.

The Place of Technology: The Laboratory and the Region

Finding the place of technology is the first step—and also the easiest. In the mid-20th century, the field of history of science, medicine, and technology emerged as a more contextually aware practice that considered more than just the "internalist" development of one technological improvement automatically leading to the next. Ever since, scholars have been noting particular sites of technological innovation, testing, marketing, diffusion, consumption, adoption, and rejection as crucial to technological change. Tracking these sites was not only a clue to where historical documentation and eyewitness testimony might be found; such sites were also historical sources in and of themselves if one knew how to look. Contextual social, cultural, political, and economic approaches to technology history leverage this situatedness to explore a larger story of how technologies develop and diffuse, change and adapt, through innovation to obsolescence.[4]

The "laboratory turn" of the 1970s and 1980s in the history and sociology of science is a key example of this contextual awareness of space. These meticulous studies treated the laboratory as a backstage area within which the contingency and contradiction of scientific exploration—imperfect equipment, recalcitrant animals, backstage assistants, and contingent funding—could be confronted and clarified before becoming packaged as "knowledge" reaching the rest of the scientific community in the form of a published paper or a paradigm shift.[5] And conceptualizing places of knowledge production as worthy of study helped legitimize the study of spaces of knowledge circulation and consumption as well.[6]

Within the subfield of technology studies, studies of modern "garages" for tinkering and invention scaled up into studies of innovation spaces at the metropolitan or regional scale—from Anna Saxenian's work on the "regional advantage" distinctions between technology firms in Silicon Valley versus Boston's Route 128, to Paul Ceruzzi's historical study of "Internet Alley" in the DC suburbs.[7] But these works identified different social aspects of the spaces they explored. For Saxenian, the networks of information and collaboration between technology professionals among different firms and universities was key; for Ceruzzi, the characteristics of greenfield suburban land which leveraged those professionals for both housing and retail development played a part. Both social and environmental differences mattered within and between spaces of technological development and information exchange.

Of course, history and geography are dialectically related concepts in that one cannot really do history without specifying geographic location, area, scale, and connections, and one cannot really do geography without specifying historical moment, period, scale, and consequences. But taking place seriously involves more than just locating historical moments on a map. All these examples point to a key insight when writing history through a geographical lens: *places matter because places differ*, and those differences can be enabling or even determinative of different historical outcomes. But finding the differences that matter at a particular historical moment to the differential development of technological change is the tricky part. Careful attention to the moments when technologies "jump context" from one place to another can be an especially productive tool.[8]

The Technology of Place: Networked Homes and Cities

But while this first strategy of seeing computing and information technology and practices as spatially situated is easy to start with, it has some limitations. Technologies, especially computing and information technologies, rarely exist in isolation—their relatedness or connectedness across space and time is often the point. This is an important opportunity to ask not "How does place shape a technology," but instead, "How does technology shape a place?" Viewed in this way, our largest, densest, and most technologically intensive places—cities—are in a very real sense communications technologies themselves, built to "overcome time with space" by bringing lots of people, products, and ideas together at a single place and at the same time in order to create, store, transmit, and share knowledge and culture.[9]

One historian who took this notion seriously with respect to information and communication technology was Joel Tarr. Noting in the late 1980s that "urban scholars generally agree that cities evolved in order to facilitate human communications," he and his colleagues pointed out that "historians have dealt with water, wastewater, and energy, as well as vehicular and pedestrian traffic, as flows in an urban network, but have paid little attention to messages. Thus, in spite of the critical nature of communications techniques for urban development, we have only a limited knowledge of their effects on the city." Tarr's work showed information studies scholars how to situate an understanding of place itself in the underlying technologies that made that place function—where "technology provides what might be called the 'sinews' of the city" to make it a feasible and attractive site of habitation, production, entertainment, and investment.[10]

Many of the formative histories of technology to incorporate a gender analysis also focused on the technology of place, in considering the way the domestic sphere of the middle-class home was a technological site for women's work—both productive and reproductive labor. In Ruth Schwartz Cowan's *More Work for Mother*, the home becomes a key locus of both consumption and labor linked to urban technologies of electrification, sanitation, and transportation.[11] Similarly, Sheila Allen and Carol Wolkowitz's *Homeworking: Myths and Realities* showed that casualized work within the domestic sphere has long been an invisible component of capitalist technological production, along with the spaces of the factory assembly line and goods distribution hub.[12]

Seeing technology in place—together with the people in that place whose lives are implicated in that technology—brings a new historical urgency to claims starting in the 1960s and 1970s that the then-nascent algorithmic information revolution would transform both homes and cities into a teleworking landscape of "electronic cottages."[13] As the rhetoric of the "information superhighway" of the 1990s morphed into that of "virtual space" and "social media" in the 2000s, despite claims from digital technology advocates such as William Mitchell that only the "city of bits" mattered in the new landscape, scholars like Stephen Graham and Simon Marvin would continue to pursue the notion that cities remained relevant in such a digital environment, as urban places which were now suffused with electronic spaces.[14]

From Place to Space: Sociotechnological Infrastructures

The nature of the information technologies that historians study, as created in place but also changing the character of place, points to a third relatively easy way to mobilize a geographical perspective: thinking about how *technologies work not just in place, but across space*. Since information technologies are often used in ways that substitute for transportation—moving information across space instantaneously after investing in the slow, painstaking process of building a material network across that space (say, telegraph wires or orbiting satellites), or consolidating information in place centrally after investing in the slow, painstaking process of building a dense material apparatus within that space (say, an urban skyscraper, or a rural data server farm).

Historians of technology and society have long explored the ways that both national governments and private enterprise committed to funding so-called "internal improvements" for the circulation of information across space, starting with roads and canals to transport printed materials and extending with electrification to telegraph, telephone, and broadcast technologies. Similarly, business historians explaining the "control challenge" of rapid economic and social growth during the second industrial revolution have studied the diffusion of point-to-point electronic communication technologies as a response to the new energy abundance provided by fossil fuels and electric motors as applied to manufacturing enterprises. Media historians explored the use of new methods of broadcast network communication to create a mass culture throughout the twentieth century, including the development of whole new industries of expert media manipulation and product marketing. And such work continues today under conditions of global economic restructuring, responding to the new challenges of transnational competition and neoliberal governance through computer-mediated communication, in what observers have labeled a new "post-industrial," "postmodern," "informational," or "network" society.[15] Janet Abbate's rich history *Inventing the Internet* is a great example linking these last two moments together—showing how private capital and public funding across extensive spatial landscapes of technological and organizational power can be mobilized to usher in the next round of change in a palimpsest of global development.[16]

Earlier histories of technologies linked across landscapes framed them as systems or networks—prioritizing questions of ownership, leadership, coordination, and command-and-control across organizational bureaucracies under the state and capital. But recent historians have framed such technologies of transportation, power generation, waterworks, and communication technologies as "infrastructure" technologies: whole systems meant to function not only profitably and reliably, but also invisibly and perpetually in order to enable other social, political, and economic activities. These *sociotechnical infrastructures* are composed of both humans and artifacts, requiring a complicated ongoing dance between both in order to persist. As such, they both are created out of particular social relations and end up incorporating particular social meanings, tied to their historical moment. We might say they are both *socially produced* (in their material geography) and *socially constructed* (in their imagined geography)—especially since we are often only forced to notice infrastructures when they break down, requiring overt social processes of assistance, repair, reproduction, and rebuilding of trust.[17]

Paul Edwards nicely summarized what we might think of as this "infrastructural turn" in the history of computing and information technology—a move away from the previous framework of technological systems and networks, credited to the work of historian Thomas Hughes—in an article written in anticipation of the Y2K computer bug crisis as the digital calendar turned to the year 2000. Noting that "computers have become the control, information storage, and information processing technology of choice in many other, pre-existing infrastructures," he argued they are thus "the infrastructure of our infrastructures."[18] He expanded on this idea in a review essay that same year, published in the history of science journal *Isis*, advocating that we transform our study of the history of "the computer" into the history of "what is fast becoming a global, infinitely reconfigurable virtual infrastructure": "A history of digital convergence and virtual infrastructures might reveal a very large iceberg, of which individual computers and even widely adopted applications are only the visible tip."[19]

In all these ways, studying the history of technology taught me that humans have long created sociotechnical information and computing systems of all sorts—from print to digital—to move knowledge across both time and space. Thus, geographical context and connection are more fundamental (and perhaps more obvious) to the history of computing and information in ways that it might be harder to find in other histories. Moving information across space and/or time requires encoding, storage, indexing/cataloging, preservation, retrieval, addressing, transmission or transportation, routing, and decoding. All these processes demand not just technologies, protocols, and labor that are present across space in order to enable the process from beginning to end, but also "containing" organizations, institutions, and infrastructures that extend and persist across space and time as well.

Seeing and Writing: Telegraph Messenger Boys

This geographically-suffused way of seeing technology in place, of place, and across space fed directly into my first book (drawn from my Ph.D. thesis), which was a historical study of the US telegraph system from the point of view of the teenaged messenger boys who served as the "last mile" connecting customer messages to the "lightning lines."[20] But although I had been searching for a dissertation project that could explore the "digital metropolis" and draw on my professional computer science training, it was actually another space-transcending technology that got me hooked in the topic of the messenger boys from a century ago: bicycles. Seeing this humble mechanical technology—and the children who used it—as part of a "high-tech" sociotechnical infrastructure for the world's most valuable economic, military, and cultural information was a leap of what geographer Derek Gregory called the "geographical imagination" that unlocked the whole story for me.[21] Suddenly the archival records of the Western Union Telegraph Company, newly available at the Smithsonian Museum of American History, took on a new meaning: photos of uniformed messenger boys waiting their turn in hidden basement rooms revealed a spatial contradiction tied to service marketing and labor organizing; totals from ledger books on the bulk prices of bicycles and shoes spoke to the overhead of transportation costs within hazardous urban environments; and census data on messenger employment across the nation revealed a nascent "temporary services industry" that is still with us today. I learned that attention to such examples of geographical imagination in the history of information technology can open up myriad ways of *seeing place and space in the archival record*: paying attention to photos, maps and blueprints as representations of place and space; noticing the material records of spatial costs and tools in economic and administrative records; and recapturing for close reading the cultural descriptions of speed and distance that historical actors leave behind from their lived experience of place and space. Once we notice such things, we as historians can ask: What kinds of spaces make what kinds of sociotechnical systems? And what kinds of sociotechnical systems make what kinds of spaces?

Analyzing Sociospatial Processes within Human Geography

Reading and Seeing: The Condition of Postmodernity

So far, this autobiographical tour of the geographical imagination in the history of computing and information might feel familiar to many readers; over my two careers I have actually met a fair number of former corporate IT professionals who, like me, found their way into the academic history of technology. But here's where my story diverges from that of many of my peers: my graduate training also involved a simultaneous course of study in a field called "human geography." If history is fundamentally about describing and explaining continuity and change over time, geography is fundamentally about describing and explaining continuity and change over space. But within this broad field of study—"how places, landscapes, and ways of life come to be, are sustained, and are eventually transformed," as one introduction put it—there are many different tools of geographical imagination.[22] And here I should pause for a confession which I hope gives some solace to any future graduate students reading this chapter: Although I write today as a successful interdisciplinary faculty member at a major US public research university who studies "the history and geography of information technology and labor," back when I started my own graduate student career almost thirty years ago, I had zero understanding of what the academic study of geography was all about. As I am sure was the case for many who were reared through the US public school system, to me "geography" meant a rather reductionist set of containers and coordinates—places where things happened, rigid borders between stark cultural similarities and differences, pinpoints on maps to which were attached events and statistics. But not a subject to study on its own, and certainly not a useful toolkit to employ in studying others.

As with my entrance into the history of technology, my gateway to human geography was also a key academic book about space, time, and technology which broke through to popular culture just at the time when I was searching for a framework for helping me contemplate my own career change out of the information industries and into the knowledge industry. What I did not realize at the time was how improbable this book's success really was: geographer David Harvey's *The Condition of Postmodernity* was a Marxian political-economy analysis of the historical and geographical workings of capitalism, especially capital's particular deepening of impact on global culture and politics during the transition from twentieth-century-style "Fordist" accumulation practices (and technologies) to the more postmodern "flexible" global accumulation strategies of the twenty-first century.[23] Fortunately I was not alone in my curiosity about what, to me, seemed to be a revolutionary way of thinking about the myriad connections and contradictions between technology, society, space, and time—Harvey's book was a prime example of the way critical academic geography was engaging with a wide range of social sciences and cultural studies disciplines during the 1990s and 2000s, in what would be called a "spatial turn" that is still playing out today.[24]

I was fortunate enough to work with not just Harvey as one of my advisers in graduate school, but also another key geographer involved in the spatial turn, Erica Schoenberger. Her book *The Cultural Crisis of the Firm* brought a similar critical geography lens to the question of why some of the most powerful economic and social organizations on the planet nevertheless stumbled when the political, cultural, and technological conditions on which they built their markets suddenly changed.[25] Her arguments are a superb complement to much of the business history literature in the tradition of Alfred Chandler which, by contrast, focuses on corporate control and rationality through homogenous technological expansion, rather than the spatial imaginary of cultural differences. My geographical training continued through a postdoctoral year where I met economic geographers Eric Sheppard and Helga Leitner, who were at that time writing critical accounts of global networking strategies employed by capitalist firms and non-governmental activist organizations alike.[26] Together my encounters with these and other geographers have convinced me of the productive parallels between technology studies and spatiotemporal studies: not only are technologies socially produced and constructed in and through places and spaces, but *place and space are themselves socially produced and constructed* (and always through technologies, as well).

Seeing Technology in Place with GIScience

To understand the centrality of the production of place and space to the "spatial turn," it may help to know that this was a time when the broad field of geography was going through a sort of spatial turn of its own—but in a different direction. I first discovered this when I was assigned by my geography advisers as a teaching assistant for an introductory undergraduate course on environmental studies, tasked with incorporating the new technology of Geographic Information Systems into the homework assignments. This seemed to be an ideal assignment for the former corporate IT student of theirs—although dating from the 1960s and increasingly tied to the Global Positioning System (GPS) satellite network of the 1970s, usable "desktop" Geographic Information Systems had only recently appeared with the GUI PC interfaces of the 1980s and 1990s. Further, the data required to make any real use of computer mapping tools was really only widely available to academics, marketers, and undergraduates once the World Wide Web emerged as a ubiquitous distribution medium around the same time. I was pleased I could use my professional computer science skills in such an academic way—and already thinking about how to start mapping the geographic data about telegraph messenger employment trends for my dissertation.

The more I explored the history of this new technology, though—as my history advisers had by this time trained me well to do—what initially seemed to be an unproblematic data visualization tool was soon revealed as a point of ideological conflict among various strands of academic geography. As Eric Sheppard pointed out in his historical analysis of the technology, several factors influenced the development of GIS to cast it within a very particular geographical epistemology, including: the high cost of "geocoded" data that can be easily imported into GIS systems; the near-monopoly dominance of one US GIS applications firm; and the initial prevalence of corporate marketing and military surveillance applications for the technology. GIS in its very design privileged a narrow, instrumental conception of "places" as definitive points on a Cartesian map, and "spaces" as consistent vector boundaries around those points. This was a reductionist and rationalizing view which served some interests in geographical thought, but not others, leading to "the transforming of space from a complex cultural category into a precise, universal, and objective metric which was highly functional to the development of trade, the cataloging of resources, and the definition of property ownership."[27]

Thus, in the mid-1990s, when I discovered it as a graduate student teaching GIS, the discipline of geography was itself going through a theoretical crisis about which direction its own "spatial turn" would actually turn. From the vantage point of the history of information and computation, the GIS critique was in large part a data critique. Much of this was played out in an excellent volume edited by geographer John Pickles, *Ground Truth: Social Implications of Geographic Information Systems*—essays that demonstrated well that geographers had absorbed the lessons from technology studies that information infrastructures are always socially produced and socially constructed.[28] For example, Sheppard argued that "data are already laden with the theories, purposes, and social norms of the agencies that collect them" in a way that echoed what scholars like Bowker and Star were arguing around the same time about the data collected behind information infrastructures of all sorts.[29] Sheppard argued that "the GIS research agenda should be broadened to incorporate questions of social embeddedness and impact of GIS. Research into these questions can draw on the complementary skills of GIS specialists and social theorists."

Fortunately, this debate did not bifurcate geography into competing factions; today, in a model of carefully considered technology adoption of the sort that information and computing scholars often advocate for, GIS is largely seen in context as a productive and limited way of knowing the world through place and space. By the second decade of the twenty-first century, geographer Stacy Warren was able to report in the *SAGE Handbook of GIS and Society* that the technology, "once scorned by critical geographers as irretrievably positivist, is now being reconceptualized as a potentially liberating discourse that situates technology in social context," pointing out that "like the telephone, the radio, and the television long before it, GIS is one of many twentieth century communications technologies that promises—and threatens—to change our understanding and experience of time and space."[30] Geographers Ian Gregory and Alistair Geddes even argued for a new form of "spatial history" or "spatial humanities" as "a field using geographical technologies [like GIS] to develop new knowledge about the geographies of human cultures past and present"—bringing GIS back to the earlier "spatial turn" in the social sciences and humanities.[31]

Seeing Technology in Space with Human Geography

That spatial turn outside of geography—providing other humanists and social scientists with a framework for thinking contextually, relationally, and theoretically about place and space—was taking part just as the concern about reductionist maps and quantitative data was roiling within geography. But while GIS affected the whole range of geographic practice across the physical and natural environment—physical geomorphology, atmospheric climate measurement, biodiversity monitoring—it was in the realm of social, cultural, political, and economic activity where the spatial turn originated. The spatial turn came in large part from the subfield of *human geography*.

Focusing on the way people inhabit, alter, and understand place and space, contemporary human geography has a long history as a critical scholarly pursuit—a reaction against both the colonialist assumptions of 19th-century geography which ignored or marginalized non-white and non-Western populations and the neoclassical assumptions of early 20th-century geography that saw humans acting in the landscape simply as rational economic value-maximizers. As Ron Johnston noted in a recent primer on the field, somewhere around the mid-20th century, human geography emerged with a new sense that "although the 'vertical' inter-relationships between people and their environments were key to human well-being and life chances, the 'horizontal' links between 'places' were much more important."[32] Today human geography draws from a wide range of inspirations and scholarly frameworks, from Marxian political economy to feminist and anti-racist studies to postmodernism and poststructuralism. (Two recent and useful primers are the 2011 *Wiley-Blackwell Companion to Human Geography* and the 2013 *Oxford Dictionary of Human Geography*.)

If there is a commonality to these pursuits, it is (a) a clear rejection of the reductionist ideas of "agents simply doing things inside of space" (similar to the rejection of data reductionism in the GIS debates) coupled with (b) an emphasis on dynamism where spatial relations are active agents within social processes (similar to the way technology scholars cast artifacts and systems as active agents within social processes). Geographer Neil Smith noted that "the tendency to see place, space, and scale simply as containers for human activity" prevalent during the field's "quantitative turn" in the 1950s and 1960s assumed (incorrectly) that "human activity does not restructure space, it simply rearranges objects in space."[33] But the history of technology shows quite clearly that human activity does restructure space—both literally and figuratively. As geographers Derek Gregory and John Urry wrote in the mid-1980s, "spatial structure is now seen not merely as an arena in which social life unfolds, but rather as a medium through which social relations are produced and reproduced."[34]

What replaced this reductionism was a new set of "epistemological and ontological commitments," as geographer Erica Schoenberger noted, based on seeing the spatiality of social processes as unfolding over time and in a reciprocal or dialectical relationship.[35] The most prominent voice in human geography for this view, David Harvey, described his view of the "geographical imagination" necessary for this turn in his 1973 collection *Social Justice and the City*: a way of viewing history that "enables the individual to recognize the role of space and place in his own biography, to relate to the spaces he sees around him, and to recognize how transactions between individuals and between organizations are affected by the space that separates them."[36] Such a dialectical viewpoint allows geographers to deal with the often confounding ability of humans not only to act as active agents of change over the natural and built environment, but also to be enabled and constrained in those very actions by natural and built environment factors—including those of their own earlier creation.

Here is where the dynamic relationships explored by historians of technology and human geographers meet: just as technologies in and across space are socially produced and social constructed, so are the new spatialities that result from (and further enable or constrain) those technologies also socially produced and constructed. For example, in *The Condition of Postmodernity*, Harvey conceptualized "time-space compression" as the moment when material changes in infrastructure connect to mental changes in perception of distance and otherness: "processes that so revolutionize the objective qualities of space and time that we are forced to alter, sometimes in quite radical ways, how we represent the world to ourselves" in such a way that "the world sometimes seems to collapse inwards upon us." To write geography into the history of technology, one must learn to see such technological spatialities—human-altered places, regions, networks, urban environments, and virtual landscapes, together with the shared or contested understandings that evolve around them—as socially produced and socially constructed themselves. Thus, another meaning of the "geographical imagination" might be the ability to see and analyze "the socially constructed spaces produced on the earth's surface."[37]

Following Harvey's work, among others, in many academic disciplines throughout the 1990s and 2000s—especially science and technology studies—there seemed to be a new attention both to digital information in humanities research and a new "spatial turn" of attention toward geographical tools and theories.[38] In summarizing this period, Ron Johnston noted that after the 1970s, "a dualism between structure and agency was identified: places provide the contexts within which people learn and behave, and as they do so, they change the places in which further learning occurs. Thus the human condition, as made, re-made and experienced in a great variety of places and at a wide range of spatial scales, became a central geographical concern—and in its exploration geographers established links with a wider range of other disciplines and sub-disciplines."[39] As economic geographers Eric Sheppard, Trevor Barnes, and Jamie Peck put it when introducing a primer on the field, this notion of ongoing, reciprocal production and reproduction of space and society—that "agency and sociospatial structure are co-constitutive"—is a particular "sociospatial ontology," or the "presumptions that practitioners of a research program share about society, space, and their interrelation."[40]

Thinking about human geography and its relationship to the history of technology in this way suggests that much of the research in the history of computing and information has already incorporated such a dialectical, spatial lens. For example, in a recent and wide-ranging *Technology and Culture* essay, historian of computing Nathan Ensmenger made a convincing case for "grounding" the history of computing by viewing it through a geographical lens in terms of environmental surveillance, control, and consequences. He points out that "when we look at the vast web of wires, cables, towers, generators, and other physical equipment that underlies the apparently virtual realm of Cyberspace—the digital present does not seem quite so discontinuous with our industrial past." Ensmenger shifts his gaze to "the material underpinnings of the digital economy—in other words, by the relationship between 'computing power' and more traditional processes of resource extraction, exchange, management, and consumption" in order to "tell a more global and inclusive history of computing."[41] In a complementary way, geographer Erica Schoenberger's book *Nature, Choice and Social Power* argued that a technology-suffused "way of life that puts unsustainable stresses on nature" does not just happen, but is historically produced "through the intricate construction of social life, through the accumulation and exercise of various forms of social power, through the always interesting ways that people live within, create, and are shaped by social structures."[42]

Technological Fixes Are Spatial Fixes

What ties such studies together for me is a recognition of the reciprocal ways that (1) technological projects always build upon and then affect spatial conditions, and (2) spatial projects always build upon and affect technological infrastructures. But this is not an arbitrary process—historians recognize that new technologies are created and promoted and adopted (or rejected) with certain actor intentions and purposes. What the dialectic between technology and geography adds is the reminder that new spatialities are also created and promoted and adopted (or rejected) with intention and purpose as well. For example, as Eric Sheppard commented in his analysis of the possible social and spatial consequences of GIS infrastructure, "GIS may make it possible to develop far more sophisticated solutions to the location of a noxious facility—solutions that take into account a greater detail and variety of information than was possible before. Yet at the same time this capability can lead analysts to become so hooked into such technical fixes that they lose sight of the larger picture altogether."[43] In other words, a specific technological tool applied to a social, political, and economic problem—a "technological fix"—might feed back into a new understanding of the spatiality for technological action in a way that forecloses other kinds of technological (or social, political, economic) solutions later. To effectively mobilize the analysis of technology together with the geographical imagination, then, it is important to look for the ways that "technological fixes" *work together with* "spatial fixes."

This concept of the "technological fix" will be familiar to computer and information historians. Contemporary proponents of the latest information technologies—the producers and marketers of that technology—often speak of their products in a teleological, deterministic sort of way, promising a systematic and universal solution to some long-standing social problem.[44] Local systems must give way to national systems, as long-distance telephone provider AT&T said in the early twentieth century, because it is a natural development. Or, digital systems will replace analog systems, as the movie industry said in the late twentieth century, because the one technology is functionally superior to the other. And especially in recent history, public systems must give way to privatized systems, as claimed critics of government funding for public broadcasting, because the public sector is inevitably inefficient. Such pseudo-historical arguments tend to coincide with a particular vision of progress as necessarily centralized, rationalized, and corporatized. Often, business managers deploy technology and labor methodically according to the bottom-line benefits of each—a "digital hand" of benevolent progress through innovation.[45] Crucially, such "fixes" build upon each other in space and over time—each new "round" of communication innovation helps alter a society's very conceptions of distance and proximity, movement and mobility, place and space. But yesterday's innovations are soon regarded as today's bottlenecks.[46]

In a similar way, the concept of a "spatial fix" has developed in human geography. The original sense of this concept, as deployed by Harvey, is a specific manifestation of a strategy within capitalism for those with surplus resources to keep those resources in circulation toward new projects which will bring ever-more accumulation of wealth. Or as Schoenberger summarized, the spatial fix is "a way to productively soak up capital by transforming the geography of capitalism."[47] Sheppard described three examples of the spatial fix: (1) the commodification of land, where land is, over time, continually and differentially valued and revalued according to new technological opportunities for accumulation; (2) the global expansion of capitalism, both in mustering wider and wider regions for production and in selling commodities and services to wider and wider consumer markets; and (3) investment into the built environment, especially when such investment is orchestrated by the state but works to benefit the competitive risks of accumulation for capital by decreasing the "friction of distance" and thus the cost of doing business.[48]

In this way the same kinds of sociotechnical infrastructure projects studied by historians of technology—networked cities, transportation and communication infrastructure, or even virtual worlds in cyberspace—represent spatial fixes as well as technological fixes. Both the spatial fix and the technological fix can appear with a certain degree of teleology or inevitability as well—spatial fixes of increased territorial appropriation, control, development, and urbanization cast as a sort of "manifest destiny" for a nation to annihilate its "frontier," while technological fixes of mechanization, automation, and now algorithmic augmentation are cast through a lens of "technological determinism" as inevitable steps toward a utopian "end of work."

Human geography and history of technology here enhance each other's understandings of these sorts of "fixes." For example, human geography points to how "technological fixes" are not just meant to solve social problems through material means, but also likely meant to solve problems of slowed capital accumulation of wealth by providing a new outlet for investment and return, production and consumption. And history of technology suggests that "spatial fixes" are not only meant as new terrains for capital accumulation and preservation of the power of capitalist actors but must point to some sort of societal hope or need or value in order to be effective sinks for investment and financial return. Put another way, technological fixes must fit within the spatial circuits of capitalism to be seen as feasible, and spatial fixes must fit within a technological imaginary of progress or the sublime to be seen as desirable.

Spatial fixes and technological fixes also relate and build upon one another historically (and somewhat paradoxically) in that new technological fixes inevitably require old space, and new spatial fixes inevitably require old technology—for example, the "new media" mimics the old and requires the old to succeed; optical fiber is laid following buried sewer conduits, disrupting vehicle travel on city streets; new cell phone towers enabling mobile telework appear first on skyscrapers meant to congregate laborers in one place throughout the working day; and all manner of communication systems meant to build a cost-free virtual space nevertheless need substations for switches, offices for managers, and fleets of vehicles for maintenance workers placed strategically throughout costly physical space. Geographer Doreen Massey uses the evocative metaphor of "sediments" to describe the geographical palimpsest of technology that results.[49]

Historians know well that the advertised purpose of new information and communication technology is almost always to make content accessible across space and across time, from one context to another, from one community of practice to the next, with as little friction of thought and required understanding as possible—a "space of flows," as Manuel Castells put it, operating in real time and without need for conscious reflection, which neatly bifurcates the world into two sides of a digital divide, where you either have ubiquitous access or virtually none at all.[50] But historians and geographers who study the details of how different physical, cultural, and political-economic geographies are linked and bypassed by each new round of information, communication, and computation technology reveal a more complex pattern—what geographer Neil Smith conceptualized as "uneven geographic development" in a way tied inextricably to the contradictions and constant innovations of global capitalism.[51]

The result is anything but a homogenous geography over space—and no matter how fine-grained we set the borders within a GIS analysis of social power and social justice over space, any boundary we might draw will hide divides and differences. While reductionist geographical research would simply identify, map, and perhaps try to explain differences based on individual characteristics of sites and social groupings, a relational, dialectical, and critical human geography sees each area of uneven development as tied to the other, being produced and reproduced over time as a result of contradictory social processes, and often working in support of particular power relations and normative goals. Thus, while we humans shape (and continually reshape) our technological environments for a myriad of social, political, and economic reasons, those technological environments in turn enable and constrain our actions and possibilities over and over again. Being able to envision and reveal this archaeology of "fixes" is a key skill in bringing a geographical imagination to the history of computing and information technology.

Seeing and Writing: Closed Captioning

These dialectical ideas of technological fixes as tied to spatial fixes were crucially important for me as I searched for an appropriate technological, geographical, and labor infrastructure to study in the follow-up to my book on the telegraph messenger boys. I had just been hired as an assistant professor, but not in either of my graduate fields of history of technology or human geography. Instead, I was employed in an information school and a journalism school. There was a strong tradition of studying the "technological fix" in service to societal norms of progress and democracy in each profession—and a strong fear in both professions at the time that the current "spatial fix" of new media technology in cyberspace was going to cause a geographical apocalypse for local libraries and local newspapers. And I must be honest (with apologies to any historians of the telegraph reading this essay) that I was worried at this early point in my career about being typecast by my peers as an historian of the telegraph. But it was more than that: Did my scholarly interests center on the same mid-19th-century-to-mid-20th-century period as my first book, or was I willing to be one of those historians who studied more recent or even contemporary technological developments? Was I interested in another study so clearly centered on one or two large corporate actors like Western Union and AT&T, or did I want to focus on smaller activist organizations instead? And if I was serious about exploring the "hidden labor" behind information infrastructures, how was I supposed to start looking for those secret spaces?

All these concerns and questions were in my thoughts as I watched television one night and saw a commercial appear for "high tech" training in the field of "realtime stenography" where one could enter a lucrative trade as a courtroom reporter, a work-at-home television closed captioner, or even an itinerant, on-call real-time speech interpreter for a deaf or hard-of-hearing student. Here was a set of intertwined technological fixes purporting to address a set of long-standing social and political problems (representing truth in the legal system, providing media accessibility and justice, and offering educational accommodations and opportunity). But each "fix" had different spatial implications for those doing the work (working within the highly regimented location of a local justice system, working at home during all hours of the day, or working wherever the client went using portable technology). Technological fixes or spatial fixes? Using both lenses to explore them simultaneously revealed the many contradictions, connections, and wider implications of both. The same hybrid human-computing practice was at the heart of all three careers—but as that common practice "jumped context" into different social and political and economic situations, those workers who performed it took on wholly different values and meanings.[52]

Conclusion: Spatial Turns and Digital Turns

As we approach the second quarter of the twenty-first century, within the history of computing and information, the "spatial turn" is still unfolding. It was not too long ago that computing historian Nathan Ensmenger effectively noted that within computing history, "our treatment of these topics is too narrow, too focused on machines rather than data, representations, and processes," arguing that "it is the activity of digitization, not computerization, that is occurring repeatedly throughout our laboratories, factories, studios, schools, shopping malls, and living rooms." Without overtly stating it, Ensmenger was making a case for situating a new history of "digitalization" both socially and spatially, in order to better understand the ongoing and transformative relations between all three.[53] One example of this can be seen in the 2009 special issue of the journal *Information Society* on the "geographies of the information society," inviting collaborative research into "the actual geographies that evolve on the surface of the earth in the information age, the virtual geographies that are the direct products of modern ICT, or the conceptual geographies gradually developed in individual and social consciousness through the representations of earth by these technologies."[54]

What makes this an especially exciting time to take a spatial turn as a historian of the digital world is the recognition that many geographers are simultaneously taking a "digital turn" of their own, reckoning with the ongoing production, repair, reproduction, and (sometimes) dissolution of our still-new global and virtual cyberspace for commerce and communication. Of course, right from the debut of the World Wide Web in the early 1990s, geographers such as Michael Batty realized, "We need to begin to map this space, to visualize its architecture, and to show how it connects to and transforms our traditional geographies."[55] The late 1990s and early 2000s saw the first geographical articles and special journal issues on what geographer Rob Kitchin called "geographies of cyberspace" and Martin Dodge called "cybergeography": "the study of the nature of the Net through the spatial perspectives of geography and cartography." Dodge suggested that new cybergeographers had an obligation to counter "the popular, casual reporting that suggests that the Internet inevitably makes geography redundant, that distance, location, and place no longer matter in peoples' daily lives, and that cyberspace enables action and interaction between anything, anytime, anywhere." And Kitchin agreed that "cyberspaces are dependent upon spatial fixity, they are embodied spaces and access is unevenly distributed" such that "cyberspaces do not replace geographic spaces, nor do they destroy space and time."[56]

But the current digital turn in geography takes these ideas further. Writing in the journal *Ethics, Place, and Environment* for a special issue on technological change, geographer Paul Adams introduced the concept of "technicity," or the notion that "space/place and technology are intertwined in dialectical fashion." He elaborated, "The task of exposing the social and spatial dimensions of technology is therefore a necessary part of any effective political project."[57] Geographer Aaron Kellerman recently proposed mobilizing this "technicity" concept by shifting away from the theoretical study of "virtual spaces" and instead operationalizing the study of specific "Internet screen spaces," or "the visual interface between the Internet and its users," building upon the long-standing ethnographic tradition from human-computer interface studies.[58] And according to a recent review in *Progress in Human Geography*, these new geographical studies "have grounded metaphors, such as those of 'cyberspace' and 'the cloud,' by tracing the actual spatialities of internet infrastructures at both local and global scales"—and most recently, "begun to trace the generation and flows of big data and algorithms."[59] Today's geographic atlases of the internet still focus on issues of e-commerce and digital divides (though materially and socially constructed very differently than twenty years ago), but also explore new geographical materialities of cyberspace through "augmented reality" and "the Internet of things"—further ways that web-connected information technology promises both a "technological fix" for persistent social problems (e.g., smart sensors and vivid visualizations to confront global climate change) and a "spatial fix" for new capitalist services that extend both information access and consumer surveillance into more areas of private life.[60]

This crossroads where the spatial turn and the digital turn meet is the place to find interdisciplinary dialogue and collaboration of the same sort that I was fortunate enough to experience as a graduate student; as my doctoral adviser argued, "Sitting down with someone trained in a completely different discipline forces us to explain why we study science and technology in the way we do, to preach to someone other than the converted or the antagonistic."[61] Recently more and more geographers have "sat down" with science and technology studies because "with increasingly opaque code shaping and delimiting the virtual realities and opportunities to be found in cyber space, and the operations of bioscience increasingly re-defining the realms of nature and humanity at the nano-scale, it is becoming increasingly difficult for geographers to see, let alone understand, the socio-technological networks they seek to chart."[62] And as the guest editor of a special issue of *Environment and Planning B: Planning and Design* on "space, sociality, and pervasive computing" noted, "As these investigations about the complex overlaying of spatial design, social interactions, and digital technologies continue, there is a need to cultivate a dialogue between those operating at the boundaries between disciplines."[63]

In my own path through academia, I first followed the spatial turn in history throughout my interdisciplinary graduate school training, and then followed the digital turn in geography throughout my interdisciplinary faculty position. These turns of imagination have not just been academic—the students we graduate and the stakeholders we serve continue to wrestle with difficult questions about the death of local journalism and the dematerialization of the local library, where new technologies and new spatialities are fundamentally entwined. Nowhere is this more evident than in the newly-emerging cross-disciplinary field of "critical data studies."[64] Such work often focuses on the corporate "platform economy" services like Uber, Airbnb, or Amazon Mechanical Turk which today mediate the provision of transportation, housing, and labor in cities around the globe—a ubiquitous geographical presence when it comes to finding opportunities for investment and profit, coupled with an apparent universal geographic absence when it comes to legal jurisdiction or local democratic accountability. Geographer Mark Graham called this a strategy of producing "conjunctural geographies" on the part of platform firms—"a way of being simultaneously embedded and disembedded from the space-times they mediate."[65] Critical data studies problematizes the latest technological fixes as spatial fixes, and vice versa, as when geographer Graham Pickren argues that "one of the goals of a critical data studies is to situate the phenomenon of big data as political, historical, and geographical as opposed to simply an inevitable outcome of technological progress."[66]

One thread of this research that I am especially excited about recognizes that the "big data" that fuels our global economy, society, and polity also depends on new categories of human labor that seem to develop and transform more quickly than ever before—but for reasons of corporate branding, state secrecy, or plain old social isolation, such work is often hidden, obscured, or invisible. This may seem counterintuitive given the prominence of "remote work" conversations in the media since the onset of the global COVID-19 pandemic. But looking closely at technological labor in and through geographical relationships reveals a more complicated picture. Consider content moderators: those who enforce the boundaries set by online social networks to screen, review, and redact content like personal attacks, images of traumatic violence, and hate speech out of supposed "free speech" platforms. In her groundbreaking book *Behind the Screen*, Sarah Roberts investigated these workers and noted "this invisibility is by design," not only because the brands employing these moderators do not want their images tarnished by any revelation of the toxic material the content moderators screen out, but also because keeping the content moderators invisible to each other—often working remotely from their private homes, under legal nondisclosure agreements—keeps them from organizing for better working conditions in what is a high-burnout, high-turnover job.[67] Similarly, in what the nonprofit Partnership on AI has called "data enrichment work," ChatGPT vendor OpenAI formally employed only 375 employees in 2023, but relied on a global labor force of about 1,000 additional contract workers to help label and train its large language model (LLM) AI system.[68] The global outsourcing of tagging and labeling tasks necessary to effectively train machine learning systems on unstructured data maps to low-resource and low-wage developing countries.[69] This is precisely the kind of mapping that we must continue to create if we are to have a fuller understanding—as consumers, citizens, and even capitalists—of how our information ecosystem works, and what kinds of critical skills and skepticism we need to practice in order to make it work for us. As anthropologist Nick Seaver, in calling for a new "anthropology of algorithms," recently argued: "If you cannot see a human in the loop, you just need to look for a bigger loop."[70]

Looking for humans in the digital loop brings us back to what I hope is a useful heuristic for finding your own path through these spatial and digital turns, no matter what disciplinary terrain you call home: *a way of seeing technology in place, place in technology, and the socio-spatial processes that unite the two*. As sociologist Thomas Gieryn observed in commenting about the disciplinary connections between geography and sociology, "Maybe a place-sensitive sociology is not a set of empirical findings at all or even a distinctive kind of explanatory model, but rather a way to do sociology in a different key—a visual key."[71] In trying to apply this visual key to my own work, there are three strategies I find myself returning to again and again: (1) looking for the *connections* between places and spaces that are outsourced, obscured, or unacknowledged (such as the ecological circuits of raw material extraction, production, distribution, use, and waste through which every modern consumer electronic device flows); (2) looking for the *cultural moments* when places and spaces are redefined or reimagined or reinhabited in new ways (such as when home becomes a workplace, school becomes a screen, or a phone becomes mobile in an "everything" digital convergence media tool); and finally, (3) looking for the *concealment* of activity in places and spaces that are overlooked, invisible, or deliberately hidden behind the most prominent of new innovations (such as the global labor force behind the training of labor-saving machine learning systems).

But perhaps the best heuristic for computing and information historians endeavoring to better "see" the socio-spatial lives of their human and machine subjects of study is to simply remember that the very concept of information itself, as well as any assessment of information's value, is meaningless without some sort of dual spatial and social context—an organizational location, a community of practice, an end-user market, a shared public purpose within which to construct that information as "data" to be manipulated, "content" to be enjoyed, or "knowledge" to be utilized. The value of information can continue to be realized only to the degree that people circulate that information, not just from one place to another, or between two nodes in a network at a time, but from one spatial, social, political, and economic context to the next—especially across sites and communities which are diverse and disparate. These are the moments of change and connection when scholars can best reveal those who keep that information circulating across time, space, and culture in order to realize and redefine its value—and its purpose.[72]

Notes

1. Scott Frickel, Mathieu Albert, and Barbara Prainsack, eds., *Investigating Interdisciplinary Collaboration: Theory and Practice across Disciplines*, The American Campus (New Brunswick, NJ: Rutgers University Press, 2017).

2. William Cronon, *Nature's Metropolis: Chicago and the Great West* (New York: Norton, 1992).

3. Stuart W. Leslie, *The Cold War and American Science: The Military-Industrial-Academic Complex at MIT and Stanford* (Columbia University Press, 1993), https://hdl.handle.net/2027/heb00988.0001.001; Stuart W. Leslie and Robert H. Kargon, "Selling Silicon Valley: Frederick Terman's Model for Regional Advantage," *Business History Review* 70, no. 4 (Winter 1996): 435–72.

4. Wiebe E. Bijker, Thomas P. Hughes, and Trevor Pinch, *The Social Construction of Technological Systems: New Directions in the Sociology and History of Technology* (The MIT Press, 2012), https://muse-jhu-edu.ezproxy.library.wisc.edu/book/19813/.

5. Peter Louis Galison et al., eds., *The Architecture of Science: Papers Presented at a Conference That Met 21–22 May at Harvard University* (Cambridge, MA: MIT Press, 1999).

6. Allan R. Pred, *Urban Growth and the Circulation of Information: The United States System of Cities, 1790–1840*, Harvard Studies in Urban History (Cambridge, MA: Harvard University Press, 1973).

7. AnnaLee Saxenian, *Regional Advantage: Culture and Competition in Silicon Valley and Route 128* (Harvard University Press, 1996), https://hdl.handle.net/2027/heb00993.0001.001; Paul E. Ceruzzi, *Internet Alley: High Technology in Tysons Corner, 1945–2005* (Cambridge, MA: MIT Press, 2008).

8. Greg Downey, "Jumping Contexts of Space and Time," *IEEE Annals of the History of Computing* 26, no. 2 (April 2004): 94–95, https://doi.org/10.1109/MAHC.2004.1299668.

9. Greg Downey, *Technology and Communication in American History*, Historical Perspectives on Technology, Society, and Culture (Washington, DC: American Historical Association, 2011).

10. Joel A. Tarr, Thomas Finholt, and David Goodman, "The City and the Telegraph: Urban Telecommunications in the Pre-Telephone Era," *Journal of Urban History* 14, no. 1 (November 1, 1987): 38–80, https://doi.org/10.1177/009614428701400103.

11. Ruth Schwartz Cowan, *More Work for Mother: The Ironies of Household Technology from the Open Hearth to the Microwave*, Repr. (New York, NY: Basic Books, 1983).

12. Sheila Allen and Carol Wolkowitz, *Homeworking: Myths and Realities* (London: Macmillan Education UK, 1987), https://doi.org/10.1007/978-1-349-18879-6.

13. Alvin Toffler, *Future Shock* (New York: Bantam Books, 1970).

14. William J. Mitchell, *City of Bits: Space, Place, and the Infobahn* (Cambridge, MA: MIT Press, 1995); Steve Graham and Simon Marvin, *Splintering Urbanism: Networked Infrastructures, Technological Mobilities and the Urban Condition*, 1st ed. (London: Taylor and Francis, 2002).

15. Manuel Castells, *The Rise of the Network Society*, Information Age, v. 1 (Malden, MA: Blackwell Publishers, 1996).

16. Janet Abbate, *Inventing the Internet*, Inside Technology (Cambridge, MA: MIT Press, 1999).

17. Geoffrey C. Bowker and Susan Leigh Star, *Sorting Things Out: Classification and Its Consequences* (The MIT Press, 2000), https://doi.org/10.7551/mitpress/6352.001.0001.

18. Paul N. Edwards, "Y2K: Millennial Reflections on Computers as Infrastructure," *History and Technology* 15, no. 1–2 (September 1998): 7–29, https://doi.org/10.1080/07341519808581939.

19. Paul Edwards, "Virtual Machines, Virtual Infrastructures: The New Historiography of Information Technology," *Isis* 89 (1998): 93–99.

20. Greg Downey, *Telegraph Messenger Boys: Labor, Technology, and Geography, 1850–1950* (New York: Routledge, 2002).

21. Derek Gregory, *Geographical Imaginations* (Cambridge, MA: Blackwell, ©1994), https://search.library.wisc.edu/catalog/999699133102121.

22. Michael Batty and Sam Cole, "Time and Space: Geographic Perspectives on the Future," *Futures*, Time and Space Geographic Perspectives on the Future, 29, no. 4 (May 1, 1997): 277–89, https://doi.org/10.1016/S0016-3287(97)00016–5.

23. David Harvey, *The Condition of Postmodernity: An Enquiry into the Origins of Cultural Change* (Oxford [England]; Cambridge, MA: Blackwell, 1989).

24. Diarmid A. Finnegan, "The Spatial Turn: Geographical Approaches in the History of Science," *Journal of the History of Biology* 41, no. 2 (June 2008): 369–88, https://doi.org/10.1007/s10739-007-9136-6.

25. Erica J. Schoenberger, *The Cultural Crisis of the Firm* (Cambridge, MA: Blackwell, 1997).

26. H. Leitner and E. Sheppard, "'The City Is Dead, Long Live the Net': Harnessing European Interurban Networks for a Neoliberal Agenda," *Antipode* 34, no. 3 (June 2002): 495–518, https://doi.org/10.1111/1467–8330.00252.

27. Eric Sheppard, "GIS and Society: Towards a Research Agenda," *Cartography and Geographic Information Systems* 22, no. 1 (January 1, 1995): 5–16, https://doi.org/10.1559/152304095782540555.

28. John Pickles, ed., *Ground Truth: The Social Implications of Geographic Information Systems*, Mappings (New York: Guilford Press, 1995).

29. Bowker and Star, *Sorting Things Out*.

30. Timothy L. Nyerges, Helen Couclelis, and Robert McMaster, *The SAGE Handbook of GIS and Society* (SAGE Publications, Inc., 2011), https://doi.org/10.4135/9781446201046.

31. Alistair Geddes and Ian N. Gregory, eds., *Toward Spatial Humanities: Historical GIS and Spatial History* (Bloomington: Indiana University Press, 2014), https://muse-jhu-edu.ezproxy.library.wisc.edu/book/29406.

32. James S. Duncan, *The Wiley-Blackwell Companion to Human Geography* (Chichester, West Sussex; Malden, MA: Wiley-Blackwell, 2011).

33. Neil Smith, *Uneven Development: Nature, Capital, and the Production of Space* (University of Georgia Press, 2008), https://hdl.handle.net/2027/heb34639.0001.001.

34. Derek Gregory and John Urry, *Social Relations and Spatial Structures* (New York: St. Martin's Press, 1985), https://search.library.wisc.edu/catalog/999558802002121.

35. Erica Schoenberger, "Interdisciplinarity and Social Power," *Progress in Human Geography* 25, no. 3 (September 2001): 365–82, https://doi.org/10.1191/030913201680191727.

36. David Harvey, *Social Justice and the City* (revised edition; Athens: University of Georgia Press, 2009), https://muse-jhu-edu.ezproxy.library.wisc.edu/book/13205.

37. Yuko Aoyama and Eric Sheppard, "The Dialectics of Geographic and Virtual Space," *Environment and Planning A: Economy and Space* 35, no. 7 (July 2003): 1151–56, https://doi.org/10.1068/a3623.

38. Charles W. J. Withers, "Place and the 'Spatial Turn' in Geography and in History," *Journal of the History of Ideas* 70, no. 4 (2009): 637–58, https://doi.org/10.1353/jhi.0.0054; Finnegan, "The Spatial Turn."

39. Duncan, *The Wiley-Blackwell Companion to Human Geography*.

40. Trevor J. Barnes, Jamie Peck, and Eric S. Sheppard, *The Wiley-Blackwell Companion to Economic Geography*, Wiley-Blackwell Companions to Geography (Chichester, West Sussex; Malden, MA: Wiley-Blackwell, 2012).

41. Nathan Ensmenger, "The Environmental History of Computing," *Technology and Culture* 59, no. 4S (2018): S7–S33, https://doi.org/10.1353/tech.2018.0148.

42. Erica Schoenberger, *Nature, Choice and Social Power* (Abingdon: Routledge, 2015).

43. Sheppard, "GIS and Society."

44. Sean F. Johnston, "Alvin Weinberg and the Promotion of the Technological Fix," *Technology and Culture* 59, no. 3 (2018): 620–51, https://doi.org/10.1353/tech.2018.0061.

45. James W. Cortada, *The Digital Hand: How Computers Changed the Work of American Manufacturing, Transportation, and Retail Industries* (Oxford; New York: Oxford University Press, 2004).

46. Downey, *Technology and Communication in American History.*

47. Erica Schoenberger, "The Spatial Fix Revisited," *Antipode* 36, no. 3 (June 2004): 427–33, https://doi.org/10.1111/j.1467–8330.2004.00422.x.

48. Eric Sheppard, "The Spatiality of The Limits to Capital," *Antipode* 36, no. 3 (June 2004): 470–79, https://doi.org/10.1111/j.1467–8330.2004.00426.x.

49. Doreen B. Massey, *Spatial Divisions of Labor: Social Structures and the Geography of Production* (New York: Methuen, 1984).

50. Castells, *The Rise of the Network Society.*

51. Smith, *Uneven Development.*

52. Downey, "Jumping Contexts of Space and Time"; Greg Downey, *Closed Captioning: Subtitling, Stenography, and the Digital Convergence of Text with Television* (Baltimore: Johns Hopkins University Press, 2008), https://doi.org/10.1353/book.3337.

53. Nathan Ensmenger, "The Digital Construction of Technology: Rethinking the History of Computers in Society," *Technology and Culture* 53, no. 4 (2012): 753–76, https://doi.org/10.1353/tech.2012.0126.

54. Hamid R. Ekbia and Nadine Schuurman, "Introduction to the Special Issue on Geographies of Information Society," *Information Society* 25, no. 5 (October 2009): 289–90, https://doi.org/10.1080/01972240903212425.

55. M. Batty, "The Geography of Cyberspace," *Environment and Planning B: Planning and Design* 20, no. 6 (December 1, 1993): 615–16, https://doi.org/10.1068/b200615.

56. Rob Kitchin, *Cyberspace: The World in the Wires* (Chichester, England; New York: J. Wiley, 1998); Martin Dodge, "Cybergeography," *Environment and Planning B: Planning and Design* 28, no. 1 (February 2001): 1–2, https://doi.org/10.1068/b2801ed.

57. Paul C. Adams, "Introduction to 'Technological Change': A Special Issue of Ethics, Place & Environment," *Ethics, Place & Environment* 10, no. 1 (March 1, 2007): 1–6, https://doi.org/10.1080/13668790601149968.

58. Aharon Kellerman, "Image Spaces and the Geography of Internet Screen-Space," *GeoJournal* 81, no. 4 (2016): 503–17, https://doi.org/10.1007/s10708-015-9639-1.

59. James Ash, Rob Kitchin, and Agnieszka Leszczynski, "Digital Turn, Digital Geographies?," *Progress in Human Geography* 42, no. 1 (February 2018): 25–43, https://doi.org/10.1177/0309132516664800.

60. Barney Warf, *Geographies of the Internet*, 1st ed. (Milton Park, Abingdon, Oxon; New York, NY: Routledge, 2020. | Series: Routledge studies in human geography: Routledge, 2020), https://doi.org/10.4324/9780367817534.

61. Stuart W. Leslie, "Reestablishing a Conversation in STS: Who's Talking? Who's Listening? Who Cares?," *Bulletin of Science, Technology & Society* 19, no. 4 (August 1999): 271–80, https://doi.org/10.1177/027046769901900403.

62. Deborah P. Dixon and Mark Whitehead, "Technological Trajectories: Old and New Dialogues in Geography and Technology Studies," *Social & Cultural Geography* 9, no. 6 (September 2008): 601–11, https://doi.org/10.1080/14649360802320560.

63. Bharat Dave, "Space, Sociality, and Pervasive Computing," *Environment and Planning B: Planning and Design* 34, no. 3 (June 1, 2007): 381–82, https://doi.org/10.1068/b3403ed.

64. Craig M. Dalton, Linnet Taylor, and Jim Thatcher (alphabetical), "Critical Data Studies: A Dialog on Data and Space," *Big Data & Society* 3, no. 1 (June 1, 2016): 2053951716648346, https://doi.org/10.1177/2053951716648346.

65. Mark Graham, "Regulate, Replicate, and Resist—the Conjunctural Geographies of Platform Urbanism," *Urban Geography* 41, no. 3 (March 15, 2020): 453–57, https://doi.org/10.1080/02723638.2020.1717028.

66. Graham Pickren, "'The Global Assemblage of Digital Flow': Critical Data Studies and the Infrastructures of Computing," *Progress in Human Geography* 42, no. 2 (April 2018): 225–43, https://doi.org/10.1177/0309132516673241.

67. Sarah T. Roberts, *Behind the Screen: Content Moderation in the Shadows of Social Media, Behind the Screen* (Yale University Press, 2019), https://doi.org/10.12987/9780300245318.

68. David Ingram, "The AI Revolution Is Powered by These Contractors Making $15 an Hour," NBC News, May 6, 2023, https://www.nbcnews.com/tech/innovation/openai-chatgpt-ai-jobs-contractors-talk-shadow-workforce-powers-rcna81892.

69. Billy Perrigo, "OpenAI Used Kenyan Workers on Less Than $2 Per Hour to Make ChatGPT Less Toxic," *Time*, January 18, 2023, https://time.com/6247678/openai-chatgpt-kenya-workers/.

70. Nick Seaver, "What Should an Anthropology of Algorithms Do?," *Cultural Anthropology* 33, no. 3 (August 21, 2018): 375–85, https://doi.org/10.14506/ca33.3.04.

71. Thomas F. Gieryn, "A Space for Place in Sociology," *Annual Review of Sociology* 26, no. 1 (August 2000): 463–96, https://doi.org/10.1146/annurev.soc.26.1.463.

72. Greg Downey, "Making Media Work: Time, Space, Identity, and Labor in the Analysis of Information and Communication Infrastructures," in T. Gillespie, P. J. Boczkowski, and K. Foot, eds., *Media Technologies: Essays on Communication, Materiality, and Society* (Cambridge, MA: MIT Press, 2014).

Chapter 14

Traveling Across the Lands of History, Philosophy, and Practices of Computing

Confessions of a Nomad

Liesbeth De Mol

When I was invited to write a chapter for this book, I initially had significant doubts. Of course there was the usual problem of being already overworked but the real problem for me was the topic proposed to me: *Writing Computing History from a Philosophical Perspective*. The chapter that ultimately resulted is written as a personal and self-reflective essay.[1]

The reader should not expect a "conventional" historiographical or historical academic paper, which assumes a clear divide between what one could call the historian scientist and historian author. Instead, I challenge that divide here by making explicit the intertwinement between my personal history and the methodological and topical choices I have developed in my past work. My writing here is not focused first on convincing the reader of a particular historiographical approach or a specific reading of history. Instead, I want to share an experience of someone who has often felt themselves to be an outsider to the field of the history of computing, and what are and have been, in that experience, the pitfalls and opportunities encountered and observed.

My style is then one of semi-autobiographical history writing, which has been called the experimental style in historiography.[2] In particular, I show how and why I steadily realized that I do not belong and do not want to belong to one discipline only. That is, I show how I became an academic nomad traveling across a variety of different methodological lands (first section). This travel has allowed me to identify in my own work, but also in that of others, what I call epistemological obstacles (second section), which are used throughout the remainder of the chapter as an analytical tool to discuss the limitations in my own work and that of others to develop my *personal* view on the history of computing, which seeks out opportunities for broader collaborations across methodological divides aimed towards the future. Based on an obstacle I inherited from my philosophy education, I provide a first opportunity of where I believe historians of computing have something to contribute to philosophy and beyond (third section). In the fourth section, I describe my initial methodologies for doing historical research, which focus on the technical discourse and the epistemological assumptions it unveils. This, I argue, resulted in an obstacle for myself but also enables me to argue how any approach that focuses on one level of historical analysis is bound to create its own basic blind spots. On that basis I argue for a more pluralist historiography of computing based on two concrete cases. Given my background in philosophy, it is probably not too surprising that I am rather presentist in my historical research. While acknowledging that this is not an obvious if not subversive position for the historian, I provide several arguments why I believe we should dare and have a discussion on this within history of computing (fifth section). In the next section, I give a brief description of my own initiatives towards developing a nomadic approach in history and philosophy of computing (sixth section). I end with a short concluding section revisiting the potential of the style chosen (seventh section).

Becoming an Academic Nomad

Let me now return to the original problem I had with accepting to write this chapter, which is anchored in my own personal history as a researcher. After a master's degree in art history[3] (1999), I believed I had found my one and true love: philosophy. I embarked on a second master's in philosophy (2002), followed by a PhD in philosophy (2003–2007). But during my doctoral years, engaging with the history of computability and computing became unavoidable, and I steadily realized that historical research too was a love, only forgotten for some years.

At my philosophy department, the tendency was towards analytical philosophy. During my years as a philosophy student, I was introduced quite heavily into logic and the idea that there is only one reliable method, namely the scientific method. Underlying that was a strong belief that truth and the good are achievable as long as we are rational; and so the question of what rationality is was and is a basic one. But there were also some other courses that introduced me to authors such as Levinas, Foucault, Derrida, Heidegger, Jankelevich, Bloch, Arendt, and Sartre—philosophers who were identified as "continental," where that adjective was used by the analytical philosophers in a deprecatory manner. It was from some of those lectures that I would get to a basic source for my later research. This was the essay entitled "The Question Concerning Technology" by Martin Heidegger.[4] At my department, Heidegger was considered by most members as one of the most obscure philosophers, one to be avoided. That view is historically anchored in how earlier analytical philosophers, most notably Rudolf Carnap, rejected Heidegger's work as an example of metaphysical pseudostatements.[5] Anecdotal but telling here is perhaps the way one professor introduced Heidegger: they opened *Sein und Zeit* at an arbitrary page and read an arbitrary passage as an argument to show that we could all be happy to ignore such nonsense. These kinds of experiences as a philosophy student taught me at quite a young age that one should not take for granted what some authority is stating for a fact and dare to question it. It was probably there that I started to feel like an outsider to the field in which I was being trained. I refused to accept without questioning the main disciplinary assumptions required to become an analytical philosopher. I refused in a sense what I was being taught.

At that time I had a rather difficult relation with computers and tried to avoid them as much as I could. In retrospect, I see this as a case of alienation in the sense of the French philosopher of technology Gilbert Simondon,[6] flirting with neo-Luddism. But my reading of the Heidegger essay changed my perspective fundamentally and brought me onto the path of computing and programming. Computational technologies can be read as the perfect embodiment of what Heidegger called "*das Bestand*," the "standing reserve." Roughly speaking, this refers to a particular mode of existence in which humans "order" the world as a "standing reserve." This is the essence of modern technology, the particular manner in which we bring things into our world. While on the basis of that analysis one might easily become a technopessimist,[7] the message I took from that work was that there is a way out of this: instead of avoiding such technologies out of fear of becoming part of *das Bestand*, what I should be doing is to try and *understand* those very same technologies: how they function, how they came to be, what is their foundation. If I could have an improved understanding of what happens "beyond" the graphical user interfaces I was using, alienation could become emancipatory, enabling me to see a way out of *das Bestand* rather than a way into it.[8] In a sense I was going through a process of generalized Heideggerian breakdown as developed in *Sein und Zeit*.[9] To put it very roughly, this comes down to the idea that, as long as we are *using* a particular technology, we are unaware of the complex network of meanings and practices that make it possible for that technology to be ready to hand. We need to have this state of unawareness to use it as intended. However, in order to understand and see it *as* a technology, we need to become aware of the complexities that are hidden by its use. This can be achieved when something breaks down. In my case then, going against the user interface also meant that I had to somehow deconstruct it.

But what could that mean, to break through the interface? At the time it resulted in two related but different decisions.

First, I was going to teach myself how to program. That experience had quite a profound effect on me—my first program, the typical "Print 'Hello world,'" was a revelation, and later I would use that to develop an experimental attitude towards computation and programming leading me to the idea of recasting human-computer interactions as conversations.[10]

Second, I decided to go back to some basic historical sources for modern computing. Given the opacity of computing technologies of that time—an opacity that has only increased since then—perhaps going back to the sources would give me some sense, some understanding of our current situation. Steadily, I developed a view of how historical research can help us to make our current situation comprehensible, to see the historical roots and traces of particular problems related to computational technologies and to try and see new opportunities to turn them around, at least locally. To give a very simple but hypothetical example, we could ask ourselves how most of us became a particular type of user of such computational technologies, the so-called end-user, and what this entails. To answer that question what is needed is a longer, careful, and diversified history of the conception*s* of "user" to contextualize historically the most common view of the User as commercialized by Apple and later by Microsoft and Google, anchored in visions of personal computing of the late 1970s and early 1980s.

After having graduated in philosophy, I decided to try and continue along the path of computing and programming, combining my renewed interest in historical thinking and my steadily developing experiences in programming. My PhD project was aimed at the theoretical formalisms of logicians such as Emil Post, Alonzo Church, and Alan Turing. They developed their formal devices in the 1920s and 1930s with the aim of showing that there does not exist a uniform finite method or procedure to solve certain problems relevant to the foundations of mathematics (known as *decision problems*).[11] One speaks here of (recursive) unsolvability results. One of the more familiar formalisms to the reader is probably that of the Turing machine, a formalism aimed at formalizing the intuitive notion of computability. These formalisms were later recast as formal foundations for the modern computer and programming.[12] These impossibility results were, however, quite abstract; and in the proofs one could not find any real indication of the consequences of these impossibility results on actual mathematical practices—and, most important to me, concrete computational practices. The aim then of my PhD dissertation was to understand the connection between these abstract results and the concrete computations of real computing machines.[13] The latter resulted in two different research problems. The first one asked the extent to which the emergence of the first computers could be historically related to these so-called logical foundations. Second, in order to understand "unsolvability" as a phenomenon of concrete computations, I focused in part II of my dissertation on a particular formalism known as tag systems, developed by Emil Post.[14] I was able to prove a number of theoretical results related to solvability and unsolvability of tag systems and conducted a number of so-called computer experiments to explore a series of characteristics of tag systems, which could be interpreted as phenomena of unsolvability. These experiments were much inspired by my reading of Stephen Wolfram's book *A New Kind of Science*.[15]

Today, I look back at that dissertation as a somewhat naive enterprise containing a number of mistakes and even conclusions I no longer support. Nevertheless, much of what I learned and experienced during my doctoral years has been quite basic to my later work. In fact, when I hesitantly went back to parts of this dissertation for the writing of this chapter, I was surprised to find there a preemption of my current stance with respect to disciplines and their boundaries and which I develop in this chapter in later sections (see the sixth section):

As is clear from the outline of this dissertation I have not restricted myself to

> one specific research domain or methodology. [. . .] It is a current trend both of the humanities as of the exact sciences that one needs to specialize into one sub-sub- . . . -sub domain to get anywhere. This is even becoming a harder reality for the younger researchers who seem to have no other choice but to specialize. This evolution I regret. Although I am the last to say that one should not know his or her subject well-enough, otherwise one can only make mistakes, I think being able to cross the borderlines of domains, trying to link them up, is at least as important as this specialization. In my personal opinion, one of the ways to make progress is to have the freedom to mix up domains.[16]

My publications and talks during my postdoc years were varied but did not make a real effort to really open up towards one another the different disciplines in which I had been working. Still, it was unavoidable that my style was not really "pure." I steadily started to feel more like an outsider at whatever conference I attended. That subjective impression was sometimes confirmed by the public. I might meet historians and they would tell me that my work was not *really* historical, driven as it was by concerns with current developments in computing or was being too technical; or I might be attending a philosophy conference and people would perceive me as a historian and not as a *real* philosopher, or would blame me for not following the *proper* philosophical method taking distance from the practice itself.[17] I return to these points in more detail later. It is then perhaps no coincidence that I was hired as a permanent researcher in 2013 by one of the interdisciplinary committees of the Centre National de la Recherche Scientifique.[18]

But while I surely felt and often still feel like an "outsider," I steadily realized that this does not correspond to my reality (anymore): if I really would have been an outsider, how could I present my results in these different communities, how could it be that these different communities accepted my submissions to their journals and invited me to their events? A more appropriate description of my own way of being an academic is that of an academic nomad:[19] someone who travels across different academic communities and is usually welcomed and accepted *as* a nomad.[20]

To conclude this section, I would like to return to the problem I had with accepting to write this chapter: I simply could not write a chapter on the history of computing from the perspective of philosophy given my steadily developed nomadic position with respect to both fields. Instead, I chose the more radical personal style of writing from the perspective of someone who does not subscribe any longer to one particular discipline. In the remainder of this chapter, I would like to develop how this nomadic way of being has steadily resulted in specific methodological stances and their implications for my *personal* view on the field history of computing.

Overcoming Obstacles

As I explained, as a nomad I have travelled across a variety of different academic and sometimes less academic worlds. During that travel, I have met many different people and had to learn how to open up towards and engage with the methods, knowledge, and views of people from different communities. In that process, I learned to detect and uncover my own biases, limitations, and hidden assumptions—an unending process. But I also steadily learned about others, in particular the methodological assumptions and habits as implicitly or explicitly defined by a particular field of research. Such assumptions are mostly productive and are even necessary for the foundation of a discipline. But these assumptions can also become counterproductive when the context in which these were originally put forward has changed significantly.

It is in this setting that I would like to introduce the notion of obstacles inspired by the French historical epistemologist Gaston Bachelard in his book *The Formation of the Scientific Mind*.[21] He introduced the notion of *epistemological obstacles* to achieve what he called a *psychoanalysis of reason*,[22] which aims at revealing epistemological obstacles in the history of scientific thought. Roughly speaking, an epistemological obstacle is a particular way of thinking, of using words, of being educated, which might in the longer term become obstacles to seeing things differently. To put it in Bachelard's words:[23] "An epistemological obstacle will encrust any knowledge that is not questioned. Intellectual habits that were once useful and healthy can, in the long run, hamper research."

I do not develop Bachelard's notion here in full detail[24] but introduce it in a general manner to indicate how a nomadic perspective on *both* history and philosophy of computing enabled me to identify and see more clearly such obstacles and their consequences in my own work and that of others. In what follows, I consider a number of struggles and obstacles I personally encountered during my travels within philosophy and history and indicate how they led me to my current position.

Computing as an Object of Study

At the time when I decided to have computing, programming, and computers as my objects of study, that topic was not a common one within philosophy; and I had to find my own communities in which to present my work. Anecdotal was my accidental meeting with a colleague philosopher in the coffee room of my department. We started to chat about our topics and when I told him about mine the immediate reaction was: "Well, that is quite a special topic especially for a woman."

At the time I was working at the Centre for Logic and Philosophy of Science. In that group, there were indeed connections to computing.[25] First, the topic of logic related quite naturally to the aforementioned foundational work of Church, Post, and Turing. Second, at the time, there was a new subcommunity being established around the philosophy of mathematical practices, a community that explicitly engaged in more historical work, with one key actor being affiliated to my research center.[26] I presented some of my work in these two communities. There was a general interest in the topic but never a strong engagement. Often, I was the only one at conferences who was doing something in connection to computing. In retrospect, I would say that, for many analytical philosophers, computers and computing then belonged to the field of technology and less to that of science, whereas the latter was regarded as superior to the former.[27]

A recurring question then was if and how such a technology could really have a basic impact on, say, mathematics. One could, for instance, wonder how far computer-assisted proofs are really fundamentally different from other, more traditional proofs. As an example, I take a paper by Jeremy Avigad, a well-known philosopher of mathematics who has also actively engaged with computing. In a paper titled "Computers in Mathematical Inquiry," the conclusion could not have been any clearer:[28]

> Ask not what the use of computers in mathematics can do for philosophy; ask what philosophy can do for the use of computers in mathematics. [. . .] What we need now is not a philosophy of computers in mathematics; what we need is simply a better philosophy of mathematics.

To a contemporary historian of technology, such a view might appear alien since it is not uncommon for a historian to look at the effect of a particular technology on discourse and practice without being bothered by essentialist questions such as: Has the nature of proof changed because of the computer? But for many a philosopher, this was not a given, and one needs to develop means to overcome what I would like to call the obstacle of historical contingency. That is, the assumption that the effects of technological developments *are* contingent and can be bracketed when raising what I would call foundational questions such as that of the nature of mathematical understanding. So, one of the objectives I put for myself *after* the PhD was to write on how and why the computer as a technology did not just have a contingent but a basic impact on mathematics.

There, of course, a focus on *practices* was key; and my tentative conclusion was that of a bottom-up approach tracing down characteristics of computer-assisted practices that are considered as specific to the use of the machine.[29] I would soon give up on the topic of computer-assisted mathematics as inscribed within philosophy of mathematics altogether—realizing that I would probably never be able to provide a satisfactory answer to such foundationalist endeavors. Today, I would probably see it as the struggle of someone who, philosophically, subscribed more to existentialism than to essentialism and, by consequence, the rejection of the idea of *a priori* truths altogether.[30]

My approach to this was anchored, on the one hand, in historical research of particular cases in the history of computer-assisted mathematics and, on the other, in my own experiences with computer experiments in the setting of computer science and mentioned in the previous section. The mix of these two approaches was not always appreciated by my colleague philosophers. My supervisor once told me that, we, the philosophers of science, have the task of telling the scientists how they *should* be doing their work. At the time, I was rather surprised by that remark because of the implicit assumption of a basic distinction between, on the one hand the practitioner, and on the other hand the philosopher—where the latter knows what is best. The reason for this was clear: I was no longer myself a pure philosopher since I had engaged with the practice and so, by consequence, could not have that distance anymore. This, I learned steadily, was considered a methodological no-go not only to some philosophers of science but also to historians.

In my ambition to argue for the epistemological impact of the computer and computing in mathematics, I steadily also built an obstacle for myself—one which I call that of the historical singularity. In its more extreme form, it comes down to the idea of a fundamental historical disruption between the times before and after the first computers. We can find an instance of this, for example, in the work of the Dutch computer scientist Edsger Dijkstra when he wrote:[31] "Automatic computers represent a radical novelty and that only by identifying them as such can we identify all the nonsense, the misconceptions and the mythology that surround them."

By the time I was working on computer-assisted mathematics, I had done some significant work on what is known as "one of the first" computers, the ENIAC (see also the next section). My study of that machine and the discourse that developed around it in the work of practitioners such as John von Neumann, Derrick H. Lehmer, and Douglas Hartree provided significant historical arguments for why the first computers were considered disruptive from an epistemological viewpoint. But I also know now that I was misled in taking the claims from the discourse too much for granted and, by consequence, assigned too much historical value to "one of the firsts." It was my collaboration with a historian of mathematics in the setting of a larger historical project on numerical tables in the history of mathematics that helped me to realize that many of the ideas associated with "the firsts," most notably that of a machine being general-purpose, had to be situated in a much longer history of developing more general devices—even when viewed only from the perspective of the history of mathematics.[32] Over the years, my encounters (in person and in writing) with historians of science, technology, and computing in particular,[33] have made this obstacle of the singularity more visible and problematic. Indeed, in the historical discourse, claims about "revolutionary ideas and technologies" are considered suspicious—and rightfully so. The consequence of this encounter with historians thus resulted in a more nuanced view: while I do support the view that computing technologies have and are changing our ways of organizing and living our lives, extreme care is required when critically examining their effects, avoiding a discourse of disruption and singularities.

It is this historical technique of seeking continuations and parallel histories in what appear to be disruptions that might offer something not only to philosophers, but most notably to practitioners, and by extension to non-informed citizens, who tend to blindly accept such discourse of disruption without further investigation. A recent example here is when ChatGPT and other similar machine learning technologies were launched in the programming discourse as being the end of programming (as we know it);[34] and, in the general discourse, as a disruptive technology threatening a wide range of human activities including teaching, journalism, and art.[35] It is in such contexts that I believe historians could play a key role in helping to demystify the discourse of disruption, by pointing, for instance, at the larger business and labor histories underlying such claims or the recurrence of such claims throughout the history of the discourse.

Already during my PhD student years, I realized that analytical philosophy was not the only possible community for me, and I actively sought out others. At the time, I was not too aware of SIGCIS or SHOT,[36] I only subscribed to the SIGCIS mailing list in 2010. I also attended several conferences that focused on Philosophy and Computing, but by far the most important community I joined then was that of the conference series "Computability in Europe," which later resulted in the founding of the association Computability in Europe. A key figure in that setting was Barry S. Cooper. My first CiE conference was held in Swansea, Wales, in 2006; and the welcoming words by Cooper deeply impressed me: the vision was that of an interdisciplinary conference bringing together researchers to think and discuss topics related to computability, whatever their background. That is, *the community was not defined by a discipline but by the topic*. Most participants did have a background in mathematics, logic, or computer science; but they also explicitly welcomed historical and philosophical contributions. Since then, I have attended thirteen CiE conferences and steadily became more and more involved with the association and its conferences. At the time, it was the community with which I felt most at home even if most of the talks were quite far removed from my comfort zone. The reasons why I felt at home there were threefold. First, it was a community that, very early on in my career, showed an honest interest in my work, even if more historical. They had a clear interest in involving what they considered to be "professional" historians and philosophers in their community; and they appreciated my technical knowledge of the topics I spoke about (see the fourth section, below). Second, it was a conference that had the intention of being truly interdisciplinary. That resonated very well at the time with my own stance, which was one of bringing together a variety of methods and disciplines (see the sixth section, below). Third, unlike the other conferences I had attended up to that point, the atmosphere was friendly, welcoming, and generous. It was a key concern of the people behind the conference to be as diverse as possible,[37] a situation that was not the one I was used to in philosophy.[38] Each of these dimensions became key in my own later explorations into community-building (see the sixth section, below). The one in particular I would like to focus on in the next (fourth) section is the first of these reasons mentioned above, which can be seen both as an obstacle to myself and as a means to overcome others: the attention to and insistence on (technical) details.

The Pertinence of Details

As explained earlier, my way into computing was anchored in my interpretation of Heidegger's essay "The Question Concerning Technology."[39] I understood the computer as a technique that embodied the very idea of the "standing reserve." Thus, to work against a mode of existence in which not only the world around us, but also humans would be "ordered" into the "standing reserve," I considered it key to try and understand the historically constructed systems that enabled and at the same time were hidden away in my experience as an end-user. In a certain sense, the process I went through when reading and studying Heidegger's text was one of Heideggerian breakdown: once I understood that my reluctance towards computers was anchored in the fact that I was only using them as something at hand, the need to see the larger system in which they were and are embedded became clear. The GUI itself became an obstacle for me.

While this is not a standard interpretation of that essay, I learned I was not the only one to have developed such a view. Most important here was perhaps Friedrich Kittler, a well-known media theorist who was very familiar with the Heidegger essay and had insisted in some of his work on the illusions created by software and the consequences of those for the so-called end-user. In particular, I was most impressed by one of Kittler's conclusions to the essay "*Hardware, das unbekannte Wesen*" in which he appealed to the philosophers as follows:[40]

> In this way, computers are sold whose architecture is not so much defined by the state of the art but by a pre-history or firm bureaucracy that crystallises into hardware right away. And if the ideal of software

> [. . .] would ever triumph, the bureaucratisation would be perfect: The hardware, in spite of its programmability, would irrevocably be obscured under its packaging. To stop this coincidence from happening seems to be an eminent political goal. If computers are the first machines to reduce the contingency or incomputability of some, though not all futures to a finite degree, its own contingency should remain as open as possible. [. . .] If somebody went and wrote all the programmes hitherto running under the name of philosophy into hardware, that would be the goal itself.

It became a basic source of inspiration, among others, for my later work on ENIAC. For the younger me, the only way to break through these created illusions of the familiar meant to engage actively with the techniques and technology underlying the interfaces with which I interacted as user. Since I had no training as a software designer or programmer, I assumed that the best way into this was to go back to some of the historical origins of the modern computer, assuming that these would be readable and understandable even for a simple end-user like me. As explained before, since I had had some training in logic, I was driven towards the basic papers by Church, Post, and Turing, which at that time, were often viewed as providing the logical foundations to the modern computer. These texts are quite technical and the reading was not easy, but they allowed me to develop my own interpretation of what is today known as the Church-Turing thesis.[41] That interpretation was heavily anchored in my reading of Emil Post's work. That work was essentially a combination of historical and philosophical methods: archival work; understanding the historical relations between these different authors and interpreting their work against the logico-mathematical background of that time to understand their views on the "Church-Turing thesis," And, based on that, to understand its possible relevance (historical and epistemological) for the modern computer and computer science. Clearly, that work required me to engage actively with mathematics and logic.

Shortly after the PhD, I also went more deeply into what was presented as one of the first computers, the ENIAC. That research was largely due to an accidental meeting with another media theorist, Martin Carlé, who was working on an art project around ENIAC in which the simulated sounds of ENIAC's vacuum tubes (and other sounds) would order dancers in a choreography. But to achieve that, reconstructions of ENIAC programs would be required. To me, the idea of understanding how programmed computations can be achieved on a hardware level resonated very well with the aforementioned Kittler quote; and so, together with Maarten Bullynck, I started to study ENIAC. That work was quite "technical" in nature; our aim was to investigate a multiplicity of practices that developed around ENIAC, focusing on the work of several actors who had worked with the machine. The idea was to reconstruct the programs prepared by those actors. Based on those, we wanted to understand the evolution in their practices to show the epistemological and technical effects of using a high-speed programmable machine. Since our aim was to understand the practices that developed around ENIAC, we had to engage actively with the details of the programming techniques and mathematical methods and problems to be tackled with the help of ENIAC. One series of papers focused on the work by Derrick H. Lehmer and contained detailed paper reconstructions of the number-theoretic problem that was set up on the machine. We also did a reconstruction of two conditionals—in current terminology these are "If . . . then . . . else" constructs—on the early ENIAC.[42] These detailed investigations not only provided us with an improved understanding of what it means to run a program *physically*, but also gave insight into how and why a practice is affected by the particularities of a machine, resulting, ultimately, in new machine and programming techniques. One example here are the reflections by Haskell B. Curry—a logician—who, based on his ENIAC work and anchored in his logical background, developed a so-called theory of program composition for the IAS machine, which was quite different from the more famous flow diagram approach by von Neumann and Goldstine.[43]

So, roughly speaking, my earlier historical work can be divided into two clusters: one that focused on the early history of mathematical logic (1920s–1940s) and one that focused on programming practices around one of the first computers. These two clusters were methodologically related through an approach that went deeply into the mathematical and programming techniques involved. The main connection between them was anchored in a question of my PhD, left largely unresolved or underexplored: What, exactly, are the historical and epistemological connections between the theoretical techniques and methods developed in logic and mathematics on the one hand and actual concrete programming practices on the other?

The particularities of that work confronted me with an important obstacle that I had to overcome: the above question already presupposes that logic *did* play a role in the shaping of the computing field and, in particular, in the rise of the first computing and programming practices. I know now that during my PhD I was too biased by a common narrative that claimed the modern computer can be (partially) anchored in logic and mathematics. While I still believe this to be true to a certain extent—roughly speaking, the first computers were built to solve mathematical problems—I started to question that narrative. The technical work on ENIAC was key here: it allowed me to realize that these so-called logical foundations and, in particular, the notion of the universal Turing machine, was in no way "essential" to the conception of the first computers or the notion of programs.[44] More important perhaps was the insight that technical histories *alone* cannot provide a proper explanation for the origin of such narratives. If we want to understand, for instance, where the idea of logic being foundational to computing comes from, one not only needs to study the technical discourse but also the social dimension of a new academic field as it struggles to find a disciplinary identity. The study of the technical discourse enables an understanding of when and how certain logical insights were effective for the technique itself. One could call this the epistemological level of analysis. An example is the introduction of the universal Turing machine concept in the 1950s discourse to tinker with the idea of building the smallest possible computers (see the fifth section, below, for more details); another is the use of constructive type theory to build proof assistants like COQ. But the study of the general discourse in terms of disputes, community-shaping, etc. also enable one to see when and how certain logical insights were introduced to shape and support a particular social identity. A generic example here comes from the so-called "software crisis" in which people like Perlis and Dijkstra relied on a mathematical view on programming to give the field (academic) legitimacy. The realization that I was initially too biased by what appeared to be standard narratives contributed to a general skepticism towards accepted narratives and a conviction that any historiographic method, ultimately, has its own blind spots and biases.

My earlier historical papers were not published in "traditional" venues for historians of computing (e.g., *IEEE Annals for the History of Computing*). Some were published in edited volumes on foundations of computing/philosophy of science; most of the ENIAC papers appeared in special issues resulting from the CiE conferences; and some ended up in venues for history and philosophy of (mathematical) logic. I had regular interactions with historians, but these were mostly European historians of mathematics who have quite a different style from US-based historians of technology. In all of these contexts, it was considered fairly unproblematic to draw historical conclusions based on a detailed study of particular practices.

It was thus quite confronting when I started to meet some "professional" historians of computing and engaged more actively with the standard work in that community. I was reading up on historiography and especially on Michael Mahoney's work. An influential insight from that work was that we should move away from machine-centered histories of computing and consider instead historie*s* that focus on communities of computing, social groups of people who conceive of and envision computing in particular manners.[45] I was also following closely some of the discussions on the SIGCIS mailing list; and I also had some personal encounters that, initially, were quite alienating for me. It appeared that to the "professional" community of historians of computing, the kind of work I was doing was considered "old school," "internalist," "too much history of ideas," "not for historians but for computer scientists," "machine-centered," "lacking social and cultural context," etc. Steadily, I learned about the reasons why my work was classified by some in this manner: apparently, I had missed out on the history of the history of computing, which developed, so goes one official narrative, from an internalist, technical, and machine-centered and hero-centered kind of history writing, written by computer scientists, to more externalist histories that focus on topics such as the programmer's profession; the problem of gender in computer science and programming; the business of programming, to name but a few. All of these topics are basic and each require their specific approaches.

But that need not be a reason for being deprecatory about histories that *do* focus on the practices of programming and machine use itself nor to assume that somehow these more technical developments can be safely ignored when writing, for instance, about the profession of the programmer. A clear argument here was given by the historian of mathematics Karine Chemla. She argued that historians of science also need to acquire a certain level of literacy when dealing with historical sources to understand the different skills that are required by different scholarly milieux.[46] That is, having as a historian a certain level of literacy is intrinsically important to the field of history of science itself.

The deprecation of technical histories was, of course, but a personal experience and should not be generalized. However, it was a fact that, at the time, the majority of history of computing books being published hardly engaged at all with the more technical aspects of computing, programming, and software design with some even explicitly opposing them to highlight the novelty of the work. A confirmation and clarification for this arrived, for me, when in 2014 a discussion ensued on the SIGCIS list around a recent talk by Donald Knuth, a well-known computer scientist and important advocate for the history of computing. The talk was titled "Let's Not Dumb Down the History of Computing." As is clear from that title, Knuth was not feeling very happy about the current state of affairs in professional history of computing as it appeared to him. The feeling was based on his reading of a paper by Martin Campbell-Kelly entitled "The History of the History of Software" and published in one of Knuth's favorite publications, *IEEE Annals for the History of Computing*. In that paper, Campbell-Kelly sketched a development in the history of the history of software whereby earlier publications were mostly technology-focused, identified as being of the "low-hanging fruit variety"; and it was celebrated that, steadily, these kinds of more technical publications disappeared from the field of history of software. Knuth, who is known to have drawn insights from the history of computing and mathematics for his own work, was of course quite upset since, in his reading, the paper implied that we should be happy that technical histories are no longer being written. One conclusion from the discussion on the SIGCIS mailing list was that, of course, current history writing is not in the style Knuth prefers, since it is not realistic for someone who seeks a position at a (US) history department to write in that style *and* find the necessary appreciation for that kind of work at a humanities department. If computer scientists prefer more technical history, then they should make sure there would be funding for such positions at their departments. The problem was thus regarded largely as an institutional and social problem.[47]

From my own experience at the time, however, technical histories are not just relevant from a computer science perspective; and the debate that ensued hinted for me at a number of methodological blind spots within the field of the history of computing itself. Of course, it is true that the "job market" forces researchers to do their research in a particular manner, but that need not mean we should agree to ignore a particular kind of method simply because it does not fit the agenda of hiring committees. What was missing for me in the Knuth passage was a deeper reflection on the potential of a history of computing which, rather than being defensive about particular methods and offensive towards others, attempts to consider the value of a multiplicity of approaches for the field *proper*. Let me explain this from the more technical perspective that is mine, keeping in mind that this is an illustration of the plurality point and not a generalization of the necessity of technical histories. I discuss two cases exemplifying two separate points.

First, there is a significant risk of blind spots and simplifications if one relies on just one level of analysis. I take the example of Nathan Ensmenger's book *The Computer Boys Take Over: Computers, Programmers, and the Politics of Technical Expertise* to illustrate this point.[48] The book's approach is quite clearly inspired by Michael Mahoney's historiographical work mentioned above and explicitly focused *not* on the machines, but instead on the people that built and maintain the software. That decision to focus on the people, however, comes with a method that inscribes the history of (the profession of) the programmer in a social history whereby, "[w]hat might on the surface appear to be disagreements about the particular technical challenges associated with software development were in reality local disputes about organizational power and authority and [. . .] about the peculiar character of the people involved with software development."[49] In other words, technical discussions are recast as discussions about social identity and professional status. This is of course a valid and valuable approach; but in its focus *only* on the social dimension, there is a risk of no longer being able to differentiate between what I called earlier the epistemological level of analysis and the social level of analysis. Indeed, if developments in programming are recast as social phenomena then it becomes difficult to perceive, for instance, the finer-grained differences between *why* different people used or introduced a particular programming technique like object-orientation;[50] or, the realization that the dichotomy between programming as an art or as a science, while perhaps true from an externalist view, breaks down once one engages with the details of the work of some of those supporting the "scientific" view.[51] Moreover, in reference to the above literacy argument of Karine Chemla, a focus only on the social level of analysis without engaging with the more technical discourse makes it hard if not impossible to diversify between the different skills required or assumed by different (historical) conceptions of the profession of the programmer. To be clear, this is not intended as a criticism on this book *per se*, but rather the expression of a general worry: if one sticks to just one level of analysis *and* assumes that other levels of analysis can be reduced to it or can be considered irrelevant to the research questions one wants to resolve, one might end up with a narrative that orders and represents concepts and techniques in a particular manner with the risk of ignoring other possible orderings that would contribute to a more completed picture.

Second, if one's level of analysis largely ignores the technical discourse, then it is quite unavoidable that you might have to rely on central concepts and notions that were historically developed in a more technical and historically or geographically distant discourse, which you simply take for granted in their more contemporary meaning without further study. Examples for the history of computing include: "algorithm," "code," "software system," "computability," "A.I.," and "automatic programming." I would like to briefly focus here on the notion of algorithm. Historians, when using that notion, in fact rely on a very recent and particular interpretation of that concept that essentially comes down to the idea of having a finite step-by-step method to solve some problem, inherited essentially from the way Donald Knuth and the ALGOL community introduced it in the US programming discourse. The comparison with recipes is never far removed. While perhaps one could then safely ignore the complex semantic and technical history of that notion when it is used in passing only, this becomes more problematic if the concept is a key notion in the narrative being developed. I illustrate this here with one example.

Common semantic associations with the commonsense understanding of algorithm is that of control, formality, and solvability; and this is also the manner in which the notion is introduced in the book *How Reason Almost Lost its Mind: The Strange Career of Cold War Rationality*.[52] In that work, the notion of algorithm is given a central role to describe the rise of so-called Cold War rationality in the US context.[53] The elaboration of the notion in the monograph *Theory of Algorithms* by the Russian mathematician A. A. Markov is mentioned in passing only. During my PhD years I had explored, to some extent, writings from Russian mathematicians and logicians from the 1940s to 1960s and noticed the prominence of the notion without really understanding where it was coming from. This was even more interesting to me since it was largely absent from the Western logical discourse: neither Post nor Turing really make it a topic and use other terms, while Church used it merely in reference to mathematics. Moreover, in the first one or two decades of the Anglo-Saxon history of the modern computer, "algorithm" is also largely absent from discourse and replaced by the more common "method." It would be many years later that I would finally have an answer to the question of the introduction and use of the notion in the Russian-speaking context thanks to a collaboration with two historians, Ksenia Tatarchenko and Anya Yermakova. A key outcome of that collaboration was to learn that an explanation of "algorithm" in its original Russian-speaking context requires a longer-term perspective that does not fit with the US perception of Cold War rationality. To the contrary, it fits much more with what one could call a humanist ideal of computer literacy.[54] That outcome was only possible because of our integrating of institutional, social, cultural, and more technical histories focusing on the content of certain logical and mathematical work. In other words, the US narrative of Cold War rationality was counterbalanced by an alternative history of *algorithm*, which is more about humanistic values and visible only through the integration of a number of different approaches.

Let me now briefly return to my own biases and the Knuth passage. When I made my first steps into the history of computing, my preference for a more technical history writing went hand-in-hand with a certain bias that required correction by integrating different levels of analysis and, in the case of logic, to try and carefully differentiate between what is socio-historically constructed and what is epistemologically relevant as much as possible. The latter level of analysis is one that is heavily anchored in my personal background in analytical philosophy but also in my regular meetings with computer scientists in my first years. The former is one I steadily learned to accept thanks to my meetings and discussions with a variety of historians, but also to my meetings with computing people who simply do not care much about logic when it concerns their practice. The Knuth passage then for me was both a confirmation of a situation I had experienced myself but also a disappointment: instead of having a serious debate about the possibility and desirability of integrating or, at least, complementing of different levels of analysis within the field of history of computing *proper*, the discussion resulted in a conclusion of an academic division of methods due to institutional and social realities. The need for such a methodological debate is, I believe, partially anchored in the diversity of computing as a practice and as a concept to which I return below. It is one towards which the current book is making a significant contribution.

Presentism or Why Are We Writing Histories of Computing?

In the SIGCIS discussion on Knuth's lament about the lack of technical histories in the history of computing, Bill Aspray wrote:

> This discussion is not new. When the history of science began to professionalize after the Second World War, at first the scientists (mostly

> physicists) welcomed them with open arms. But the honeymoon soon ended when they found that the historians were not going to simply be apologists for the scientists but might have their own set of interests and interpretations that did not square with those of the scientists.[55]

Indeed, the professionalization of the history of science was quite heavily anchored in sometimes virulent fights with the scientists whose fields professional historians were attempting to study.[56] In a recent paper by the historian of science Kostas Gavroglu titled *The Sisyphean Fate of History of Science,*[57] Gavroglu describes this professionalization of the history of science and the disciplinary fights that were part of that process in mythological terms referring to Sisyphus. Sisyphus was punished by the gods because he had cheated death not once but twice. The professionalization of the history of science, despite the resistance from scientists, is then described as history of science having tricked death a first time.

This historical passage, of course, could not have happened to the history of computing in the 1970s since the field did not exist yet. That is, at the time, it was mostly still computer scientists who were writing their own histories.[58] But, as suggested by Aspray, the Knuth passage could be read as an instantiation of such passage in the history of computing. It is in such settings that historians of science and technology developed a rather high sensitivity against what is known as anachronisms or, more specifically, Whiggish history. In its more extreme form this comes down to rewriting the history of field *x* in terms of field *x*'s contemporary terminology. In general, it is the idea of writing history with reference to the present. Anti-Whiggism then is about "treating the past on its own terms as much as possible, and not simply as a runway aimed at the present," as Michael Gordin described it in his lengthy review of Hasok Chang's presentist *Is Water H2O?: Evidence, Realism and Pluralism.*[59] Such opposition against Whiggism is almost natural to historians of science and technology, that is, "we have wired it into the central core of our field *as a discipline.*"[60] Chang, in the reviewed book, proposed a presentist view on the history of science under the banner of *complementary science.* Roughly speaking, it is the idea of learning from our pasts for current scientific practices, as a means to ask questions that are excluded from current practice; to encourage re-examining accepted wisdoms; and for showing how certain truths and methods have been accepted as common sense. Chang, however, is not a pure historian of science but part of the school known as integrated history and philosophy of science. The inclusion then of a more presentist view can be seen as a normative dimension anchored in certain philosophical traditions.

Not knowing too much about all this several years ago, I was not very conscious that, in fact, I was, to a certain extent, a presentist too nor of the reasons why this might be a problem for the "pure" historian. But then I had an encounter with a well-known historian of computing who confronted me with this "issue," telling me that it is bad practice to write history in function of the present. Initially, that comment came as a surprise to me; and I was upset if not offended by it. However, I understand now that (1) they were in fact right but only to a certain extent and (2) that this was perhaps an issue for the historian but not for me, a nomad, who clearly cannot and does not want to escape fully her own education as a philosopher.

So, what does my presentism entail? It surely has certain aspects in common with Chang's presentism: a deep motivation for doing history of computing to critically think about our current digital condition. In that sense, one could say I give purpose to a history that it surely did not have at the time it was happening. Emil Post, when writing about solvability, surely was not thinking about the limits of the modern computer; and Lehmer, when working around ENIAC, was surely not using a notion of "program" that can be equated to our current notion. By doing this kind of history writing in function of today, it is unavoidable that I have sinned against the hard-wired anti-Whiggism that is part of the general fields of history of science and technology and, by consequence, also history of computing. I have chosen particular topics and methods and ignored others because of my presentism and probably have sometimes fallen into the trap of describing things from the past in current terminologies.

Let me elaborate a bit more on this sin. As described before, my motivation for doing history of computing is anchored in the belief that our current situation is problematic and that we need a deeper, critical reflection to change that situation. It is a normative dimension comfortable for philosophers, less so for historians. For me, it is about investigating the conditions which have led to most of us having become illiterate end-users, an investigation that, rather naively perhaps, started with a more technical history writing on the so-called logical foundations and the first computers. To put it differently, for me, doing history serves another purpose that I prefer to call *political*. In general, it is the idea to render transparent computing by showing how it is shaped by its material, economic, social, notational, and other practices and conditions to go against certain "ideologies" that I consider to be dangerous. It implies also that, when I am writing about the history of computing, I do not have in mind only a readership of fellow historians but people who share with me similar concerns, whatever their background, thus also including, but not exclusively writing for, programmers and computer scientists.

In most of my historical work, the "sin" manifests itself most clearly when introducing certain new concepts to study a longer historical development. One example is given in the paper "Less Is More in the Fifties: Encounters between Logical Minimalism and Computer Design during the 1950s,"[61] that was briefly mentioned above. *My* co-authors and I introduced the notion of logical minimalism to tie together and compare certain developments in mathematical logic from the 1920s to 1940s with engineering practices from the 1940s to early 1960s. This was to relate developments in mathematical logic to tinkering about the minimal number of components to build circuits and computers. One important methodological challenge of that work was to understand the transformations and transpositions that certain logical techniques had to undergo when introduced in the discourse of engineers.

A more recent example is given in a forthcoming chapter, "Notations. There is no escape," written by the PROGRAMme collective.[62] In that chapter, we provide a framework to think about the meaning of programming beyond the common "programming in a programming language" to include also, for instance, programming with visual icons or prompt engineering. To that end, we introduced the notion of *notational programmability* as a means to compare different programming notations relative to their programmability, keeping in mind that any such notation always needs to be seen in its own local historical context. It is used as a positive explanation for the explosion of programming notations instead of the more negative "Tower of Babel" metaphor that is often used in the programming discourse. It enables us to conclude that any attempt at fully automating the programmer has significant implications for the user of the notation. That chapter spans a history that starts around 1800 and ends in our current times. It clearly makes the sin of presentism despite the rather extended historical scholarship on which we have relied for the development and elaboration of the concept. But it is one that had to be made for the purpose of the chapter, since it allowed us to evaluate developments in the past and current discourse from a very different angle, one that is historically motivated and informed. That is, while we do use a new concept anchored in concerns with present developments, it is that very same concept that also allows us to take distance from the current discourse to place it in a non-linear and complex historical development of programming notations.

These two examples hopefully give the reader an impression of my presentism. I know from experience that "pure" historians might struggle with it. Typical replies to presentations of these works at conferences with historians in the public are: But are you saying that practitioners themselves were guided by notational programmability in their choice for or reliance on a certain notation? How can that be true since they never used such a concept explicitly? Is it not misleading to speak about logical minimalism with reference to two very different practices with neither of them using that concept explicitly? These are criticisms that, by now, I respect and understand, but at the same time, given their frequency, have also made me wonder about more hidden Whiggisms in the works of "pure" historians. I think here, for instance, of the examples from the previous section, that is, the use of current terminologies from the technical discourse (often that of Big Tech) without taking into account the rich and more diverse histories of these very same terminologies.

Besides the perhaps too obvious argument that no one can wash their hands in complete innocence here,[63] the more important question that I want to raise here is in how far our anti-Whiggism has become too much of an obstacle, anchored too much in our own pasts and our struggles for disciplinary independence. It is here that I would like to briefly return to Gavroglu's paper and the possible second death of the history of science and, in general, the social sciences and humanities. He argues that our field is under attack again, this time perhaps less by the scientists but more by the politics of academic funding. More than ever, decisions about how to spend research and education budgets are guided by instrumentalist short-term visions in terms of job markets and economic impact. We might call these short-term presentist views. The humanities and social sciences are the first victims to fall: they have to survive by focusing only on the immediate societal returns from investments in science and technology.[64] This is of course a presentism that should be avoided since it does not consider or take into account our fields as being fields *on their own*. Interestingly, Gavroglu sees a way to trick death a second time by refocusing the social sciences and humanities on our digital condition and the threats and challenges it poses to our identities, our freedoms, our thinking, our educational systems, our democratic political systems:

> In the past 200 years democracy has been threatened by wars, dictatorships, coups d'état, famine, pandemics, and civil strife. But never before has it been threatened by unseen, immaterial, and mathematical entities called algorithms that dominate our everydayness. We need to acquire a deep understanding of this historically unique state of affairs. What other arguments do we need in order to let the social sciences and humanities chart their own research strategies instead of constraining them to produce "useful" results?[65]

While one might well argue that this should not be the *only* raison d'être for our fields, it surely is a viewpoint that could be picked up by historians of computing to contribute to the kind of "deep understanding" Gavroglu was referring to.[66] It would mean that perhaps we should dare and critically question our old defenses, anti-Whiggism included, which were effective perhaps in the 1970s and 1980s but might be less effective under this new attack. This is not to say that we should get rid of all our previous standards, but rather to have a serious discussion on what could be an acceptable presentism from a historian's viewpoint, if any.[67] This is not to imply that my personal presentism is on the right track and there are certainly other options to explore. One interesting path was offered quite recently in the monograph by Thomas Haigh and Paul Cerruzi.[68] In an extended methodological reflection on the use of grand narratives, Haigh explicitly considers the risks of Whig history in a setting that describes the history of the computer as that of the computer steadily becoming (practically) universal and indicates how its unacceptable form can be avoided.[69]

The Computer Is Not One Thing but Many Different Things

As explained earlier, already during the writing of my PhD thesis I was quite convinced that computing requires a multi-disciplinary approach. Clearly, my presentism results in an approach that combines historical research with philosophically inspired views and insights from the practices of computing and programming. But these should be seen against the background of a much broader multidisicplinarity, anchored in two well-known and related views:

1. computing and/or programming itself cannot be reduced to any particular existing field of research (e.g., mathematics, literature, engineering, logic, etc.). This is a view that has received some acceptance in the history of computing field itself. It was most famously captured in the following quote by Michael Mahoney: "The computer is not one thing but many different things and the same holds true of computing."[70] As is known, in the history of the computing field proper, there have been quite some discussions about the nature of the computing field as a discipline going from more mathematical conceptions to engineering views and multidisciplinary views.[71] But there have also been other views such as Donald Knuth's conception of programming as a literary activity;[72] or Warren Sack's historical contextualization of computing and programming from an arts and humanities perspective.[73]

2. the technology of computing has steadily become ubiquitous.[74] This generally refers to the idea that computational technologies have steadily pervaded our lives (scientific, personal, social, etc.). It implies that a study of computing and programming should not and cannot belong to just one field of study.

These two perspectives naturally lead to a study of computing and programming that is open to a variety of disciplinary and methodological perspectives. Among others, it provides an additional argument for a broader methodological debate within the history of computing proper about the need and potential of bringing together a variety of historiographic methods, which is one of the aims of this volume.

But how can one achieve such diversified perspective on computing and programming as an individual? Even as a nomad, one must be realistic and know that no individual can ever achieve that on their own, determined as we are by that path and the limits of time. One possible path then towards such a methodological and disciplinary diversified perspective is through collaborative community building. In the last ten years I have dedicated much of my professional time exactly to this kind of activity, to the detriment of others.

A first basic step in that direction was in collaboration with Giuseppe Primiero, a philosopher and logician who worked at the same philosophy department as I did as a postdoc. Just like me, he felt that we needed a space in which we could combine and mix different views and approaches. During a meeting in the coffee room of the philosophy department, we started to talk about some shared frustrations of that time—this was in 2009—and we developed the idea of organizing a conference on the history and philosophy of computing that would explicitly involve also more technical talks and views by trained computer scientists and programmers. The aim was to think together about the computing field without assuming that one field would be superior to another. The first HAPOC conference was a fact in 2011 and, to our feeling at least, it was also a success. The leading quote of that conference was the above-mentioned Mahoney quote.[75] HAPOC soon developed into an organization under the umbrella of the IUHPST (International

Union for History and Philosophy of Science and Technology) and in particular its two divisions DHST and DLMPST.[76] We steadily built up a number of recurring events including the biennial HAPOC conference (odd years) and the biennial symposium on the History and Philosophy of Programming (HAPOP, even years). In the main HAPOC conference we have invited well-known historians of computing[77] alongside philosophers, media theorists, artists, and computer scientists. One key feature of this "community" is that people who identify with it often have rather clear disciplinary and methodological stances but enjoy and see a need for opening their minds to those of others from other milieux.

But HAPOC also has its shortcomings: besides the fact that sometimes talks might be so far removed from someone's interests and competencies that they become incomprehensible to some, the conference really is about *pluri*disciplinarity. That is: what we achieved was the creation of a space where people with diverse backgrounds feel relaxed enough to share their views and research results with a diverse public, hoping that this would be enriching for the different participants. But going from there to a collaboration that integrates the different fields and methods is quite a different step. This has happened occasionally in the wake of HAPOC but was not its core business.

A step in that direction was made with the PROGRAMme project, which brings together a variety of researchers (programmers, historians, logicians, artists, media theorists, computer scientists, philosophers, etc.) to try and work together *as a collective* on one and the same question: What is a computer program? The rationale behind that project has been from the start the idea that conceptions and understandings of programs really depend on the historically contextualized practice in which one is designing, programming, studying, and using programs. To put it quite bluntly: a user of phone apps is not the same user as that of ENIAC, and they will have very different conceptions and understandings of programs. To capture these differences, the project has been defined around four core modalities of programs. As abstract objects, programs are open to scientific research; as textual objects, they are notational; in the sense that programs are stored and executed on a computing device, programs are physical; and to the extent that programs are made, shared, and used in our human world, they are socio-technical. These modalities enable a range of different views and ways of handling programs. They are reflected in several different fields that focus on programs, both within the humanities and the computing field, each of which tend to prioritize one of the modalities. For instance, to view programs as text is the approach favored in the broader domain of critical code studies but also corresponds to the way programmers often view programs when "coding" them.[78] In the project, rather than favoring just one of these modalities and the methods that come with it, the idea is to look at programs from the four different modalities *together*. This is rooted in the idea of breaking through illusions created by one perspective only and to open the possibility of other configurations. It implies, among others, that standard narratives are not taken for granted but critically examined. We call this an anti-disciplinary approach, not in the sense of going *against* disciplines but rather to develop a way of thinking that is anchored in a variety of disciplines and approaches but also transcends them by challenging accepted methods and views by confronting one with the other. A good example here is the chapter that was mentioned earlier, "Notations. There is no escape." For some in the group it was initially considered unconventional that this chapter would not have "programming languages" as a main topic. But we did not want to start from such a standard notion or conception that history has given us, but instead challenge it by focusing instead on the very different notion of notations and their programmability. That chapter was written and conceived by a PROGRAMme subgroup, which included three historians (or four, if I count myself as a historian), two computer scientists with different backgrounds, a media theorist, and a manager turned philosopher. It took around three years, and numerous online and in-person meetings, to finally arrive at a text that was acceptable to all. This is to indicate that such collaborations are not evident or straightforward but assume significant commitment of those involved.

From a history of computing viewpoint, one might of course wonder what there is to gain from such a mixed approach, given the time and commitment it requires. Here I can only speak from a personal perspective. For me, the gain is in the confrontation not only with other methods but also other insights, other problems, other facts that I simply would not have known about or have been able to perceive. It has opened up my view on the literature and the sources, allowing me to see certain passages in the history of computing in a different light; it has forced me to focus more on the bigger picture than on the particular details of a number of specific cases; and it has made me realize quite explicitly some of my own obstacles and biases of which I described some in this chapter.

Concluding—Opportunities as a Nomad

As announced at the start of this chapter, I develop here my personal views on the history of computing based on my academic nomadism. The choice for a (semi-)autobiographical style was a logical choice: in order to convey to the reader *why* I am writing and performing history in a particular way, it was necessary to be explicit about its relation to my own travels across the lands of history, philosophy, and practices of computing. But that choice was also a difficult one: the style requires a certain honesty towards one's own choices, which one can easily hide in "conventional" texts, making me, as an author, much more vulnerable. That is: I had to make the effort to understand how my own arguments, how my own work, is also shaped by my own personal experiences in academia. It means that I had to try and identify my own epistemological obstacles and make them explicit to the reader. These are my weak points and my blind spots as a researcher.

It is then my hope that some readers of this text can gain from such perspective to see the potential, especially in teaching, of contextualizing one's own methodological and other assumptions. This, to be clear, does not result in a relativist, or worse, nihilistic historiographical position in which *anything goes*.[79] To the contrary, the appeal lies in the identification of obstacles, among others, by trying to work together, despite what might appear as fundamental methodological divides that cannot be overcome. In particular, I argued for the significance of bringing together a diversity of methodological approaches in history of computing and questioned the anti-Whiggish stance that we often find in the more general field of the history of science and technology.

I also briefly sketched my own alternative approach which tries to bring together several different fields and methods and which requires significant efforts into community-building. The experience with such cross-disciplinary collaborations has given me the courage to write this rather personal self-reflective chapter; to be explicit about some of my own obstacles and biases; and to be honest about and share my personal views on the history and philosophy of computing. It has enabled me to challenge conventions about what it means to be an academic and, instead, to experiment with new formats, other styles of writing, new ways of doing research despite all the fears I might have had and still have for being rejected because of that stance. Most importantly perhaps, it has led me to the conviction that what we need today is not an intensification or confirmation of disciplinary and methodological boundaries by working against one another. That would only confirm and encourage a political system that celebrates a naive and ideological vision of neo-Darwinism applied to research and education.[80] Instead, I hope that in the future there will be more opportunities to explore ways of working across such boundaries together not only to achieve more on a scientific level but also to achieve a change locally to our current and historically developed academic situation.

Notes

1. Supported by the ANR PROGRAMme project ANR-17-CE38-0003–01.

2. The potential of this style of writing has been discussed within historiography. I refer for instance to Jaume Aurell (2006): "Autobiography as Unconventional History: Constructing the Author," *Rethinking History: The Journal of Theory and Practice*, 10:3, 433–449. Aurell distinguishes between two styles, the constructivists and the experimentalists. The current chapter then would belong to this second style of historical autobiographies. As Aurell writes about this style: "The role that the historians assume as authors of their own lives challenged the norms of objectivity that rule their profession, and provide them—if they are engaged with the epistemological issues—a privileged platform for renewing and revitalising their conception of history" (Aurell, idem, p. 444).

3. The official title of my study was "Kunstwetenschappen," "Art sciences" in English. The main focus, however, was on art history. The first two years were a mix of art sciences, history, and archeology; during the last two years we had to choose a specialty. I chose Visual Arts and Architecture and did a master's thesis on linear perspective.

4. Martin Heidegger is a well-known German philosopher who has been very influential amongst others on some basic French thinkers like Derrida or Foucault. The essay mentioned is considered a key source in the philosophy of technology. It is known that

Heidegger supported the Nazi regime. In April 1933, shortly after Hitler had become chancellor of Germany, he was elected as the rector of the University of Freiburg. He accepted that position and soon after became a member of the Nazi Party. About one year later he offered his resignation but remained a member of the Nazi Party until the Second World War ended. This chapter is of course not the place to discuss how to handle an influential philosopher who was also a Nazi and, in particular, the question of the relation between Heidegger's Nazi politics and his thinking. There have been recurring and heated discussions on this amongst renowned philosophers like Derrida and Jankelevich and I refer the reader to these.

5. Carnap famously called metaphysicians like Heidegger "musicians without musical ability" in his most significant critique: Rudolf Carnap (1931), "Überwindung der Metaphysik durch logische Analyse der Sprache," Erkenntnis 2, pp. 219–241 (English translation: Rudolf Carnap); and "The Elimination of Metaphysics Through the Logical Analysis of Language," in A. Ayer (ed.), *Logical Positivism*, New York: The Free Press, 1959. The critique of Carnap on Heidegger went quite far and is often seen in the history of philosophy as indicative of the larger divide between "continental" and analytical philosophy. See, for instance, chapter 6, "A Case Study in Misunderstanding: Heidegger and Carnap," in Simon Critchley, *Continental Philosophy: A Very Short Introduction*, Oxford University Press, 2001.

6. Alienation is a well-known term in Marxist and Hegelian philosophy. Simondon, a well-known French philosopher of technology used it in his *On the Modes of Existence of Technical Objects* (Gilbert Simondon, *De mode d'existence des objets techniques*, Editions Aubier, 1958; translated to English as: Gilbert Simondon, *On the Modes of Existence of Technical Objects*, translated from the French by Ninian Mellamphy with a preface by John Hart, University of Western Ontario, 1980. A more recent translation is: Gilbert Simondon, *On the Modes of Existence of Technical Objects,* translated by Cecile Malaspina and John Rogove, University of Minnesota Press, 2016. In that work he speaks of how humans have become alienated from the machine caused by, amongst others, their lack of understanding of the machine. As he wrote: "Culture behaves towards the technical object much in the same way as a man caught up in primitive xenophobia behaves towards a stranger. This kind of misoneism directed against machines does not so much represent a hatred of the new as a refusal to come to terms with an unfamiliar reality. Now, however strange this reality may be, it is still human, and a complete culture is one that enables us to discover that this stranger is indeed human. Still, the machine is a stranger to us; it is a stranger in which what is human is locked in, unrecognized, materialized and enslaved, but human nonetheless. The most powerful cause of alienation in the world of today is based on misunderstanding of the machine" (Idem, p. 1).

7. This is how the Heidegger essay is often interpreted in philosophy of technology.

8. Essential in my interpretation of the Heidegger essay is his use of a quote from the German poet Hölderlin: "*Wo aber Gefahr ist. Wächst das Rettende auch*" (English translation: "But where the danger is, also grows the saving power") from the poem "Patmos."

9. Ironically, I was doing the exact opposite of how Heidegger's earlier work, and in particular *Sein und Zeit*, has been interpreted in the setting of human-computer interaction. For instance, Mark Weiser of Xerox PARC, who developed the notion of ubiquitous

computing in the late 1980s, was heavily inspired by Heidegger's earlier work when arguing for the need to completely hide the computer for the user to turn it into a silent and invisible servant, that is, "only when things disappear [. . .] are we freed to use them without thinking and so to focus beyond them on new goals." That view, however, entirely ignored the essential notion of Heideggerian breakdown. The result was the concept of a technology that ides itself fully and went hand-in-hand with the concept of proprietary hardware and software. [Mark Weiser, The Computer for the 21st Century, *Scientific American*, 265, 3 (September 1991), pp. 94-104. The citation is from p. 94.]

10. Liesbeth De Mol, "How to Talk with a Computer? An Essay on Computability and Man-Computer Conversations," *Off Topic: Zeitschrift für Medienkunst der KHM*, vol. 1, 2008, pp. 80–89; and Liesbeth De Mol, "How to Talk with a Computer? An Essay on Human-Computer Conversations Continued," unpublished, available at: https://hal.univ-lille.fr/hal-01305968/document

11. A decision problem is, essentially, the problem of finding a uniform and effective procedure for solving every instance of that problem. An example of a solvable problem is the problem to decide for any two integers x and y whether or not y is a divisor of x. Indeed, there exists at least one finite procedures (we would call these algorithms today) to solve every instance of that problem. An example of an unsolvable problem is to decide for any enunciation in first-order logic whether or not that enunciation is derivable or not in first-order logic. This latter problem is the most famous version of the Entscheidungsproblem.

12. The Turing machine model was introduced in the computing discourse in the 1950s and early 1960s as a formal model of the modern computer; the formalism of Post and Church were more relevant in theoretical reflections on programming. Church's formalism known as lambda-calculus is today still a key formalism for people working in formal verification; Post's production systems were picked up in the foundations of compilers as well as in certain early developments of AI (most notably the work of Newell and Simon). However, as has been shown elsewhere (see, e.g., L. De Mol, M. Bullynck, and E. G. Daylight, "Less is More in the Fifties: Encounters Between Logical Minimalism and Computer Design during the 1950s," *IEEE Annals for the History of Computing*, vol. 40, no. 1, 2018, pp. 19–45), extreme care is required from the historian here to understand that the use of such formal devices in a different discourse (e.g., programming) involves a complex process of transpositions often resulting in a rather weak connection between the original discourse and the target discourse.

13. The title of the PhD thesis was: "Tracing Unsolvability: A Mathematical, Historical and Philosophical Analysis with a Special Focus on Tag Systems." It was defended on May 23, 2007 at Ghent University. My jury consisted of two analytical philosophers (Erik Weber and Jean-Paul van Bendegem), one logician (Diderik Batens), and two theoretical computer scientists (Maurice Margenstern and Martin Davis).

14. Initially, I hoped to rely on tools of visualizations using fractal geometry but realized quite early that these would not get me any further. That initial research did result in my first published paper: L. De Mol, "Study of Fractals Derived from IFS-Fractals by Metric Procedures," *Fractals*, vol. 13, no. 3, 2005, pp. 237–244.

15. Stephen Wolfram, *A New Kind of Science*, Champaign, IL: Wolfram Media, 2001.

16. Liesbeth De Mol, "Tracing Unsolvability: A Mathematical, Historical and Philosophical Analysis with a Special Focus on Tag Systems," PhD thesis, Ghent University, 2007, p. 19.

17. The only public where I ever really felt at home was a public of artists who think less in terms of disciplines. My meetings with that community showed me the way of understanding and looking at research in a very different manner.

18. In particular this is Committee 53, *Sciences en société: production, circulation et usages des savoirs et des technologies* (Sciences in society: Production, circulation and uses of knowledge and technologies). It consists of researchers from different disciplines, from the humanities as well as from the "exact" sciences. Unlike other commissions, the "auditions" (interviews) were in front of the full committee.

19. The philosophically informed reader might see here an implicit reference to the concept of nomad science as developed by Deleuze and Guattari in their *A thousand plateaus* and, in particular, the *Treatise on Nomadology—The War Machine.* Andrew Pickering used that concept to describe cybernetics (Andrew Pickering, *The Cybernetic Brain: Sketches of Another Future*, Chicago: University of Chicago Press, 2010). In that sense, there are probably some similarities to be found. Since I am not a specialist, however, on this work, I did not want to fully engage with that concept and rather use the notion of nomad in a common sense manner.

20. The above description of my personal journey might create an impression that it was always my intention to be a nomad. That, of course, is not true and but a projection in retrospect. Reality is that it took me many years to realize that my "home" is not and should not be in any particular field of research and that there are opportunities in the academic system for researchers like me.

21. Gaston Bachelard, *The Formation of the Scientific Mind: A Contribution to a Psychoanalysis of Objective Knowledge*, Manchester: Clinamen Press, 2002. Originally published in French as: Gaston Bachelard, *La formation de l'esprit scientifique: Contribution à une psychanalyse de la connaissance objective*, Paris: Vrin, 1938.

22. Ibidem, p. 29.

23. Ibidem, p. 25.

24. Bachelard's analysis was focused first on science and not on history and philosophy *per se*. Moreover, one should keep in mind the times when this work was being written (1930s), when many philosophers and epistemologists had a rather positivist and progressivist view on the sciences, which I do not share.

25. Initially, I was affiliated to another group (Laboratory for Applied Epistemology). However, over the course of my first year, I encountered several basic and unresolvable problems with my then supervisor and was happy when someone else at the department was willing to take over as a supervisor.

26. This was the philosopher of mathematics Jean-Paul van Bendegem. There were other topics that naturally related computing to logic and philosophy of science. For instance, philosophy of language is a classical field in philosophy of science which has some clear connections to programming languages; the question of valid reasoning had strong ties with certain developments in AI, etc. These topics, however, were not really a theme in my research environment of that time.

27. Of course, there were some exceptions. Most notably here was Peter Galison's work which, among others, showed how technology and science cannot be strictly separated from one another. The interest in philosophy of technology has increased over the years. The so-called Delft school of the philosophy of technology has been quite influential here in its study of technology from a more analytical perspective. See, for instance: https://plato.stanford.edu/entries/technology/. Note that there has been a similar development in history of science and history of technology. The latter was initially treated rather deprecatory by the community of historians of science. On an international level it is telling that the Division for the History of Science added "Technology" to its name only in 2013 to become the DHST.

28. J. Avigad, "Computers in Mathematical Inquiry," in P. Mancosu (ed.), *Philosophy and the Many Faces of Science*, Oxford University Press, 2008, pp. 302–316.

29. L. De Mol, "The Proof is in the Process: A Preamble for a Philosophy of Computer-Assisted Mathematics," in M. C. Galavotti, D. Dieks, W. J. Gonzalez, S. Hartmann, T. Uebel, and M. Weber (eds.), *New Directions in the Philosophy of Science*, Cham: Springer, 2014.

30. A well-known phrase in this setting, due to Jean-Paul Sartre, is "existence precedes essence." For me it comes down to the idea that truth is always relative to one's own historical and cultural context.

31. E. W. Dijkstra, "On the Cruelty of Really Teaching Computing Science, *Communications of the ACM*, 32, 1989, pp. 1398–1404.

32. L. De Mol and M.-J. Durand-Richard, "Calculating Machines and Numerical Tables—A Reciprocal History," draft version available here: https://hal.univ-lille.fr/hal-01396846. Note that this chapter will be published in a volume edited by D. Tournès but that it was written in the period 2011–2013.

33. I think here, for instance, of the work by historians like John Agar: *The Government Machine: A Revolutionary History of the Computer*, MIT Press, 2003; and James R. Beniger: *The Control Revolution: Technological and Economic Origins of the Information Society*, Cambridge: Harvard University Press, 1986, which contextualize the rise of the modern computer in a longer and broader socio-historical development that relates to the reorganization and systematization of information processing. Another important source for me was Michael S. Mahoney, "The Histories of Computing(s)," *Interdisciplinary Science Reviews*, 30:2, 2005, pp. 119–135. In that paper Mahoney considers two different models of viewing ENIAC. The first is the machine-centered view assuming a more singular view on ENIAC, the other is a model from the perspective of communities of computing in which ENIAC becomes but one of many parallel developments.

34. Matt Welsh, "The End of Programming," *Communications of the ACM*, vol. 66, no. 1, 2023, pp. 34–35.

35. See, for instance: https://www.theguardian.com/commentisfree/2020/sep/08/robot-wrote-this-article-gpt-3

36. SHOT is the US-based Society for the History of Technology, founded in 1958. SHOT has a tradition of Special Interest Groups, with SIGCIS being one of them. SIGCIS stands for Special Interest Group in Computing, Information, and Society and was founded in 1987 as a more informal Special Interest Group on the History of

Computing. Today it is probably the most important community of historians of computing. Central is the SIGCIS mailing list and the annual meetings of the SIG at the SHOT conference.

37. To give just one clear example: the people behind the series have made significant and successful efforts to have a good gender balance at the conference by aiming for a fair number of female keynotes and a fair number of female PC members. When, for instance, during PC discussions an imbalance seems probable, people have no reluctance to interfere and make sure the balance is kept. Since 2007, they also have an annual Women in Computability workshop, providing a space for senior women to speak about and share their own experiences; there is a mentoring program and a grants program.

38. Indeed, philosophy in particular is reputed not to create such friendly atmospheres and was, until a couple of years ago, known for its sexism with, among others, plenty of conferences without any female keynotes. Luckily, that situation has steadily changed.

39. The title of this section, "The Pertinence of Details," is an implicit reference to a paper by the ethnomethodologist Michael Lynch ("Ethnométhodologie et pratique scientifique: la pertinence du détail," *Cahiers de recherche sociologique* 5, pp. 45–62).

40. Author's translation from: Friedrich Kittler, "Hardware—Das Unbekannte Wesen," in Sybille Krämer (ed.), *Medien, Computer, Realität: Wirklichkeitsvorstellungen Und Neue Medien*," Frankfurt Am Main: Suhrkamp, 1998. Original German quote: "*Wenn Computer die ersten Maschinen sind, die die Kontingenz oder Unberechenbarkeit einiger, nicht aller Zukünfte um endliche Grade vermindern können, sollte ihre eigene Kontingenz so offen wie möglich bleiben. Die Entwicklung möglichst komplexer und möglichst vielfältiger Hardwareplattformen ist daher eine Aufgabe, an der zum Beispiel Europa, von der bewundernswerten Ausnahme des Transputers abgesehen, bislang noch kaum teilgenommen hat. Wenn mein Beitrag zur Ringvorlesung es vermocht hätte, diese Aufgabe wenigstens wahrnehmbar zu machen, wäre ein Teilziel schon erreicht. Wenn jemand hinginge und all jene Programme, die bislang unterm Titel Philosophiestudium liefen, in Hardware gießen würde, das Ziel selber.*"

41. The Church-Turing thesis in fact does not really exist as has been argued elsewhere by Edgar Daylight: E. G., Daylight and E. Schüttpelz (2022), "The Turing Machine as a Boundary Object: Sorting Out American Science and European Engineering." Filmed July 2022 at WAI22: Wittgenstein and AI, London. Video, https://www.dijkstrascry.com/Daylight_Schuettpelz_BoundaryObject, and, to a certain extent, myself. Rather, what we have, are a variety of recast versions that fit to particular discourses. One well-known version of it is that anything that can be computed can also be computed by a Turing machine.

42. It is well-known that ENIAC had two lives: in its first life it was a highly parallel and modular machine that was hard to program (it required, essentially, a recabling of the entire machine for every new problem to be set up); in its second and much longer life, it was, essentially, rewired into a serial machine that could be "coded" by using the machine's order code. A main historical source for understanding these transitions is: T. Haigh, M. Priestley, and C. Rope, *ENIAC in Action: Making and Remaking the Modern Computer*, MIT Press, 2018. They see the rewired ENIAC as the first computer within the so-called modern code paradigm.

43. In fact, Curry provided a partial compiler on paper. However, he soon left computing again to get back to working in logic and so his work was never picked up and, by consequence, had no influence on further developments.

44. See, for instance: L. De Mol and M. Bullynck, "Roots of Program Revisited," *Communications of the ACM*, vol. 64, no. 4, 2021, pp. 35–37. In that short paper we showed that the notion of "program" as used in the ENIAC discourse is anchored in a decades-old engineering discourse around so-called program clocks and program devices. The original conception of programs in ENIAC was thus completely disconnected from any logical discourse.

45. See, for instance: Michael Mahoney, "The Histories of Computing(s)," *Interdisciplinary Science Reviews*, vol. 30, no. 2, pp. 119–135.

46. As she writes there: "The fact that historians feel aware in some respect of how they are illiterate, when they tackle a new set of documents, is the best indicator for the existence of the specific skills that the actors possessed. In my view, without describing the skills and attempting to acquire them, historians of science cannot adequately deal with their sources" (Karine Chemla, "Literacy and the History of Science: Reflections Based on Chinese and Other Sources," in David R. Olson and Nancy Torrance, *The Cambridge Handbook of Literacy*, Cambridge University Press, 2009, pp. 253–270).

47. Thomas Haigh, "The Tears of Donald Knuth," *Communications of the ACM*, vol. 58, no. 1, pp. 40–44.

48. Nathan Ensmenger, *The Computer Boys Take Over: Computers, Programmers, and the Politics of Technical Expertise*, Cambridge, Massachusetts: MIT Press, 2010.

49. Ibidem, p. 11.

50. In Ensmenger's book object-orientation is presented as a means to reorganizing software labor. While this is of course true to a certain extent, there were others who relied on it as a means to develop a software system that could be personalized (think of the motivation to use objects instead of functions in the Smalltalk system).

51. I elaborated on this point during a talk at the SIGCIS workshop at SHOT in 2019 and which is available here: https://hal.science/hal-03287418v1/file/talk_SHOT_LDM_final.pdf

52. P. Erickson, J. Klein, L. Daston, R. Lemov, T. Sturm, and M. Gordin, eds., *How Reason Almost Lost Its Mind: The Strange Career of Cold War Rationality*, Chicago: University of Chicago Press, 2013.

53. As they write in the introduction of the work: "An ideal of cold war rationality can be described as follows: First of all, this rationality should be formal, and therefore largely independent of personality or context. It frequently took the form of algorithms—rigid rules that determine unique solutions—which were moreover supposed to provide optimal solutions to given problems, or delineate the most efficient means toward certain given goals (taken, in this instance, for granted). Second, complex tasks and episodes were analyzed into simple, sequential steps; the peculiarities of context, whether historical or cultural, gave way to across-the-board generalizations; analysis took precedence over synthesis. And finally, at least ideally, advocates hoped that the rules could be applied mechanically: computers might reason better than human minds. This ideal type shows the marks of its historical origins, as we will see in the ensuing chapters: on the one hand,

in the mathematics of algorithms, linear programming, and game theory; on the other, in the theory and practice of economic rationalization" (Idem, p. 4). I am very much indebted here to Ksenia Tatarchenko who brought up the issues around that use of algorithm in that work when I was collaborating with her and Anya Yermakova.

54. See: K. Tatarchenko, A. Yermakova, and L. De Mol (2021), "Russian Logics and the Culture of Impossible: Part II—Reinterpreting Algorithmic Rationality," *IEEE Annals for the History of Computing*, vol. 43, no. 3, pp. 57–69, and K. Tatarchenko, A. Yermakova, and L. De Mol (2021), "Russian Logics and the Culture of Impossible: Part I—Recovering Intelligentsia Logics," *IEEE Annals for the History of Computing*, vol. 43, no. 3, pp. 43–56. It should be pointed out here also that my contributions to these two papers were modest compared to the contributions from Ksenia Tatarchenko and Anya Yermakova.

55. Bill Aspray, message on the SIGCIS mailing list, thread *Using History to Teach Computer Science,* May 18, 2014.

56. One well-known case here are the Unguru debates in the history of mathematics. In 1975 Sabbetai Unguru published a paper in the *Archive for History of Exact Sciences* where he argued that Greek (or Babylonian) mathematics is not algebra; and it should thus not be put in modern mathematical terms but has to be considered as a discourse in its own. Unguru's paper resulted in a controversy during which a number of famous mathematicians would send angry letters to the editors of the *Archive.* It was symptomatic of the growing divide between mathematicians and historians vis-à-vis history of mathematics, which eventually led to a greater intellectual independence of the historians from mathematicians and to the slow recognition of the history of mathematics as a specialization in its own right. For a discussion of these debates and its relevance for the history of the history of computing, see: L. De Mol and M. Bullynck, "Making the History of Computing. The History of Computing in the History of Technology and the History of Mathematics, *Revue de Synthèse*, vol. 139, Issue 3–4, 2018, pp. 361–380.

57. K. Gavroglu, "The Sisyphean Fate of History of Science," *Centaurus*, vol. 64, no. 2, 2022, pp. 809–828.The paper is based on Gavroglu's Neuenschwander prize lecture presented at the 2022 ESHS conference held in Brussels and with the same title.

58. For a short history of the history of computing see, for instance: Thomas Haigh, "The History of Computing: An Introduction for the Computer Scientist," in: A. Akera and W. Aspray, *Using History to Teach Computer Science and Related Disciplines*, Computing Research Association, 2004, pp. 5–26.

59. Michael D. Gordin, "Book Review: The Tory Interpretation of History," *Historical Studies in the Natural Sciences*, vol. 44, no. 4, 2014, pp. 421–422.

60. Ibidem, p. 320.

61. L. De Mol, M. Bullynck, and E. Daylight, 2018, idem.

62. This is one of the chapters of the forthcoming book written by the PROGRAMme collective and discussed in more detail in section 6.

63. For an argument, see Hasok Chang's reply to Gordin's review and in particular section 2.2.: Hasok Chang, "Presentist History for Pluralist Science," *Journal for General Philosophy of Science*, vol. 52, 2021, pp. 97–114.

64. One example discussed by Gavrogli concerns the European Horizons program where the social sciences and humanities used to have their own niche. But since Horizon 2020 this is no longer the case—instead they have been "fully integrated" into the program "to maximize the returns to society from investment in science and technology" (Ibidem, p. 818).

65. Idem, p. 824.

66. Ibidem.

67. There is an implicit reference here to Latour's critical reflections on critical theory of which he is supposed to be one of the founders: B. Latour, "Why Has Critique Run out of Steam? From Matters of Fact to Matters of Concern," *Critical Inquiry*, vol. 30, 2004, pp. 225–248. In that text he uses the military metaphor of the officer having to consider revising their military strategies when faced with a new situation: "I simply want to do what every good military officer, at regular periods, would do: retest the linkages between the new threats he or she has to face and the equipment and training he or she should have in order to meet them—and, if necessary, to revise from scratch the whole paraphernalia. This does not mean for us any more than it does for the officer that we were wrong, but simply that history changes quickly and that there is no greater intellectual crime than to address with the equipment of an older period the challenges of the present one" (Idem, p. 231).

68. Thomas Haigh and Paul Ceruzzi, *A New History of Modern Computing*, MIT Press, 2021.

69. Thomas Haigh, "Finding a Story for the History of Computing," Collaborative Research Center 1187 Media of Cooperation, Working paper series, no. 3, July 2018.

70. Michael Mahoney, "The History of Computing in the History of Technology," *Annals of the History of Computing* 10, 1988, pp. 113–125

71. For a detailed account see: Matti Tedre, *The Science of Computing: Shaping a Discipline*, CRC Press, 2014. Peter Denning is known to have frequently launched debates on the nature of the field. Most well-known is probably the 1989 report: D. Comer et al., "Computing as a Discipline," *Communications of the ACM*, vol. 32, no. 1, 1989, pp. 9–23. In that report a multi-disciplinary view is proposed.

72. Donald E. Knuth, "Literate Programming," The Computer Journal (British Computer Society) vol. 27, no. 2, 1984, pp. 97–111.

73. Warren Sack, *The Software Arts*, MIT Press, 2019.

74. In the recent book by Tom Haigh and Paul Cerruzi (*A New History of Modern Computing*, MIT Press, 2021) the notion of ubiquity is replaced by that of universality but the two seem quite related if not identical. The technical notion of ubiquity is of course a rather common one in the computing field itself (see, for instance, the ACM UbiComp conference or the ACM journal *Ubiquity*). The more technical notion of ubiquitous computing was launched in the late 1980s by the computer scientist Mark Weiser working at Xerox PARC, where ubiquitousness required that the technology itself should disappear from sight hidden behind interfaces that are "human-centered," familiar, intuitive. Interestingly, he was inspired by Heidegger's philosophy too.

75. In the call for papers we wrote: "The number of researchers working in fields related to computing is growing rapidly in many different directions. As Mahoney once

stated, 'the computer is not one thing but many different things, and the same holds true of computing.' As a consequence, the computing sciences collect the most diverse complex of experts: philosophers, logicians, historians, mathematicians, computer scientits, programmers, engineers. The number of involved subjects grows accordingly: from the foundational issues to their applications; from the philosophical questions to problems of realizability and design of specifications; from the theoretical studies of computational barriers to the relevance of machines for educational purposes. Given the significance of computing for modern society, the relevance of its history and philosophy can hardly be overestimated. The aim of this conference is to bring together these two streams: we are strongly convinced that an interplay between researchers with an interest in the history and philosophy of computing can crucially add to the maturity of the field."

76. DHST stands for Division for the History of Science and Technology; DLMPST stands for Division for Logic, Methodology and Philosophy of Science and Technology. Each of these divisions has its main conference every four years. While sister divisions, their inner ways of working are not entirely the same. For instance, the DHST has a long-standing tradition of commissions which are specialized in particular themes such as history of physics or history of mathematics, whereas DLMPST has fewer commissions.

77. Among others: Janet Abbate, Bill Aspray, Nathan Ensmenger, Con Diaz, David Gugerli, and Tom Haigh.

78. Other examples: programs as formal objects fits well with a more logical approach one finds in analytical philosophy but also with work on formal verification; programs as socio-technical objects fits well with the approach of Science and Technology Studies but also with particular strands of software engineering or ethical analyses of computer programs; programs as physical objects fits well with certain approaches of media studies but also with the views of hardware engineers; etc.

79. This is of course an implicit reference to the well-known work by Paul Feyerabend, *Against Method: Outline of an Anarchistic Theory of Knowledge*, New Left Books, 1975. The *anything goes* approach developed by Feyerabend is often interpreted as a kind of relativism if not nihilism. But the core of the argument is in fact resonant with part of what was argued here: the idea that one should always be very wary of methodological dogmatism.

80. I refer here to the view that was defended by the president of the CNRS, Antoine Petit, who argued in an interview published November 26, 2019, in *l'Echo*, the need for a reform of research and education in France, proposing a new law which he described as follows: "*Il faut une loi ambitieuse, inégalitaire—oui, inégalitaire, une loi vertueuse et darwinienne, qui encourage les scientifiques, équipes, laboratoires, établissements les plus performants à l'échelle internationale, une loi qui mobilise les énergies.*" (English translation: "We need a law that is ambitious, unequal—yes, unequal, a law that is virtuous and Darwinian, that encourages the best-performing scientists, teams, laboratories and establishments on an international scale, a law that mobilizes energies.")

Index

Page references for figures or tables are italicized*

About the Editor

William Aspray is senior research fellow at the Charles Babbage Institute, a research center for the study of Computing, Information, and Culture at the University of Minnesota Twin Cities. Recent publications include *Computer: A History of the Information Machine* (4th ed., Routledge, 2023, co-author) and *Understanding Information History* (Springer, 2023). He is a former editor of *Information & Culture: A Journal of History*.

About the Contributors

Geoffrey C. Bowker is Emeritus Donald Bren Chair at the School of Information and Computer Sciences, University of California at Irvine. Together with Leigh Star he wrote *Sorting Things Out: Classification and Its Consequences* (MIT Press, 2000). He is currently working on time and computing under the rubric "life at the femtosecond."

Martin Campbell-Kelly is professor emeritus at the University of Warwick. He has held visiting appointments at the Smithsonian Institution, the MIT Sloan School of Management, and the National Archive for the History of Computing at Manchester University. He specializes in the business and economic history of the information-technology industries. His books include *From Mainframes to Smart Phones: A History of the International Computer Industry* (Harvard University Press, 2015) and *Cellular: An Economic and Business History of the Mobile-Phone Industry* (MIT Press, 2022), both co-authored with Daniel D. Garcia-Swartz. He is co-author of *Computer: A History of the Information Machine* (4th ed., Routledge, 2023).

James W. Cortada is senior research fellow at the Charles Babbage Institute, a research center for the study of Computing, Information, and Culture at the University of Minnesota Twin Cities. He writes extensively on business and economic history of information and information technologies, including *IBM: The Rise and Fall and Reinvention of a Global Icon* (MIT Press, 2019), *All the Facts: A History of Information in the United States since 1870* (Oxford University Press, 2016), and *Birth* *of Modern Facts: How the Information Revolution Transformed Academic Research, Governments, and Businesses* (Rowman & Littlefield, 2023).

Liesbeth De Mol is researcher of the French National Research Foundation (CNRS), based in Lille. She is a co-founder of the Commission for the History and Philosophy of Computing and member of the PROGRAMme collective, the author of the forthcoming monograph *What is a Computer Program?* She writes extensively on the history and epistemology of logic, programming, and computer science.

Philip Doty, emeritus faculty member in the School of Information at the University of Texas–Austin, focuses on information policy, including surveillance, privacy, copyright, and IT. He is editor of *Privacy in America* (Scarecrow, 2011). Recent publications appeared in the *Journal of the Association for Information Science & Technology*, the Information Ethics Round Table, and proceedings of the annual conferences of the Association for Information Science and Technology and the Canadian Association for Information Science.

Greg Downey is Evjue-Bascom Professor in the School of Journalism and Mass Communication and the Information School at the University of Wisconsin, Madison. He writes on the history and geography of information/communication technology and labor, including *Telegraph Messenger Boys: Labor, Technology, and Geography 1850–1950* (Routledge, 2002) and *Closed Captioning: Subtitling, Stenography, and the Digital Convergence of Text with Television* (Johns Hopkins University Press, 2008).

Daniel D. Garcia-Swartz is economist with Charles River Associates in Chicago. His publications include many peer-reviewed articles and book chapters on the economics of high-technology industries. He has co-authored two books with Martin Campbell-Kelly: *From Mainframes to Smartphones: A History of the International Computer Industry* (Harvard University Press, 2015), and *Cellular: An Economic and Business History of the International Mobile-Phone Industry* (MIT Press, 2022).

Shane Greenstein is the Martin Marshall Professor of Business Administration at Harvard Business School. His most recent book, *How the Internet Became Commercial* (Princeton University Press, 2015), won the 2016 Schumpeter Prize. Professor Greenstein previously taught at the Kellogg School of Management, Northwestern University, and in the economics department at the University of Illinois, Urbana/Champaign. He continues to receive a daily education in life from his wife and children.

David Gugerli is professor of history of technology at the ETH Zurich. He is the author, among many other works, of *Simulation for All: The Politics of Supercomputing in Stuttgart* (together with Ricky Wichum; Chronos, 2021) and *How the World Got into the Computer* (Chronos, 2022)

Thomas Haigh is professor of history at the University of Wisconsin, Milwaukee. His numerous writings on the history of computing include *A New History of Modern Computing* (MIT Press, 2021) and *ENIAC in Action* (MIT Press, 2018).

Matthew L. Jones is the Smith Family Professor of History at Princeton University. In 2023, Norton published his *How Data Happened: A History from the Age of Reason to the Age of Algorithms*, written with Chris Wiggins. He is completing a book, *Great Exploitations*, on state surveillance of communications and information warfare. He has published two books previously, *Reckoning with Matter: Calculating Machines, Innovation, and Thinking about Thinking from Pascal to Babbage* (University of Chicago Press, 2016) and *The Good Life in the Scientific Revolution: Descartes, Pascal, Leibniz and the Cultivation of Virtue* (University of Chicago Press, 2006).

John Leslie King is Professor Emeritus of Information at the University of Michigan. He studies the relationship between social and technical change. For 20 years he was on the faculties of management and information and computer science at UC Irvine. For another 20 years he was on the faculty of information at the University of Michigan, Ann

Arbor. Along the way he was department chair, dean, vice provost, and university librarian.

Howard Rosenbaum is professor of information science at Indiana University, Bloomington. He is the author of *Understanding and Communicating Social Informatics* (Information Today, 2005) and *Social Informatics: Past, Present, and Future* (Cambridge Scholars Publishing, 2014), among other works.

Valérie Schafer is professor in contemporary European history at the University of Luxembourg. She writes extensively on the history of computing, telecommunications, and data networks, including *En construction: La fabrique française d'Internet et du Web* (INA, 2018) and she co-edited *Digital Roots: Historicizing Media and Communication Concepts of the Digital Age* (De Gruyter Oldenbourg, 2021). She is a founder of the journal *Internet Histories*.

Laura Skouvig is associate professor in the Department of Communication at the University of Copenhagen. She writes extensively on information history, history of surveillance, and history of knowledge, including editing *Histories of Surveillance from Antiquity to the Digital Era* (Routledge, 2021) and editing *The Organization of Knowledge* (Emerald, 2017).

Toni Weller is research fellow in history at De Montfort University, UK. She is the editor of *Information History in the Modern World* (Bloomsbury, 2017), author of *Information History—An Introduction: Exploring an Emergent Field* (Chandos, 2008), and lead editor of the forthcoming Routledge *Handbook of Information History*. She is also a former editor of the international journal *Library & Information History*.

Daniela Zetti is historian of technology. Since April 2020 she has been vice head of the Ethical Innovation Hub and a postdoctoral researcher at the Institute for the History of Medicine and Science Studies, both at the University of Lübeck, Germany. Her publications include *Digital Federalism. Information, Institutions, Infrastructures (1950–2000)*, Itinera 49/2022, DOI: 10.24894/978-3-7965-4509-2 (with Paolo Bory).

www.ingramcontent.com/pod-product-compliance
Ingram Content Group UK Ltd.
Pitfield, Milton Keynes, MK11 3LW, UK
UKHW020222290926
15470UKWH00001B/335

9 781538 183816